Report of Investigations 9689

Dust Control Handbook for Industrial Minerals Mining and Processing

Andrew B. Cecala, Andrew D. O'Brien, Joseph Schall, Jay F. Colinet, William R. Fox, Robert J. Franta, Jerry Joy, Wm. Randolph Reed, Patrick W. Reeser, John R. Rounds, Mark J. Schultz

DEPARTMENT OF HEALTH AND HUMAN SERVICES
Centers for Disease Control and Prevention
National Institute for Occupational Safety and Health
Office of Mine Safety and Health Research
Pittsburgh, PA • Spokane, WA

January 2012

This document is in the public domain and may be freely copied or reprinted.

Disclaimer

Mention of any company or product does not constitute endorsement by the National Institute for Occupational Safety and Health (NIOSH). In addition, citations to Web sites external to NIOSH do not constitute NIOSH endorsement of the sponsoring organizations or their programs or products. Furthermore, NIOSH is not responsible for the content of these Web sites. All Web addresses referenced in this document were accessible as of the publication date.

Ordering Information

To receive documents or other information about occupational safety and health topics, contact NIOSH at

> Telephone: **1–800–CDC–INFO** (1–800–232–4636)
> TTY: 1–888–232–6348
> e-mail: cdcinfo@cdc.gov
>
> or visit the NIOSH Web site at **www.cdc.gov/niosh**.

For a monthly update on news at NIOSH, subscribe to NIOSH *eNews* by visiting **www.cdc.gov/niosh/eNews**.

DHHS (NIOSH) Publication No. 2012–112

January 2012

SAFER • HEALTHIER • PEOPLE™

CHAPTER CONTENTS

- FUNDAMENTALS OF DUST COLLECTION SYSTEMS
- WET SPRAY SYSTEMS
- DRILLING AND BLASTING
- CRUSHING, MILLING, AND SCREENING
- CONVEYING AND TRANSPORT
- BAGGING
- BULK LOADING
- CONTROLS FOR SECONDARY SOURCES
- OPERATOR BOOTHS, CONTROL ROOMS, AND ENCLOSED CABS
- HAUL ROADS, STOCKPILES, AND OPEN AREAS

TABLE OF CONTENTS

ACRONYMS AND ABBREVIATIONS USED IN THIS HANDBOOK xxiii
UNIT OF MEASURE ABBREVIATIONS USED IN THIS HANDBOOK xxiv
ACKNOWLEDGMENTS xxv
ABOUT THIS HANDBOOK xxvi
    TASK FORCE COMMITTEE COCHAIRS xxvi
    TASK FORCE COMMITTEE MEMBERS xxvi
    EDITOR xxvii
    HANDBOOK LAYOUT COORDINATOR xxvii
    IMA-NA COORDINATOR xxvii
INTRODUCTION 1
    SILICA EXPOSURE AND POTENTIAL HEALTH IMPACTS 2
    REFERENCES 4
CHAPTER 1: FUNDAMENTALS OF DUST COLLECTION SYSTEMS 7
    BASICS OF DUST COLLECTOR SYSTEMS 8
    AIRFLOW AND DUST CONTROL 9
    EXHAUST SYSTEMS DESIGN 11
    HOODS 11
        Hoods and Blowing versus Exhausting Ventilation 11
        Hood Types 12
        Hood Design 14
            Air Induction 15
            Capture Velocity 16
        Other Hood Considerations 18
        Checklist for Hood Effectiveness 19
    DUCTWORK AND AIR VELOCITIES 19
        High-Velocity Systems 20
        Low-Velocity Systems 21
            Airflow Target of 1,800 fpm 23
            Avoidance of Horizontal Ductwork 23
            The Use of Main Duct Trunk Lines 24

       Minimizing Field Fits and Welds 24

       Avoiding Mitered Elbows Greater Than 90 Degrees 25

       Sizing and Locating Orifice Plates... 25

       Minimizing the Use of Flexible Hoses 26

   Modified Low-Velocity (MLV) System 26

AIR CLEANING DEVICES 28

   Gravity Separators (Drop-Out Boxes) 28

   Centrifugal Collectors or Cyclones 29

   Baghouse Collectors 31

       Mechanical Shaker Collectors 34

       Reverse Air Collectors 36

       Reverse Jet (Pulse Jet) Collectors 37

       Cartridge Collectors 38

   Wet Scrubbers 40

       Venturi Scrubbers 41

       Impingement Plate Scrubbers 41

       Spray Tower Scrubbers 42

       Wet Cyclone Scrubbers 42

   Electrostatic Precipitators (ESPs) 42

COLLECTOR DISCHARGE DEVICES 43

   Rotary Airlock Valves 43

   Double Dump Valves 43

   Tilt Valves 44

   Vacuum or Dribble Valves 44

FILTER FABRICS 44

   Particulate Collection of Fabrics 46

DESIGNING MINE/PLANT DUST COLLECTION SYSTEMS 48

   Clay Application 49

FANS 50

   Fan Operating Characteristics 50

   Fan Laws 52

   Fan Types... 53

       Axial-Flow Fans 53

      Centrifugal Fans 54

    Other Fan Types 56

    Roof Ventilators 56

  REFERENCES 57

CHAPTER 2: WET SPRAY SYSTEMS 61

  PRINCIPLES OF WET SPRAY SYSTEMS 61

    Water Application 61

    Nozzle Location 62

    Controlling Droplet Size 63

    Methods of Atomization 64

    Chemical Additives to Control Droplets 64

  NOZZLE TYPES AND SPRAY PATTERNS 65

    Air Atomizing Nozzles 66

    Hydraulic Full Cone Nozzles 67

    Hydraulic Hollow Cone Nozzles 67

    Hydraulic Flat Fan Nozzles 68

  MAINTENANCE ISSUES WITH WET SPRAY SYSTEMS 70

    Water Quality.. 70

    Nozzle Maintenance 72

    Erosion and Wear 72

    Corrosion 72

    Clogging 73

    Caking 73

    Temperature Damage.. 73

    Improper Reassembly 73

    Accidental Damage 74

    Checking Spray Nozzle Performance 74

    Performance Testing of Equipment.. 75

  REFERENCES 75

CHAPTER 3: DRILLING AND BLASTING 79

  SURFACE DRILLING DUST CONTROL... 79

    Wet Drilling.. 80

      Advantages to Wet Drilling.. 81

- Recommendations for Proper Wet Drilling 82
- Disadvantages to Wet Drilling 82
- Water Separator Sub 83
- Dry Drilling 84
  - Medium- to Large-Diameter Drill Dust Collection Systems 84
  - Sources of Dust Emissions from Collection Systems 84
  - Deck Shroud Leakage Solutions 85
    - Rectangular Shroud 85
    - Circular Shroud 86
    - Collector to Bailing Airflow Ratio 86
    - Airflow Maintenance for Dust Control 87
    - Use of Collector to Bailing Airflow Ratios for Dust Control 88
  - Air-Blocking Shelf 90
  - Collector Dump Discharge 93
    - Preventing Collector Dump Dust Entrainment 93
  - Drill Stem/Deck Leakage 95
    - Preventing Drill Stem/Deck Leakage 95
  - Small Diameter Drill Dust Collection Systems 96
- UNDERGROUND DRILLING DUST CONTROL 97
  - Wet Drilling 98
    - Water Mists and Foams 99
    - Wetting Agents 99
    - External Water Sprays 100
- RESPIRABLE DUST CONTROL FOR BLASTING 100
  - Blasting Dust Control Measures 101
    - Wetting Down Blasting Area 101
    - Water Cartridges 102
    - Fogger Sprays 102
    - Air Filtration Systems 103
    - Ventilation System Dispersal 105
- REFERENCES 105
- *CHAPTER 4: CRUSHING, MILLING, AND SCREENING* 111
  - PREVENTION AND SUPPRESSION APPLICATIONS 112

Wet Control Methods	112
Dry Control Methods	112
CRUSHING	113
Work Practices to Minimize Dust Exposure from Crushers	122
MILLING	122
Design and Work Practices to Minimize Dust Exposure from Milling	124
SCREENING	125
Work Practices to Minimize Dust Exposure from Screens	127
MAINTENANCE	128
REFERENCES	129

CHAPTER 5: CONVEYING AND TRANSPORT 133

BELT CONVEYORS	133
Controlling Material Spillage	134
Carryback	141
Belt Scrapers	141
Belt Washing	142
CONVEYOR DESIGN AND MAINTENANCE ISSUES	143
Belt Splices	143
DESIGN CONSIDERATIONS FOR TRANSFER POINTS	144
WATER SPRAYS FOR PREVENTION OF AIRBORNE DUST	146
SCREW CONVEYORS	147
BUCKET ELEVATORS	148
PNEUMATIC CONVEYANCE	150
REFERENCES	153

CHAPTER 6: BAGGING 157

BAGS AS DUST SOURCES	157
BAG CONSIDERATIONS	159
Bag Construction	159
Bag Perforations	160
Overall Perforations	161
Undervalve Perforations	161
Bag Fill Type	162
Effects of Bag Failures	165

CONTROL TECHNOLOGIES FOR FILLING 50- TO 100-POUND BAGS 166
   Exhaust Hood Placed around Bag Loading Nozzles 166
   Exhaust Hoppers (Below Entire Fill Station) 167
   Dual-Nozzle Bagging System 168
   Overhead Air Supply Island System 170
   Automated Bag Placer and Filling Systems 172

CONTROL TECHNOLOGIES FOR CONVEYING AND PALLETIZING 50- to 100-POUND BAGS 174
   Bag and Belt Cleaning Device 174
   Semi-Automated Bag Palletizing Systems 175
   Automated Bag Palletizing Systems 177
   Plastic Pallet Wrap 180

FLEXIBLE INTERMEDIATE BULK CONTAINERS 181

MAINTAINING CLEAN FLOORS IN BAG LOADING, PALLETIZING, AND WAREHOUSE LOCATIONS 182

ENCLOSING DUST-LADEN AREAS WITH PLASTIC STRIPPING AND USING AN LEV SYSTEM 183

WORK PRACTICES TO MINIMIZE DUST EXPOSURES FROM BAGGING AND PALLETIZING 184

REFERENCES 185

CHAPTER 7: BULK LOADING 189

LOADING SPOUTS 189
   Dust Collection Systems and Loading Spouts 192
   Cascading Loading Spout 193

DUST SUPPRESSION HOPPER 194

ENCLOSURES 194
   Enclosed Loadout 194
   Operator Booths/Control Rooms 196

REFERENCES 197

CHAPTER 8: CONTROLS FOR SECONDARY SOURCES 201

CLEANING DUST FROM SOILED WORK CLOTHES 201

NEW TECHNOLOGY FOR CLEANING DUST FROM WORK CLOTHING 202
   Clothes Cleaning Booth 203
   Air Reservoir 203

Air Spray Manifold	203
Exhaust Ventilation	204
Safety Issues	205
EFFECTIVENESS OF CLOTHES CLEANING TECHNOLOGY	205
CURRENT STATUS OF CLOTHES CLEANING REGULATIONS	206
HOUSEKEEPING PRACTICES	207
TOTAL STRUCTURE VENTILATION DESIGN	210
PRINCIPLES OF TOTAL STRUCTURE VENTILATION	210
Clean Makeup Air Supply	211
Effective Upward Airflow Pattern	212
Competent Shell of Structure	212
VARIABLES AFFECTING TOTAL STRUCTURE VENTILATION	212
Costs	213
OPEN STRUCTURE DESIGN	213
BACKGROUND DUST SOURCES	214
EXAMPLES OF SECONDARY DUST SOURCES	215
Dusts from Outside Sources Traveling Inside Structures	215
Personal Dust Exposure from Differences in Job Function Work Practices	215
Broken Bags of Product	217
Cloth Seat Material/Worn Acoustical Material	218
Dirty Work Boots	218
MAINTENANCE	220
Work Practices to Reduce Dust Exposure during Maintenance Activities	220
REFERENCES	221
CHAPTER 9: OPERATOR BOOTHS, CONTROL ROOMS, AND ENCLOSED CABS EFFECTIVENESS TERMINOLOGY	225
FILTRATION AND PRESSURIZATION SYSTEMS	227
Field Studies	227
Laboratory Study	228
Mathematical Model to Determine Enclosure Protection Factor	229
RECOMMENDATIONS FOR FILTRATION/PRESSURIZATION SYSTEMS	230
Ensuring Enclosure Integrity by Achieving Positive Pressurization against Wind Penetration	230
Keeping Doors and Windows Closed	231

Effective Filtration	232
Effective Outside Air Filtration and Enclosure Pressurization	232
Outside Air Inlet Location	233
PreCleaner	233
Effective Recirculation Filtration	233
Floor Heaters	234
Good Housekeeping (Enclosure Cleanliness)	235
Mechanical Filter Media and Visual Indicator for Filter Changing	235
Ease of Filter Change	236
Uni-Directional Design	236
Generalized Parameters for an Effective System	237
OTHER METHODS TO LOWER RESPIRABLE DUST IN ENCLOSURES	238
Small In-Unit HEPA Filters	238
REFERENCES	239
CHAPTER 10: HAUL ROADS, STOCKPILES, AND OPEN AREAS	243
HAUL ROADS	244
Basics of Road Construction	244
Haul Road Dust Control Measures	249
Road Preparation	250
Water	251
Surfactants	253
Salts	254
Petroleum Emulsions	255
Polymers	255
Adhesives	256
Other Dust Control Methods	257
Speed Control	257
Traffic Control	257
Load Covers	257
Cab Maintenance	258
STOCKPILES AND OPEN AREAS	258
Surface Roughness	259
Water	262

    Coatings 263
    Wind Barriers 265
  REFERENCES 267
GLOSSARY 275

ILLUSTRATIONS

Figure 1.1. Relationship between static, velocity, and total pressure 10

Figure 1.2. A basic depiction of a simple exhaust system with the major components being the hood, duct, air cleaning device, and fan 11

Figure 1.3. Comparison of ventilation characteristics for blowing versus exhausting system 12

Figure 1.4. Simple design of basic push-pull ventilation system 13

Figure 1.5. Basic setup for a type of receiving hood 14

Figure 1.6. Demonstration of air induction as material falls from a conveyor 15

Figure 1.7. Hood entry loss calculation 17

Figure 1.8. Hood entry loss coefficients for different hood types 18

Figure 1.9. Schematics of high- and low-velocity systems 20

Figure 1.10. Sawtooth design of a low-velocity system with a dropout duct back into the process at the base of each leg 22

Figure 1.11. Duct design angles for low-velocity systems 24

Figure 1.12. Demonstration of poor orifice plate placement 25

Figure 1.13. Depiction of the horizontal and vertical velocity relationship in a modified low-velocity system 27

Figure 1.14. Typical design of gravity separator (drop-out box) 29

Figure 1.15. Typical design of a cyclone dust collector 30

Figure 1.16. Basic design of a baghouse dust collector 31

Figure 1.17. Typical design of a mechanical shaker dust collector 35

Figure 1.18. Typical design of a reverse air dust collector 36

Figure 1.19. Typical design of reverse jet (pulse jet) dust collector 37

Figure 1.20. Cartridge collector with dust filter canisters 39

Figure 1.21. Typical design of a wet scrubber dust collector 40

Figure 1.22. Typical designs of a venturi scrubber 41

Figure 1.23. Typical design of an impingement plate scrubber 42

Figure 1.24. Double dump weight-based valves 44

Figure 1.25. Demonstration of how a complex exhaust system is a combination of branches linking simple exhaust systems 49

Figure 1.26. A typical fan performance curve. Each fan curve is associated with a certain fan model at a selected RPM 51

Figure 1.27. Fan performance curve showing stalling and operating ranges 51

Figure 1.28. Wall-mounted propeller fan 53

Figure 1.29. Typical tubeaxial fan (left) and vaneaxial fan (right) 54

Figure 1.30. Typical centrifugal fan 55

Figure 1.31. Axial centrifugal fan showing airflow 56

Figure 1.32. Typical roof ventilators using axial fans (left) and centrifugal fans (right) 57

Figure 2.1. Common dust control application illustrating nozzle positioning 62

Figure 2.2. Effect of droplet size on dust particle impingement 63

Figure 2.3. Contact angle resulting from a liquid meeting a solid surface 65

Figure 2.4. Typical internal mix nozzle (top) and external mix nozzle (bottom) 66

Figure 2.5. Typical air atomizing nozzle round spray pattern (top) and fan spray pattern (bottom) 66

Figure 2.6. Typical full cone nozzle and spray pattern 67

Figure 2.7. Typical hollow cone whirl chamber nozzle and spray pattern. Right angle design shown 67

Figure 2.8. Typical hollow cone spiral nozzle and spray pattern 68

Figure 2.9. Typical flat fan nozzle and spray patterns 68

Figure 2.10. Typical loader dump dust control application 69

Figure 2.11. Typical conveyor dust control application 69

Figure 2.12. Airborne suppression performance of four types of spray nozzles and flat fan spray nozzles are hydraulically atomizing 70

Figure 2.13. Typical self-contained water delivery system 71

Figure 2.14. Typical duplex basket strainer 72

Figure 2.15. Nozzle erosion—new versus used 72

Figure 2.16. Nozzle corrosion—new versus used 73

Figure 2.17. Nozzle caking—new versus used 73

Figure 2.18. Temperature damage—new versus used 73

Figure 2.19. Nozzle damage—new versus used 74

Figure 2.20. Good spray tip showing pattern and distribution graph. Height of bar indicates distribution of water over pattern width and indicates relatively uniform flow over the width of the pattern 74

Figure 2.21. Worn spray tip showing pattern and distribution graph. Height of bar indicates distribution of water over pattern width and indicates an increase of and excessive flow in the center of the pattern due to orifice wear 74

Figure 3.1. Illustrations depicting a small surface crawler drill rig, a truck-mounted drill rig, and a large track-mounted drill rig, respectively 80

xv

Figure 3.2. The air and water flow during drilling operation to demonstrate water flushing of the drill cuttings. The water flows through the center of the drill steel and out the end of the drill bit to remove the cuttings from the drill hole 81

Figure 3.3. Internal workings of a water separator sub 83

Figure 3.4. A basic dry dust collection system on a drill 84

Figure 3.5. Reinforcing flaps added to each corner of a rectangular deck shroud to reduce leakage 85

Figure 3.6. Circular shroud design 86

Figure 3.7. Results of changing dust collector to bailing air ratio at various shroud gap heights 87

Figure 3.8. Graphs of the model that represents the severity of dust emissions based upon collector to bailing airflow ratios, showing that the respirable dust concentrations become lower as the collector to bailing airflow ratio increases 89

Figure 3.9. Qualitative models of airflow patterns underneath the shroud without the air-blocking shelf (left) and with the air-blocking shelf (right) 90

Figure 3.10. Air-blocking shelf installed on the inside perimeter of the drill shroud of a blasthole drill. Note the overlap in the rear corners to eliminate any gaps in the shelf perimeter 91

Figure 3.11. Buildup of material on the air-blocking shelf (left) and the resulting dust emissions that are created during lowering of the drill mast (right) 92

Figure 3.12. Air-blocking shelf with modifications incorporated into installation. This shelf was installed in short sections in an attempt to prevent potential damage that could occur during tramming with the drill mast up 92

Figure 3.13. Dust collector dump point prior to shroud installation (A). Two men installing the shroud onto the dust collector dump point (B). The dust collector dump point after installation of the shroud (C) 93

Figure 3.14. Comparison of dust concentrations of uncontrolled dust collector dump to shrouded dust collector dump 94

Figure 3.15. Air ring seal used to impede the movement of dust particles through an opening . 96

Figure 3.16. Typical dust collection system used by small crawler or "buggy" drills 97

Figure 3.17. Illustrations showing a typical jackleg drill, stoper drill, and 3-boom jumbo, respectively 98

Figure 3.18. Water flow during a drilling operation to demonstrate water flushing of the drill cuttings. The water flows through the drill into the center of the drill steel and out the end of the drill bit to remove the cuttings from the drill hole 98

Figure 3.19. Drill boom showing external water sprays directed onto the drill steel 100

Figure 3.20. A typical blasthole containing an explosive charge utilizing a water cartridge to suppress dust generated during blasting 102

Figure 3.21. A fogger spray used to create a mist for dust suppression in the heading where blasting will occur 103

Figure 3.22. A filtration unit, located adjacent to the blast heading, used for filtering the contaminated ventilation air from the blast heading after blasting occurs 104

Figure 3.23. A filtration unit, located in the blast heading, used for filtering the contaminated ventilation air from the blast heading after blasting occurs 104

Figure 4.1. Illustration of a jaw crusher showing material being crushed between fixed and reciprocating jaws 113

Figure 4.2. Examples of compressive crushers. Left—illustration of a cone crusher showing material being crushed between the cone mantle and bowl liner. Right—illustration of a gyratory crusher with material being crushed between the gyratory head and the frame concave 114

Figure 4.3. Examples of impactive crushers. Left—a hammermill-type crusher showing material crushed between the rotating hammers and fixed grinding plate. Center—illustration of material being crushed between rotating hammers and fixed anvil plates in an impact breaker. Right—illustration of the size reduction action of a roll crusher 114

Figure 4.4. Illustration of a wet dust control approach with partial enclosure at a crusher dump loading operation. Note the blue "fan patterns" signifying water sprays 115

Figure 4.5. Illustration of a dry (exhaust) dust control system with a partial enclosure at a crusher dump loading operation 116

Figure 4.6. Illustration of a dry (exhaust) dust control system at the transfer point of a conveyor discharge to a crusher feed hopper 118

Figure 4.7. Illustration of a dry (exhaust) dust control system on a feed chute into a transfer chute feeding a crusher 118

Figure 4.8. Illustration of a dry (exhaust) dust control system at the discharge of a jaw crusher onto a belt conveyor 119

Figure 4.9. Illustration of a dry (exhaust) dust control system at the discharge of a hammermill crusher onto a belt conveyor 119

Figure 4.10. Illustration of a wet dust control approach with a transfer chute/impact bed enclosure at a crusher loading operation 120

Figure 4.11. Illustration of a wet dust control approach on a crusher discharge/belt loading operation 121

Figure 4.12. Cut-away illustration of a ball mill showing the charge consisting of the material being processed and balls 123

Figure 4.13. Illustration of a stirring mill 123

Figure 4.14. Sampling doors located on top of a collection/transfer point to prevent leakage . 124

Figure 4.15. Illustration of a dry (exhaust) dust control system on the feed to a rotary screen with enclosed transfer chute 126

Figure 4.16. Illustration of a dry (exhaust) dust control system on a vibrating screen 127

Figure 5.1. Basic components of a conveyor belt 133

Figure 5.2. Types of fugitive dust emissions from conveyor belts 134

Figure 5.3. Basic depiction of material transfer from a top conveyor to a bottom conveyor 135

Figure 5.4. Depiction of impact bed used to control material direction and speed 135

Figure 5.5. Depiction of curved loading plate to help steer loaded material in a particular direction 136

Figure 5.6. Spillage as a result of material being improperly fed onto a belt 136

Figure 5.7. Calculations for determining material impact loading at a conveyor transfer point 137

Figure 5.8. Depiction of an impact cradle with underlying rubber layers 138

Figure 5.9. Side support cradle 138

Figure 5.10. Skirtboard used within the transfer point to help manage loaded material 139

Figure 5.11. Typical dust curtain used at the entrance and exit of the chute enclosure 140

Figure 5.12. Typical method for scraping material from a belt 141

Figure 5.13. Belt conveyor discharge chute used to return scrapings to the primary material flow 142

Figure 5.14. Illustration of a mechanical belt splice (left) and a vulcanized belt splice (right) 143

Figure 5.15. Conveyor transfer enclosure used with an exhaust ventilation system 145

Figure 5.16. Typical impact bed designed for ore-on-ore contact at transfer locations or chutes 145

Figure 5.17. Depiction of a typical screw conveyor 147

Figure 5.18. Packing gland and pillow block support bearing 148

Figure 5.19. Depiction of a bucket elevator 149

Figure 5.20. Depiction of a typical bucket elevator dust collection process 150

Figure 5.21. Depiction of two types of pneumatic conveying systems 151

Figure 5.22. HammerTek wear-resistant elbow installed in conveying ductwork 151

Figure 5.23. Rotary airlock feeding pneumatic conveying line 152

Figure 5.24. Conveyance system utilizing a venturi eductor device 152

Figure 5.25. Venturi eductor utilizing suction to draw the feed material into the conveying line 153

Figure 6.1. Product "rooster tail" shown on left, product blowback shown on top right, and dust-soiled bag shown on bottom right 158

Figure 6.2. Typical bag valve—interior view (left). Artist concept of fill nozzle and bag valve (right). Fill nozzle enters through the bag valve and delivers product into bag 159

Figure 6.3. Undervalve bag perforations (left) and overall perforations (right) 161

Figure 6.4. Undervalve perforations relieve excess pressure from bag during filling process to minimize the possibility of bag failure. Product and dust are seen escaping along with excess air 162

Figure 6.5. Open-top bag being loaded with product (normal whole-grain material, which is less dusty) 163

Figure 6.6. Extended polyethylene bag valve, interior view 164

Figure 6.7. Dust reductions with extended polyethylene and foam valves as compared with the use of the standard paper valve 164

Figure 6.8. Exhaust hood to capture and exhaust respirable dust away from each fill nozzle .. 167

Figure 6.9. Exhaust hopper below bag loading station with integrated LEV system to capture dust generated during bag loading process 168

Figure 6.10. Dual-nozzle bagging system design 169

Figure 6.11. Bag operator's dust exposure with and without the use of a dual-nozzle bagging system 170

Figure 6.12. OASIS positioned over bag operator to deliver clean filtered air down over work area 171

Figure 6.13. Bag storage shelves shown on left; a robotic arm with suction cups to place empty bags on fill nozzle shown on right 173

Figure 6.14. Bag and belt cleaner device. Note the use of a closed system under negative pressure 175

Figure 6.15. Semi-automated bag palletizing system to ergonomically improve bag stacking process with a push-pull ventilation system used to capture the dust generated 176

Figure 6.16. Semi-automated bag palletizing systems in which a worker removes bags from a conveyor and lifts and positions each layer on an air slide before automated loading onto the pallet. Both an OASIS system and an exhaust ventilation system are used to lower the worker's dust exposure during this bag palletizing process 177

Figure 6.17. Figure of the automated high-level palletizing process 178

Figure 6.18. New innovation of low-level pallet loading process using a robotic arm device .. 179

Figure 6.19. Automated spiral stretch wrap machine 180

Figure 6.20. Expandable neoprene rubber bladder used in the fill spout of the bagging unit for Flexible Intermediate Bulk Containers 181

Figure 6.21. Exhaust ventilation system used during loading of Flexible Intermediate Bulk Containers 182

Figure 7.1. Loading spouts designed with telescoping cups or pipes and dust collection capabilities 190

Figure 7.2. Loading spout discharge cones designed to seal against the loading port of closed vehicles to minimize dust liberation 191

Figure 7.3. Articulated loading spout positioner 191

Figure 7.4. Skirt (in yellow box) and tilt sensors (in red boxes) installed on the discharge end of loading spouts to help reduce dust emissions... 192

Figure 7.5. Cascade loading spout 193

Figure 7.6. Dust Suppression Hopper creating a solid stream of material during loading to reduce dust liberation 194

Figure 7.7. Rail car loadout area enclosed with plastic strips to contain dust 195

Figure 7.8. Environmentally controlled operator's booth near loadout area 196

Figure 8.1. Design of clothes cleaning system 202

Figure 8.2. Air spray manifold design with 26 nozzles spaced 2 inches apart 204

Figure 8.3. Test subject wearing polyester/cotton blend coveralls before and after using the clothes cleaning booth 206

Figure 8.4. Increase in coworker's dust exposure one floor up from floor sweeping activities. 207

Figure 8.5. Various types of floor sweep units used to clean dust-laden floors 208

Figure 8.6. Portable system used to perform vacuuming 209

Figure 8.7. Basic design of total structure ventilation system 211

Figure 8.8. Open-structure design compared to standard walled structure 214

Figure 8.9. Increase in a worker's respirable dust exposure inside a mineral processing facility while bulk loading was being performed outside 215

Figure 8.10. Comparison of two workers' respirable dust exposure while performing their job function with different control techniques being implemented 216

Figure 8.11. Worker closing bag valve with his hand as he removes the bag from the fill nozzle, significantly reducing his dust exposure 217

Figure 8.12. Increase in worker's respirable dust exposure from broken bag during the conveying process 218

Figure 8.13. Product tracked onto the floor of a piece of mobile equipment from dirty work boots 219

Figure 8.14. Respirable dust concentrations inside enclosed cab of two different pieces of equipment with and without the use of a gritless, nonpetroleum sweep compound 220

Figure 9.1. General design of an effective filtration and pressurization system 226

Figure 9.2. Laboratory test setup to evaluate various operational parameters on a filtration and pressurization system for an enclosed cab 228

Figure 9.3. Positive cab pressure necessary to prevent dust-laden air from infiltrating the enclosed cab at various wind velocities 231

Figure 9.4. Respirable dust concentrations inside enclosed cab for three days of testing with cab door closed and open 232

Figure 9.5. Problem created by heater stirring up dust from the floor and blowing dust off of the worker's clothing 234

Figure 9.6. Filtration unit showing lack of maintenance and care (left side) and after cleaning and the addition of a new intake air filtration unit (right side) 236

Figure 9.7. Airflow pattern for intake and return at roof of cab and uni-directional airflow design.. 237

Figure 10.1. Example of haul road dust from a typical mine haul truck 243

Figure 10.2. Cross section of haul road 246

Figure 10.3. Use of a spreader box to lay gravel onto road surface 247

Figure 10.4. Use of end-dumping with grader to lay gravel onto road surface 248

Figure 10.5. Subbase thickness from California Bearing Ratio tests on road material 249

Figure 10.6. Water truck equipped with a front water cannon and rear water sprays 251

Figure 10.7. Various types of manufactured fan spray nozzles for use on a water truck 252

Figure 10.8. Respirable dust concentrations measured from haul road after water application occurred at 10:00 a.m. 253

Figure 10.9. Self-tarping dump truck. The blue arrow indicates the direction of movement when tarping. Section A-A shows the recommended freeboard 258

Figure 10.10. Chisel plow used for creating large clods in open areas to reduce wind erosion 260

Figure 10.11. Example of tillage (from chisel plowing, for example) exposing large clods..... 261

Figure 10.12. The armoring process for tillage rows. Blue arrow denotes wind direction perpendicular to tillage rows. Magenta arrows depict fine material being eroded from the windward side of ridges leaving large-sized particles, with the fine material being carried over ridge tops and deposited in the troughs on the leeward side of the ridges 262

Figure 10.13. Dozer being used to compact stockpile for long-term storage 264

Figure 10.14. Example of a vegetated topsoil stockpile 265

Figure 10.15. Demonstration of how fences are used as wind barriers 266

Figure 10.16. Barrier built to enclose a stockpile 266

Figure 10.17. Layout demonstrating dispersed wind barriers.. 267

TABLES

Table 1.1. Typical quantities of airflow in dust collection system components 14
Table 1.2. Emission factors for crushed stone processing operations (lb/ton)* 33
Table 1.3. Minimum Efficiency Reporting Values (MERV) according to ASHRAE Standard 52 45
Table 2.1. Particle/droplet size in comparison to common precipitation classification 64
Table 2.2. Typical applications by spray nozzle type 68
Table 4.1. Relative emission rate ratios of crushing and screening equipment 111
Table 5.1. Impact cradle ratings to be used based on impact loading 137
Table 5.2. A comparison of CEMA and Army Corps of Engineers recommendations for skirting for a belt speed of 600 fpm 140
Table 8.1. Amount of dust remaining on the coveralls after cleaning, and the cleaning time for 100 percent cotton and polyester/cotton blend coveralls 205
Table 9.1. Comparison of three different descriptors for the effectiveness at providing clean air to an enclosed cab, operator booth, or control room 227
Table 9.2. Summary of field studies evaluating upgraded cabs 227
Table 9.3. Calculated PF derived from mathematical model 238

ACRONYMS AND ABBREVIATIONS USED IN THIS HANDBOOK

AASHTO	American Association of State Highway and Transportation Officials
ACGIH	American Conference of Governmental Industrial Hygienists
AEK	airflow extensible Kraft
AIRRS	air ring seal
ASHRAE	American Society of Heating, Refrigerating and Air-Conditioning Engineers
ASTM	American Society for Testing and Materials
B&BCD	bag and belt cleaning device
BHP	brake horsepower
CBR	California Bearing Ratio
CEMA	Conveyor Equipment Manufacturers Association
CFR	Code of Federal Regulations
DSH	Dust Suppression Hopper
EPA	Environmental Protection Agency
ESP	electrostatic precipitator
FIBC	Flexible Intermediate Bulk Container
FK	flat Kraft
HEPA	high-efficiency particulate air
HVAC	heating, ventilation, and air-conditioning
LEV	local exhaust ventilation
MERV	minimum efficiency reporting value
MLV	modified low-velocity
MSD	musculoskeletal disorder
MSDS	Material Safety Data Sheet
MSHA	Mine Safety and Health Administration
NIOSH	National Institute for Occupational Safety and Health
NISA	National Industrial Sand Association
NK	natural Kraft
OASIS	overhead air supply island system
OSHA	Occupational Safety and Health Administration
PF	protection factor
PM	particulate matter
PSE	particle size efficiency
PTFE	polytetrafluoroethylene
SOP	standard operating procedure
SP	static pressure
TP	total pressure
TSP	total suspended particulates
TWA	time-weighted average
USBM	United States Bureau of Mines
VP	velocity pressure

UNIT OF MEASURE ABBREVIATIONS USED IN THIS HANDBOOK

acph	air changes per hour
cfm	cubic feet per minute
cfm/ft	cubic feet per minute per feet
cfm/ft^2	cubic feet per minute per feet squared
fpm	feet per minute
ft	feet
ft^2	feet squared
ft/min	feet per minute
gal/yd^2	gallons per square yard
gpm	gallons per minute
gr/cf	grains per cubic foot
gr/dscf	grains per dry standard cubic foot
kg/m^3	kilograms per cubic meter
lb	pounds
lbs/hour	pounds per hour
lbs/min	pounds per minute
µg/m^3	micrograms per cubic meter
µm	micrometers
mg/m^3	milligrams per cubic meter
mm	millimeters
NPT	national pipe taper
pH	potential of hydrogen
psi	pounds per square inch
psig	pound-force per square inch gauge
RPM	revolutions per minute
scfm	standard cubic feet per minute
TPH	tons per hour
wg	water gauge

ACKNOWLEDGMENTS

The developers of this handbook are indebted to the following individuals, listed alphabetically, for their review and valuable contributions to the handbook's production:

Roger Bresee, Vice President, Technical Services, Unimin Corporation, Peterborough, Ontario, Canada;

Chris Bryan (retired), C.M.S.P., Occupational Health & Safety Manager, U.S. Silica Company, Berkeley Springs, WV;

Robert M. Castellan, M.D., M.P.H., Expert—Division of Respiratory Disease Studies, National Institute for Occupational Safety and Health;

Cynthia Farrier, Graphics Specialist, National Institute for Occupational Safety and Health;

Christopher Findlay, C.I.H., C.M.S.P., U.S. Department of Labor, Mine Safety and Health Administration, Arlington, VA;

Ian Firth, Principal Adviser—Occupational Health, Rio Tinto Limited, Bundoora, Victoria, Australia;

Marion Molchen, Molchen Photography, Washington, PA;

Jan Mutmansky, Ph.D., P.E., Professor Emeritus, Pennsylvania State University, University Park, PA;

Kenneth Vorpahl (retired), C.I.H. (1973-2009), C.S.P. (1976-2009), General Manager/Safety & Health, Unimin Corporation, Winchester, VA;

Richard Wood, International Union of Operating Engineers, National HAZMAT Program, Beaver, WV.

ABOUT THIS HANDBOOK

This handbook represents a successful collaborative effort by government and industry in protecting the health of U.S. mine workers. The two principal partnerships active in creating this handbook were between the Office of Mine Safety and Health Research (OMSHR) of the National Institute for Occupational Safety and Health (NIOSH) and the Industrial Minerals Association–North America (IMA-NA). The mission of the NIOSH OMSHR is to eliminate mining fatalities, injuries, and illnesses through research and prevention, while the IMA-NA is the representative voice of companies that extract and process the raw materials known as industrial minerals.

This handbook was written by a task force of safety and health specialists, industrial hygienists, and engineers (listed below) to provide information on proven and effective control technologies that lower workers' dust exposures during all stages of minerals processing. The handbook describes both the dust-generating processes and the control strategies necessary to enable mine operations to reduce workers' dust exposure. Implementation of the engineering controls discussed can assist mine operators, health specialists, and workers in reaching the ultimate goal of eliminating pneumoconiosis and other occupational diseases caused by dust exposure in the mining industry.

Designed primarily for use by industrial minerals producers, this handbook contains detailed information on control technologies to address all stages of the minerals handling process, including drilling, crushing, screening, conveyance, bagging, loadout, and transport. The handbook's aim is to empower minerals industry personnel to apply state-of-the-art dust control technology to help reduce or eliminate mine and mill worker exposure to hazardous dust concentrations—a critical component in ensuring the health of our nation's mine workers.

TASK FORCE COMMITTEE COCHAIRS

Andrew B. Cecala, Senior Research Engineer
Dust Control, Ventilation, and Toxic Substances Branch, NIOSH

Andrew D. O'Brien, C.S.P., General Manager/Safety & Health
Unimin Corporation

TASK FORCE COMMITTEE MEMBERS

Jay F. Colinet, Senior Scientist
Dust Control, Ventilation, and Toxic Substances Branch, NIOSH

William R. Fox, Manager/Safety & Health
Unimin Corporation

Robert J. Franta, Quotation Engineer
Spraying Systems Co.

Jerry Joy, C.I.H., C.S.P., Research Scientist—Industrial Hygienist
Dust Control, Ventilation, and Toxic Substances Branch, NIOSH

Wm. Randolph (Randy) Reed, Ph.D., P.E., Research Mining Engineer
Dust Control, Ventilation, and Toxic Substances Branch, NIOSH

Patrick W. Reeser, Engineering Manager
U.S. Silica Company

John R. Rounds, Director Project Engineering
Unimin Corporation

Mark J. Schultz, P.E., Senior Mining Engineer
Pittsburgh Safety and Health Technology Center (PS&HTC), Dust Division, Mine Safety and Health Administration (MSHA)

EDITOR

Joseph Schall, Health Communications Specialist
NIOSH

HANDBOOK LAYOUT COORDINATOR

Jeanne A. Zimmer, Physical Science Technician
Dust Control, Ventilation, and Toxic Substances Branch, NIOSH

IMA-NA COORDINATOR

Mark G. Ellis, President
IMA-NA

INTRODUCTION

Throughout the mining and processing of minerals, the mined ore undergoes a number of crushing, grinding, cleaning, drying, and product sizing operations as it is processed into a marketable commodity. These operations are highly mechanized, and both individually and collectively these processes can generate large amounts of dust. If control technologies are inadequate, hazardous levels of respirable dust may be liberated into the work environment, potentially exposing workers. Accordingly, federal regulations are in place to limit the respirable dust exposure of mine workers. Engineering controls are implemented in mining operations in an effort to reduce dust generation and limit worker exposure.

For the purposes of this handbook, dust is broadly defined as small solid particles created by the breaking up of larger particles. Depending on their size, these particles can become hazardous to worker health, particularly when suspended in air. The largest size particle that can be suspended in air for long periods of time from wind velocity acting upon it is about 60 micrometers (μm), which is about the thickness of a human hair. Particles ranging from about 60 to 2,000 μm can also become suspended in air, but they only reach heights up to approximately three feet above the ground before they fall back to the surface. Particles larger than about 2,000 μm generally creep or roll along the surface due to wind velocity acting upon them [EPA 1996]. These larger particles of dust can affect the nasal passages, causing an irritated and congested nose, and might also cause an irritant cough should they deposit in the throat. Smaller airborne particles of dust, which can remain suspended in air for hours, pose a greater risk to the respiratory system when inhaled. In general, the smaller the aerodynamic diameter of the inhaled dust particle, the more likely it will be deposited more deeply in the respiratory tract.

Established sampling procedures measure various fractions of airborne dust according to aerodynamic diameter. For example, inhalable dust samplers collect dust that tends to deposit from the upper airways to the alveolar region of the lungs. Inhalable dust samplers collect approximately 97 percent of particles less than 1 μm aerodynamic diameter, but only about 50 percent of particles of 100 μm aerodynamic diameter [ACGIH 2007]. Respirable dust samplers more selectively collect dust that tends to deposit in the gas exchange region of the lungs. Respirable dust samplers collect about 50 percent of particles of 4 μm aerodynamic diameter. Their collection efficiency for other particle sizes ranges from about 1 percent of 10-μm particles to approximately 97 percent of particles less than 1 μm in diameter [ACGIH 2007]. Individually, such particles cannot be seen by the unaided eye.

The safety and health regulations for the U.S. mining industry are enforced by the U.S. Mine Safety and Health Administration (MSHA). The Federal Mine Safety and Health Act of 1977 created MSHA, giving it the authority to develop and revise improved mandatory health and safety standards to prevent injuries and protect the lives of those working in mines. These regulations are found in Chapter 1 of Title 30, Code of Federal Register (CFR).[1] Among the listed threshold limit values, this publication establishes a total airborne dust standard of 10 mg/m³ for those listed substances. However, if potential exposure to silica dust is suspected, a respirable dust sample is obtained, and x-ray diffraction following NIOSH Analytical Method 7500 [NIOSH 2003] is used to quantify crystalline silica content. The three most common forms (polymorphs) of respirable crystalline silica are quartz, tridymite, and cristobalite. If the sample contains greater than 1 percent quartz, a respirable dust standard is calculated with the following equation:

$$\text{Respirable dust standard} = \frac{10\,\text{mg/m}^3}{\%\,\text{respirable quartz} + 2}$$

Enforcement of a respirable standard based upon this equation effectively limits respirable quartz exposure to less than 100 µg/m³. NIOSH has a recommended exposure limit (REL) for respirable crystalline silica of 50 µg/m³, as a time-weighted average for up to a 10-hour day during a 40-hour week [NIOSH 1974]. The exposure limits for tridymite and cristobalite are limited to 50 percent of the value calculated from the formula for quartz.[1]

In consideration of the above standards, a critical aspect of the mining and processing of minerals is to determine when workers are exposed to unacceptable levels of respirable dust, and to implement and ensure the effectiveness of the controls described in this handbook as needed. To this end, the National Industrial Sand Association (NISA) recently published the second edition of the "Occupational Health Program for Exposure to Crystalline Silica in the Industrial Sand Industry" [NISA, 2010]. This document provides the necessary information for the proper surveillance and assessment of worker dust exposure.

SILICA EXPOSURE AND POTENTIAL HEALTH IMPACTS

As reported in "NIOSH Hazard Review: Health Effects of Occupational Exposure to Respirable Crystalline Silica," occupational exposure to respirable crystalline silica dust can have several adverse health consequences, including silicosis, tuberculosis, chronic bronchitis, emphysema, and chronic renal disease [NIOSH 2002]. In addition, NIOSH has classified crystalline silica as a potential occupational carcinogen [NIOSH 2002; 54 Fed. Reg. 2521 (1989)[2]].

Silicosis is the disease predominantly associated with crystalline silica exposure. Silicosis is an incurable and potentially fatal lung disease caused by the inhalation of respirable crystalline

---

[1] Subchapter K—Metal and Nonmetal Mine Safety and Health—details the safety and health standards for metal and nonmetal mines (Part 56 covers surface mines, while Part 57 covers underground mines). Section 56.5001 sets forth the exposure limits for airborne contaminants for surface metal and nonmetal mines, while 57.5001 sets forth the exposure limits for airborne contaminants for underground metal and nonmetal mines. Both sections state that "the exposure to airborne contaminants shall not exceed, on the basis of a time weighted average, the threshold limit values adopted by the American Conference of Governmental Industrial Hygienists, as set forth and explained in the 1973 edition of the Conference's publication entitled 'TLV's Threshold Limit Values for Chemical Substances in Workroom Air Adopted by ACGIH for 1973,' pages 1 through 54, which are hereby incorporated by reference and made part hereof" Title 30 CFR is available at the MSHA website at http://www.msha.gov/30cfr/CFRINTRO.HTM.
[2] Federal Register. See Fed. Reg. in references.

silica. Quartz is the most common single mineral in the earth's crust, and many mining operations involve direct contact with overburden and ore containing quartz. Thus, workers throughout much of the mining industry are potentially exposed to respirable crystalline silica through routine mining activities such as drilling, crushing, sizing, transporting, and loading.

When workers inhale respirable dust, the particles can penetrate the body's defense mechanisms and reach the alveolar region of the lungs. Crystalline silica particles that deposit in the alveolar region can stimulate an inflammatory and toxic process that can ultimately develop into clinically recognizable silicosis. Depending on the concentration of respirable crystalline silica and duration to which they are exposed, workers may develop any of several forms of silicosis [NIOSH 2002]:

- *chronic*—resulting from long-term excessive exposures, and first clinically apparent 10–30 years after first exposure;
- *accelerated*—resulting from exposure to higher concentrations of crystalline silica, first clinically apparent 5–10 years after the initial exposure;
- *acute*—resulting from exposure to unusually high concentrations of crystalline silica, clinically apparent within weeks to 5 years after the initial exposure.

Chronic silicosis, the most common form of the disease, results in characteristic nodular scarring in the lungs and occurs after many years of inhaling respirable crystalline silica dust. Over time, the initial small nodules can eventually coalesce into large fibrotic masses, a condition called progressive massive fibrosis. Accelerated silicosis is much less common than chronic silicosis but progresses more rapidly, and acute silicosis, a rarely occurring form of the disease, is the most serious and most rapidly fatal form. Unlike chronic and accelerated silicosis, where a chest x-ray examination typically reveals scattered discrete small (and possibly also large) opacities, the chest x-ray appearance of acute silicosis resembles that of a diffuse pneumonia. This appearance results from extensive damage to the lining of the air spaces of the lungs, causing the alveoli to become filled with an abnormal fluid containing protein, degenerating cells, and other materials [Davis 2002].

Not all workers diagnosed with silicosis will be symptomatic; some with chronic disease will have no notable symptoms despite characteristic abnormalities on their chest radiograph. However, many workers with chronic silicosis will develop symptoms over time and essentially all those with accelerated and acute forms of the disease will be symptomatic even before they are diagnosed. Though silicosis is incurable, various symptoms, including chest irritation with uncontrollable coughing and shortness of breath, can be debilitating and warrant therapeutic intervention. Additionally, those with silicosis are at substantial risk for developing tuberculosis or other mycobacterial diseases [NIOSH 2002; Davis 2002].[3]

This introduction has highlighted the need to control exposures to respirable silica dust. The control technologies discussed in this handbook for lowering dust levels below permissible or recommended occupational exposure limits are designed to control exposures not only to silica dust, but also to other types of dust. Also, while this handbook is focused on protecting workers

---

[3] Additional information on silica and silica-related diseases can be found at the National Institute for Occupational Safety and Health (NIOSH) website at http://www.cdc.gov/niosh/topics/silica/.

in the mining and mineral processing industry, the concepts and approaches presented can provide useful guidance for controlling airborne dust in other industries.

REFERENCES

ACGIH [2007]. 2007 Threshold limit values for chemical substances and physical agents and biological exposure indices. Cincinnati, OH: American Conference of Governmental Industrial Hygienists.

Davis GS [2002]. Silicosis. In: Hendrick DJ, Burge PS, Beckett WS, Churg A, eds. Occupational disorders of the lung: recognition, management, and prevention. W.B. Saunders, pp. 105–127.

EPA [1996]. Air quality criteria for particulate matter, Vol. 1. Research Triangle Park, NC: National Center for Environmental Assessment, Office of Research and Development, Environmental Protection Agency.

54 Fed. Reg. 2521 [1989]. Occupational Safety and Health Administration: air contaminants; final rule; silica, crystalline-quartz. (Codified at 29 CFR 1910.)

NIOSH [1974]. NIOSH criteria for a recommended standard: occupational exposure to crystalline silica. Cincinnati, OH: U.S. Department of Health, Education, and Welfare, Public Health Service, Center for Disease Control, National Institute for Occupational Safety and Health, DHEW (NIOSH) Publication No. 75–120.

NIOSH [2002]. NIOSH hazard review: health effects of occupational exposure to respirable crystalline silica. By Schulte PA, Rice FL, Key-Schwartz RJ, Bartley DL, Baron P, Schlecht PC, Gressel M, Echt AS. U.S. Department of Health and Human Services, Centers for Disease Control and Prevention, National Institute for Occupational Safety and Health, DHHS (NIOSH) Publication No. 2002–129.

NIOSH [2003]. NIOSH manual of analytical methods, 4th ed., 3rd supplement. Cincinnati, OH: U.S. Department of Health and Human Services, Centers for Disease Control and Prevention, National Institute for Occupational Safety and Health, DHHS (NIOSH) Publication No. 2003–154.

NISA [2010]. Occupational health program for exposure to crystalline silica in the industrial sand industry, 2nd ed. Washington, DC: National Industrial Sand Association.

CHAPTER 1: FUNDAMENTALS OF DUST COLLECTION SYSTEMS

CHAPTER 1: FUNDAMENTALS OF DUST COLLECTION SYSTEMS

Dust collection systems are the most widely used engineering control technique employed by mineral processing plants to control dust and lower workers' respirable dust exposure. A well-integrated dust collection system has multiple benefits, resulting in a dust-free environment that increases productivity and reclaims valuable product.

The most common dust control techniques at mineral processing plants utilize local exhaust ventilation systems (LEVs). These systems capture dust generated by various processes such as crushing, milling, screening, drying, bagging, and loading, and then transport this dust via ductwork to a dust collection filtering device. By capturing the dust at the source, it is prevented from becoming liberated into the processing plant and contaminating the breathing atmosphere of the workers.

LEV systems use a negative pressure exhaust ventilation technique to capture the dust before it escapes from the processing operation. Effective systems typically incorporate a capture device (enclosure, hood, chute, etc.) designed to maximize the collection potential.

As part of a dust collection system, LEVs possess a number of advantages:

- the ability to capture and eliminate very fine particles that are difficult to control using wet suppression techniques;
- the option of reintroducing the material captured back into the production process or discarding the material so that it is not a detriment later in the process; and
- consistent performance in cold weather conditions because of not being greatly impacted by low temperatures, as are wet suppression systems.

In addition, LEVs may be the only dust control option available for some operations whose product is hygroscopic or suffers serious consequences from even small percentages of moisture (e.g., clay or shale operations).

In most cases, dust is generated in obvious ways. Anytime an operation is transporting, refining, or processing a dry material, there is a great likelihood that dust will be generated. It also follows that once the dust is liberated into the plant environment, it produces a dust cloud that may threaten worker health. In addition, high dust levels can impede visibility and thus directly affect the safety of workers.

The five areas that typically produce dust that must be controlled are as follows:

1. The transfer points of conveying systems, where material falls while being transferred to another piece of equipment. Examples include the discharge of one belt conveyor to another belt conveyor, storage bin, or bucket elevator.
2. Specific processes such as crushing, drying, screening, mixing, blending, bag unloading, and truck or railcar loading.

3. Operations involving the displacement of air such as bag filling, palletizing, or pneumatic filling of silos.
4. Outdoor areas where potential dust sources are uncontrolled, such as core and blast hole drilling.
5. Outdoor areas such as haul roads, stockpiles, and miscellaneous unpaved areas where potential dust-generating material is disturbed by various mining-related activities and high-wind events.

While areas 4 and 5 can be significant sources of dust, they are generally not included in plant or mill ventilation systems design because of the vast area encompassed and the unpredictability of conditions. Therefore, dust control by methods alternative to LEVs is required, as discussed in later chapters of this handbook.

Dust control systems involve multiple engineering decisions, including the efficient use of available space, the length of duct runs, the ease of returning collected dust to the process, the necessary electrical requirements, and the selection of optimal filter and control equipment. Further, key decisions must be made about whether a centralized system or multiple systems are best for the circumstances. Critical engineering decisions involve defining the problem, selecting the best equipment for each job, and designing the best dust collection system for the particular needs of an operation.

This chapter will make a number of references to the industry handbook *Industrial Ventilation: A Manual of Recommended Practice for Design,* by the American Conference of Governmental Industrial Hygienists [ACGIH 2010], with several figures from that handbook adapted for use here. The ACGIH handbook should be considered as a primary resource for anyone interested in protecting workers from dust exposure in the mineral industry using dust collector systems, and especially for engineers who are involved in designing such systems. The material in this chapter will complement the information available in the ACGIH handbook.

BASICS OF DUST COLLECTOR SYSTEMS

Well-designed dust collection systems need to consider not only the dust as a potential contaminant, but also the attributes of the dust capturing system. In defining the nature of dust as a potential contaminant to workers, a number of issues must be examined. These include the particle size and distribution, shape, physical characteristics, and the amount of dust emitted. Particle size describes how coarse or fine particles are, and is normally defined by their upper and lower size limits. Particle sizes are measured in micrometers (μm) ($1/1,000^{th}$ millimeter). The respirable dust range harmful to workers' health is defined by those particles at, or below, the 10 μm size range. To put this size in perspective, 325 mesh is approximately 44 μm and is the smallest micrometer size that one can see with the unaided human eye. In dust collector systems, the larger particle sizes are easy to collect, often aided significantly by gravity.

The shape of particles affects how they are collected and how they are released from the collection media. *Particle shape* is a common terminology used in aerosol technology, while the term *aerodynamic diameter* is frequently used to describe particle diameters. The aerodynamic diameter of a particle is the diameter of a spherical particle that has a density of 1,000 kg/m^3 (the standard density of a water droplet) and the same settling velocity as the particle [Hinds 1999]. Aerodynamic diameter is used in many designs of filtration systems and air cleaners. Additional properties of the material that are key design considerations for dust collection systems are moisture and temperature. Moisture and temperature play a significant part in equipment selection for dust collector systems.

AIRFLOW AND DUST CONTROL

To control how air flows in a ventilation system, one must manage air velocities, air quantities, and temperature, as well as apply basic principles of static pressure (SP) and velocity pressure (VP).

Air velocity is measured in feet per minute and impacts the size of particle that can be carried by the airstream. *Air quantity* is measured in cubic feet of air per minute (cfm), which is the amount of air used in ventilating the process. Air temperature is measured in degrees Fahrenheit or degrees Celsius. It is used to determine the type of gaskets and filter media needed. Many applications where dust is being collected are thermal in nature, with examples including furnaces, kilns, and dryers.

Pressure (or head) in ventilation design is generally measured in inches of H_2O, also referred to as inches water gauge (wg). In a ventilation system, this pressure is known as the static pressure and is generally created by a fan. Static pressure is the difference between the pressure in the ductwork and the atmospheric pressure. Negative static pressure would want to collapse the walls of the duct, while positive static pressure would want to expand the walls of the duct.

Static pressure values are used to overcome the head loss (H_l) of the system, which is made up of two components: frictional resistance to airflow in the ductwork and fittings (frictional losses (H_f)) and the resistance of obstacles such as cyclones and dust collectors (shock losses (H_x)) [Hartman et al. 1997]. Static pressure is measured by inserting a pitot tube into the ductwork, perpendicular to the side walls, to determine the difference between atmospheric and duct pressures (Figure 1.1).

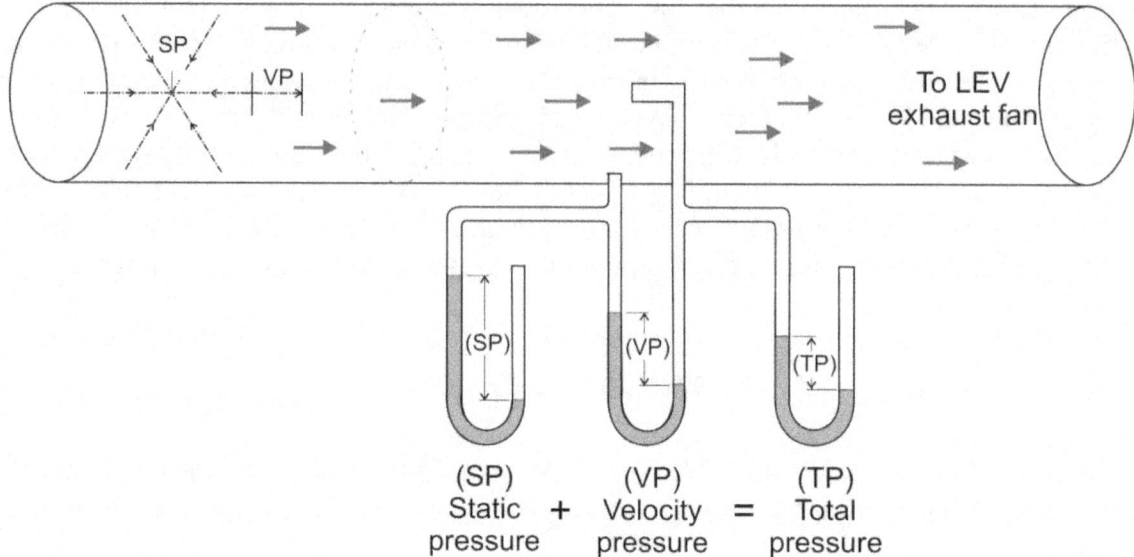

Figure 1.1. Relationship between static, velocity, and total pressure [adapted from ACGIH 2010].

Air traveling through a duct at a specific velocity will create a corresponding pressure known as the velocity pressure (VP). Velocity pressure is the pressure required to accelerate the air from rest to a particular velocity. It only exists when air is in motion, always acts in the direction of airflow, and always has a positive value. For ventilation purposes, VP is measured with a test probe facing directly into the airstream. The algebraic sum of static pressure and velocity pressure is total pressure (TP) [Hartman et al. 1997], as expressed by the following equation:

$$TP = SP + VP \tag{1.1}$$

where TP = total pressure, inches wg;
SP = static pressure, inches wg; and
VP = velocity pressure, inches wg.

The ACGIH handbook, *Industrial Ventilation: A Manual of Recommended Practice for Design*, [ACGIH 2010], gives a number of definitions and equations that are useful for describing airflow in an operation's ventilation system. The handbook also details fundamental characteristics in relation to blowing and exhausting air through a plant ventilation system. A handbook from Martin Engineering, *Foundations: The Practical Resource for Cleaner, Safer, More Productive Dust & Material Control* [Swinderman et al. 2009], also devotes a chapter to the control of air movement, including a section on effective measurement of air quantities. Finally, a recommended journal article is "Dust Control System Design: Knowing your Exhaust Airflow Limitations and Keeping Dust out of the System" [Johnson 2005].

EXHAUST SYSTEMS DESIGN

All exhaust systems, whether simple or complex, have in common the use of hoods, ductwork, and an air cleaning and collection device that leads to the exhaust fan (Figure 1.2). The ACGIH handbook, *Industrial Ventilation: A Manual of Recommended Practice for Design* [ACGIH 2010], discusses all aspects of these systems in great detail. To supplement the ACGIH handbook, the discussion below outlines some basic system design parameters and sets forth some important considerations about air velocity.

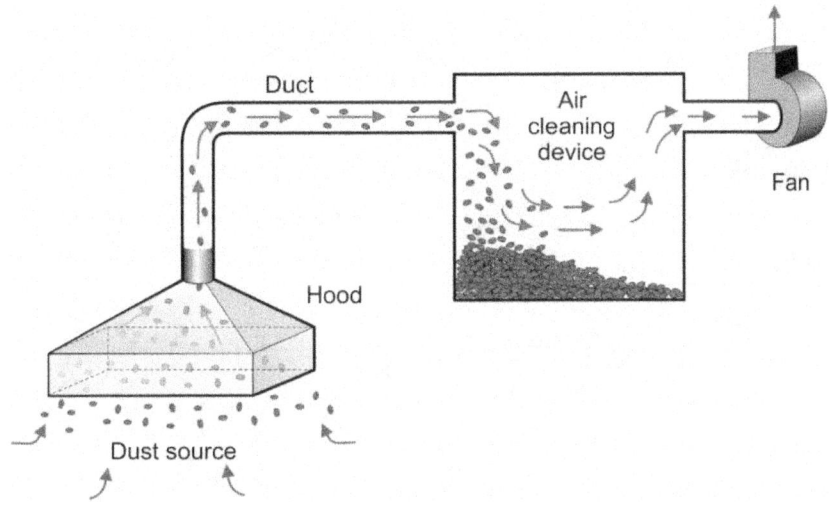

Figure 1.2. A basic depiction of a simple exhaust system with the major components being the hood, duct, air cleaning device, and fan.

HOODS

Hoods are specifically designed to meet the characteristics of the type of ore or product being processed. An effective hood is a critical part to any system because if the hood does not capture the dust, the rest of the exhaust ventilation system becomes meaningless. A properly designed hood will create an effective flow rate and airflow pattern to capture the dust and carry it into the ventilation system. The effectiveness of the hood is determined by its ability to induce an inward airflow pattern for the dust-laden air in the work environment.

Hoods and Blowing versus Exhausting Ventilation

When considering the effectiveness of a hood at capturing dust, the limitations of exhausting systems need to be considered. This issue is most evident when comparing the characteristics of blowing versus exhausting air from a duct. With a blowing system, the air delivered from the fan maintains its directional effect for a substantial distance once exiting the duct. With a blowing system, at a distance of 30 diameters (dimension of the exiting duct), the air velocity is reduced to approximately 10 percent of the exiting velocity (Figure 1.3). This blowing air tends to maintain its conical shape and actually entrains additional air, a process commonly referred to as induction. When one compares a blowing system to an exhaust system, the air velocity is at this approximate 10 percent level at only one duct diameter from the exhaust inlet.

The airflow characteristic for an exhaust system is substantially different. The air exhausted, or pulled into the duct, is captured from all directions around the duct opening and thus forms a nearly spherical shape, as opposed to the conical shape of the blowing system. Another major difference is the air velocity. The air velocity for an exhaust system is approximately 10 percent of the intake velocity at the duct opening at only 1 diameter away, as compared to 30 diameters away at the 10 percent level for the blowing system. These ventilation principles underscore how critical it is for an effective hood design to be very close to the dust generation source.

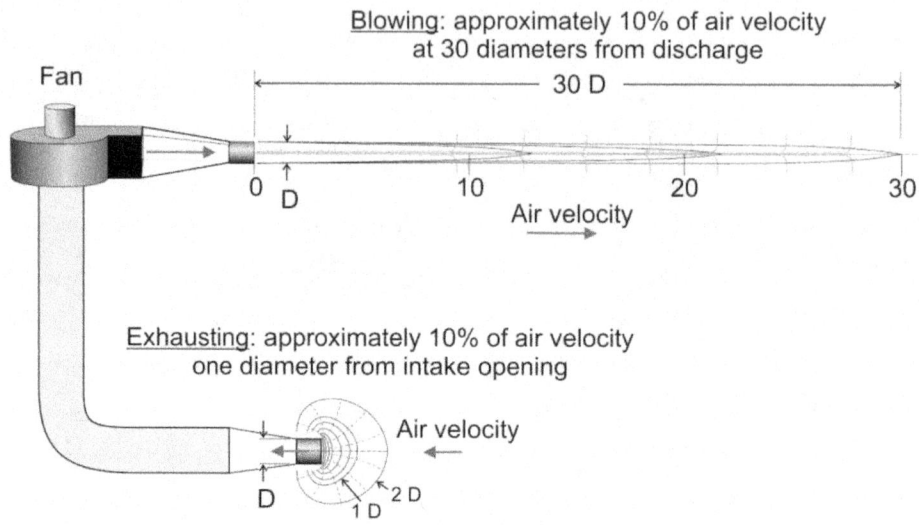

Figure 1.3. Comparison of ventilation characteristics for blowing versus exhausting system [adapted from Hartman et al. 1997].

Hood Types

Hoods have a vast range of different configurations, but usually fall into three different categories: *enclosing*, *capturing*, and *receiving*.

Enclosing hoods are those in which the source is either partially or totally enclosed to provide the required airflow to capture the dust and prevent it from contaminating the work environment. The most effective way to capture dust generated is a hood that encompasses the entire dust generation process. Openings into the enclosure (hood) are minimized with doors and access points into the contaminated work process. This situation is normally used when worker access is not necessary and openings are only necessary for the product to enter and exit a piece of machinery or a work process. These types of enclosing hoods can have numerous applications throughout the mining and minerals processing sequence, and are most often used in crushing, grinding, milling, and screening applications.

When access is necessary into the dust generation process or area, it is then common to use some type of booth or tunnel—a type of partial enclosure application. In these partial enclosure systems, the key is to provide sufficient intake airflow to eliminate, or at least minimize, any escape of dust from the enclosed area. This is best accomplished by enclosing the dust generation area or zone as much as possible. One common method to do this is with clear plastic stripping, which allows workers to have ingress and egress while maintaining an effective seal to the contaminated area. A partial booth or tunnel (hood) requires higher exhaust volumes to be effective than do totally enclosed systems.

When it is not applicable to either totally or partially enclose the dust generation source or area, *capturing hoods* are normally used and are located as near as possible to the dust source. Because the dust generation source is exterior to the hood, the ability of the hood to capture the dust-laden air is paramount to the success of the system. These types of hoods must be able to overcome any exterior air current around this area. They can be very effective when the dust is emitted in a specific area and the exhaust hood is placed in relatively close vicinity to this area. The capture velocity of the hood decreases inversely with the square of the distance from the hood. In cases where this distance becomes too great, one should consider the use of a push-pull ventilation system (Figure 1.4). In a push-pull ventilation system, a blowing jet of air provides a blast of air movement to provide the necessary quantity to overcome the distance from the hood. This air jet is normally directed across a contaminant source and towards the exhaust hood. As this jet travels towards the exhaust hood, this airflow entrains additional air with the intent to capture and move the dust-laden air. The goal is to move this total volume of air into the exhaust hood. This blowing jet coupled with an exhaust (capturing) hood provides a very effective ventilation design.

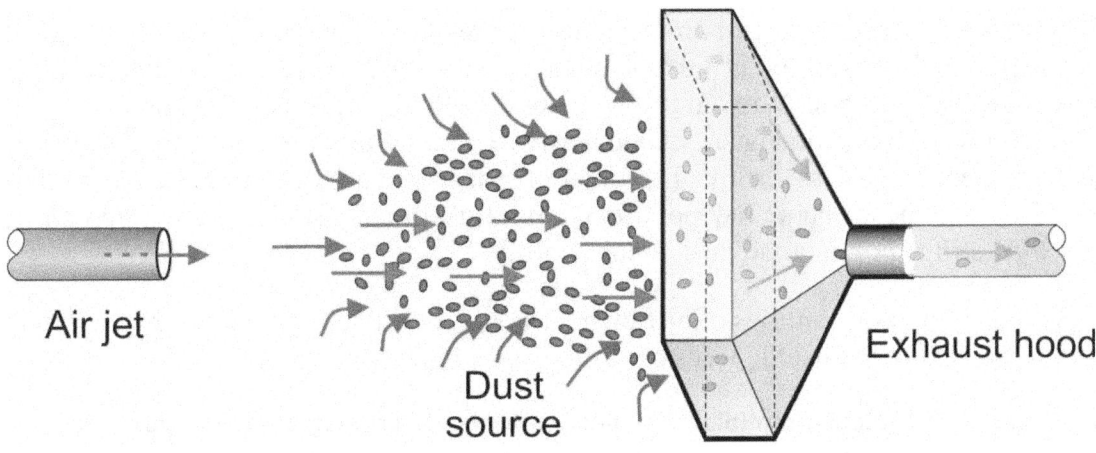

Figure 1.4. Simple design of basic push-pull ventilation system.

The third and most infrequently used type of hood is a *receiving hood* (Figure 1.5). Receiving hoods are normally located close to the point of generation to capture the dust and not allow it to escape. In most cases, these hoods are relatively small in size. The hood uses the directional inertia of the contaminant to lower the necessary capture velocity. These types of hoods have only minor applications in mining and mineral processing and are most common in small machinery and tool applications in laboratory and shop areas.

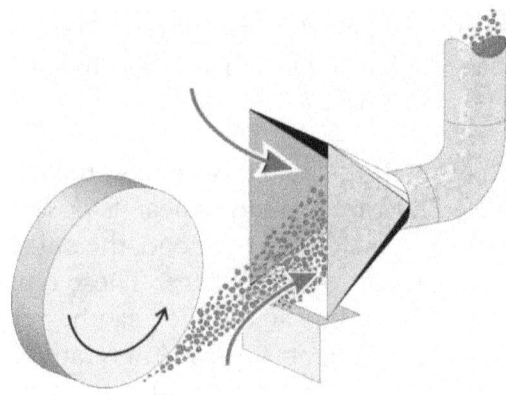

Figure 1.5. Basic setup for a type of receiving hood.

Hood Design

The most important parameters in the design of an exhaust hood are as follows:

1. The rate of airflow through the hood.
2. The location of the hood.
3. The shape of the hood.

Of these three parameters, the rate of airflow through the hood is the most important. As previously mentioned, if the hood is not able to capture the dust, the rest of the dust collector system becomes meaningless. Without an adequate air velocity, dust capture may not be sufficient. In order to maintain an acceptable negative internal pressure, new or "tightly" enclosed equipment needs less airflow than older or "loosely" enclosed equipment. Because of this, the airflow volume (in cubic feet per minute, or cfm) for similar pieces of equipment can vary widely yet still maintain good dust control ability.

As an approximation, the quantities of airflow in Table 1.1 will generally provide good dust control when applied to the listed pieces of equipment, properly hooded.

Table 1.1. Typical quantities of airflow in dust collection system components

Equipment	Airflow, cfm
Bucket elevator—sealed	400 at top and bottom
Bucket elevator—sealed	800 if top-only
Belt conveyor	1,000–1,500 per transfer point
Low-speed oscillating screens	300–500 hood
High-speed vibratory screens	Length of the hood seal x 250 cfm/ft
Screens	300–500 on each discharge chute
Loading spouts	800–1,200
Storage bins	300–400
Hoods, hoppers, and canopies	250 cfm/ft^2 of vertical curtain area around perimeter of unit

When using the air quantities in Table 1.1, it should be noted that feed rate into the bin (in tons per hour, or TPH) is to be taken into account when sizing hood air volume to allow for air that is entrained by incoming material.

Other equipment fitted with dust control hoods, such as baggers, packers, crushers, magnetic separators, palletizers, etc., will likely include a manufacturer's recommended airflow. In some instances, it may be difficult to install hoods on process equipment and maintain high collection efficiencies and easy access for operation and maintenance. In these rare cases, it may be necessary to build an enclosure for the entire piece of equipment, possibly by erecting a dust control hood above the equipment and surrounding it with flexible curtains.

There are two issues that need to be considered when determining the rate of airflow to a hood: *air induction* and *capture velocity*.

Air Induction

Air induction is based on the concept that material falling through air imparts momentum to the surrounding air (Figure 1.6). Due to this energy transfer, a stream of air always travels with the falling material. For example, a chute feeding sand to an elevator will drag air into the elevator. This air must be removed from the elevator through an exhaust hood or it will escape and carry dust with it into the plant through openings in the elevator casing, creating dust emissions.

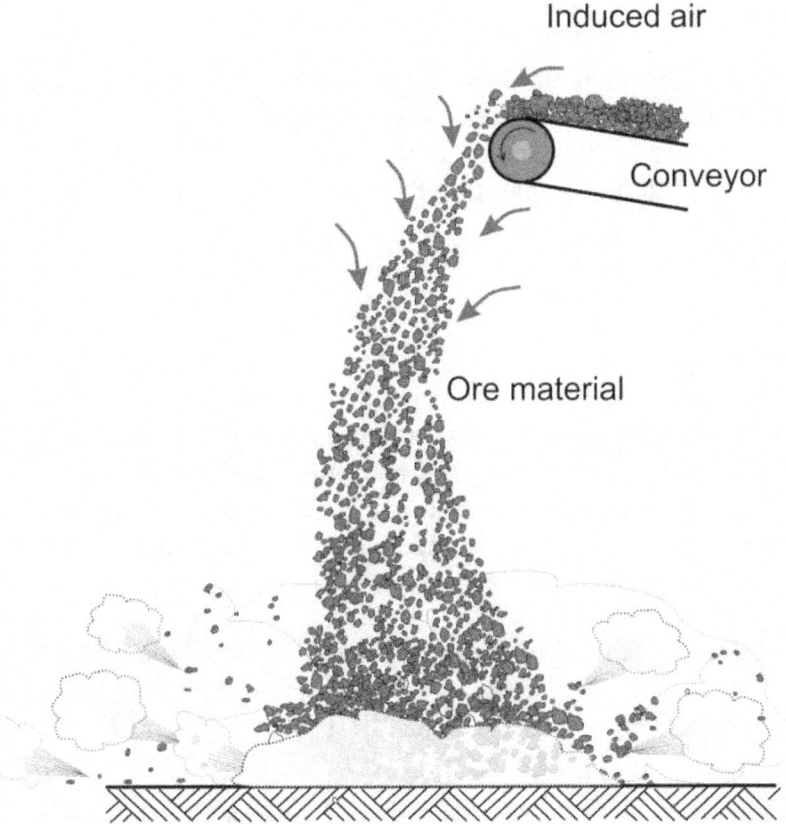

Figure 1.6. Demonstration of air induction as material falls from a conveyor.

The following air induction equation can be used to estimate exhaust volumes for hoods based on material feed rate, height of free fall, size, and feed open area:

$$Q = 10 \times A_U \sqrt[3]{\frac{RS^2}{D}} \qquad (1.2)$$

where Q = air quantity, cubic feet per minute;
A_U = enclosure upstream open area, square feet;
R = rate of material flow, tons per hour;
S = height of material fall, feet; and
D = average material size, feet.

The most important parameter in the air induction equation is A_U, the opening through which the air induction occurs (i.e., the cross-sectional area of the feed inlet). The tighter the feed enclosure, the smaller the value for A_U and the smaller the exhaust quantity required. Also, the lower the value for S the smaller the exhaust quantity required. In designing a material handling circuit, it is important to keep the values for A_U and S as low as possible to prevent excessive dust generation and to reduce the quantity of ventilation air. The values for R and D also affect exhaust quantity requirements, but are typically a constant in the mining or processing operation and cannot be altered.

The air quantity (Q) for exhaust hoods can also be estimated using the air induction approach. This concept is important because it relates various factors that affect the required air volumes, which may not be accounted for in standard tables or charts. The airflow required using air induction calculations should be compared to the published standards, with corrections made if deemed necessary.

Capture Velocity

Capture velocity is a measure of the required airflow necessary to seize the dust released at the source and then pull this dust into the exhaust hood. The capture velocity must be powerful enough to overcome all the opposing factors and air currents in the surrounding area. There are various tables available that provide a range of recommended air velocities under a variety of conditions. For mining and minerals processing, this range is normally between 100 and 200 fpm, but can increase up to 500 fpm in special cases. Most operations establish their own capture velocities for their exhaust hoods based on years of experience.

After this capture velocity is determined, the exhaust volume for the hood can be calculated. The following "DallaValle" equation is used to determine the exhaust volume needed for a basic free-standing hood arrangement (Figure 1.7) [DallaValle 1932; Fletcher 1977]:

$$Q = V_x(10X^2 + A_h) \qquad (1.3)$$

where Q = the rate of air exhausted, cubic feet per minute;
V_x = the required air velocity at the most remote point of contaminant dispersion, feet per minute;

X = distance in feet from the face of the hood to the most remote point of contaminant dispersion; and
A_h = the area of hood opening, square feet.

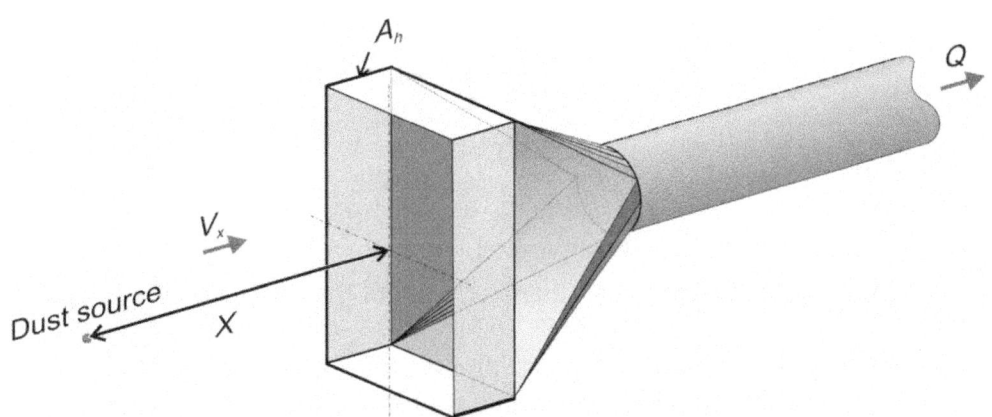

Figure 1.7. Hood entry loss calculation [adapted from ACGIH 2010].

From this equation, it becomes obvious that the air velocity, the size of the hood, and the distance from the hood to the dust source are all critical factors for the required air volume. The distance (X) from the dust source to the hood is extremely important because it is a squared relationship.

Another issue that should be noted is that for a free-standing hood, air is also being pulled from behind the hood. This lessens the hood's ability to capture and pull the dust-laden air in the source area. In order to minimize this effect, there are several approaches that need to be considered. First, if the hood is positioned on a tabletop, the airflow requirement is reduced to the following equation:

$$Q = V_x(5X^2 + A_h) \qquad (1.4)$$

Another simple technique to improve the airflow from around a hood is to place a flange around it. By doing so, the equation becomes:

$$Q = 0.75V_x(10X^2 + A_h) \qquad (1.5)$$

A flange provides a barrier that prevents unwanted air from being drawn from behind the hood and is a very simple design modification to improve the effectiveness of the hood, as well as reducing operating costs. Numerous other factors and considerations for capture velocities can be found in the ACGIH handbook, *Industrial Ventilation: A Manual of Recommended Practice for Design* [ACGIH 2010].

Other Hood Considerations

When dust is captured and pulled into a hood from a dust source, the hood converts static pressure to velocity pressure and hood entry losses. Hood entry loss is calculated according to the following equation:

$$H_e = (K)(VP) = |SP_h| = VP \tag{1.6}$$

where H_e = hood entry loss, inches wg;
K = loss coefficient;
VP = velocity pressure in the duct, inches wg; and
SP_h = absolute static pressure about 5 duct diameters down the duct from the hood, inches wg.

Air flowing through an exhaust hood will cause pressure changes that need to be calculated when evaluating the impact on each individual hood in a multiple hood system. Figure 1.8 provides the hood entry loss coefficients for three different types of hoods commonly used in mining and mineral processing facilities. The first case shows three different hood types: circular, square, and rectangular with plain openings. The second case shows loss coefficients with flanged openings, and the last case shows a bell mouth inlet for just a circular duct. This figure demonstrates the significant improvement in the design, and thus the lowering of the hood entry loss coefficient, with each improvement in the hood type.

Hood type	Description	Hood entry loss coefficient (K)
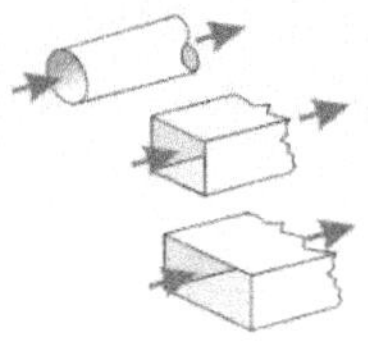	Plain opening	0.93
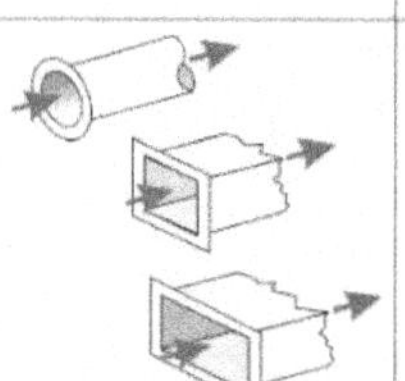	Flanged opening	0.49
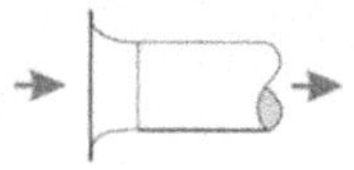	Bell mouth inlet	0.04

Figure 1.8. Hood entry loss coefficients for different hood types [adapted from ACGIH 2010].

Checklist for Hood Effectiveness

The following is a checklist of effective practices or considerations regarding the use of hoods in exhaust ventilation systems.

- The most effective hood design is one that encompasses the entire dust generation process. This virtually eliminates dust escaping and contaminating mine/plant air and the exposure of workers.
- Openings/doors/access points into the hood should be minimized as much as is reasonably possible. When access is necessary, a partial enclosure application, normally referred to as a booth or tunnel, is then recommended. When access points are not able to be closed, sealing these areas with clear plastic stripping is a common and effective technique.
- When neither total nor partial hood enclosures are possible, capture hoods should be used. The hoods need to be located as close to the dust source as feasibly possible. Remember that the distance component is a squared relationship that affects the required hood air volume.
- Hood capture velocities must be able to overcome any exterior air current between the dust source and the hood.
- The use of a push-pull ventilation system should be considered when the distance between the dust source and the hood becomes too great, or when there are significant exterior air currents in the area.
- It is critical that makeup air drawn past a worker before entering an exhaust hood be essentially dust-free. If the air is drawn from a contaminated area, the potential exists to increase a worker's respirable dust exposure. Dust-laden air from a dust source should never be pulled through a worker's breathing zone as it is being drawn into an exhaust hood.
- Flanges on hoods should be used because they significantly improve the airflow from the front of the hood area, which should be directed towards the dust source. In cases where bell shaped hoods and ducts can be used, they provide the optimum design.
- All hoods should be designed to meet the criteria established by the ACGIH handbook, *Industrial Ventilation: A Manual of Recommended Practice for Design* [ACGIH 2010].

DUCTWORK AND AIR VELOCITIES

There are three basic types of systems used to transport dust to the collector: high-velocity, low-velocity, and modified low-velocity. Most familiar to industry is the high-velocity system, where air-carrying velocities are in the range of 3,000–4,500 fpm. Fundamentally, this is a dust *collection* system, while a low-velocity system can best be described as a dust *containment* system, with duct velocities always less than 1,800 fpm. Low-velocity systems are designed so that they will not transport nonrespirable size dust particles (generally particles over 10 µm). It is important to keep in mind that low-velocity transport does not imply low airflow. Capture velocities (the air velocity at the hood opening required to capture the contaminant) and hood air volumes are the same in all designs, be they high-velocity, low-velocity, or modified low-velocity systems. Figure 1.9 depicts the basic difference between high- and low-velocity systems.

The implication of the figure is that, in a high-velocity system, the dust will be carried to the collector regardless of the slope or angle of the duct. However, in a low-velocity system, it is necessary to slope the duct to an inlet hood or discharge point, because some of the larger particles, inadvertently collected, will settle out in transport, and must be removed.

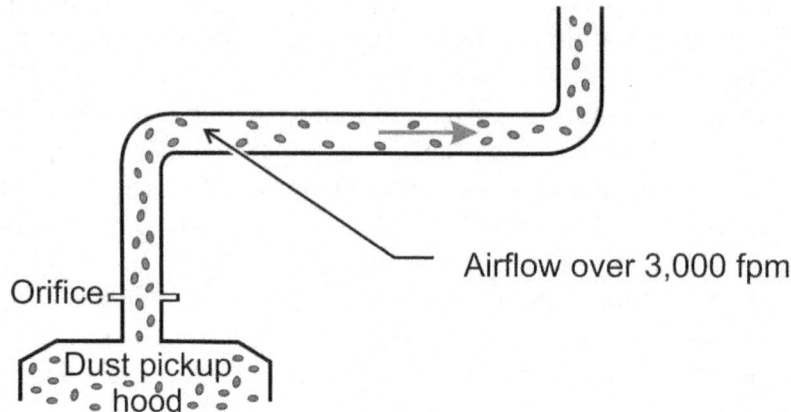

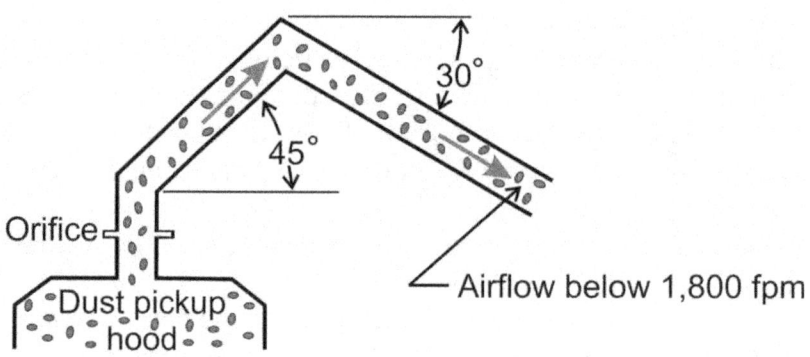

Figure 1.9. Schematics of high- and low-velocity systems.

High-Velocity Systems

The high-velocity system is characterized by its ability to carry dust-laden air (particles larger than 10 µm) from the entry points of the system to the dust collector without having the particles settle out in the duct. To keep the particles from settling, high transport velocities are required.

Because the ducting in a high-velocity system can be run both horizontally and vertically, there are few engineering restrictions. Effective ducting layouts can be easily designed, normally with a central horizontal duct and smaller collection ducts branching off to the dust-producing equipment/hood.

A major disadvantage of the high-velocity system is that the ductwork is subjected to highly abrasive "blasting" by the dust particles moving at a high rate of speed, especially when the air changes direction. Elbows and branch entries are particular areas subjected to high wear. Abrasion first shows in these areas of air direction change. This wear or abrasion, if not addressed through additional engineering design or maintenance, will result in the long-term deterioration of the collection system due to formation of holes or openings in the ductwork. Because of this abrasive wear, most ductwork and fittings have to be fabricated from heavier materials and the fittings require longer radii. The high wear rate also results in high maintenance labor and material costs. Any system with holes in the ducting loses its effectiveness at the pickup point, and even a single hole in the duct can have a strong impact on the system. Because high-velocity systems are prone to required maintenance, they may not always be operating at optimum efficiency.

High humidity, coupled with inadequate airflow, leads to further wear problems on systems with ducts that are run horizontally. Upon shutdown, the dust in the airstream settles in the horizontal runs of duct. As the air cools to the dew point, moisture can form on the surface of the duct, causing some attachment of dust particles to the wall. Over time, this buildup robs the system of airflow due to increased friction losses in the system and increases in velocity through the narrower opening. The loss of airflow in the system reduces collection efficiency, and the resulting increased velocity increases wear and can cause an even more rapid deterioration to the system than would normally be expected.

Finally, high velocity means high pressure drops throughout the system. This equates to increased horsepower and higher power consumption, raising overall operating costs. Also, although the initial installation costs for high-velocity systems are comparatively low, higher maintenance and operating costs lead to high overall lifetime costs of the system.

Low-Velocity Systems

The basis of the low-velocity collection system is to create the same negative static pressure in the area surrounding the dust source and maintain the same airflow into the collection hood, as in a high-velocity system. The difference between the systems is that, after collection, the air transport velocity in the low-velocity system is much lower. Thus the pipe is always sloped to allow the oversized particles to slide to a discharge point for easy removal. Transport velocity is designed to move only the particles in the respirable size range, generally below 10 μm. Respirable dust is carried through the ductwork by the low-velocity airflow, while heavier particles fall out into the ductwork and slide back into the process.

Ductwork in the low-velocity system cannot be run horizontally. The ductwork is designed so that the larger particles that fall from the airstream are reintegrated into the process. Therefore, a sloped sawtooth design (Figure 1.10) is used instead of long horizontal runs. In order to control the airflow, a fixed orifice plate or blast gate is positioned in the duct segment, preferably on a downward run, so as not to trap material above the high-velocity flow through the orifice.

Figure 1.10. Sawtooth design of a low-velocity system with a dropout duct back into the process at the base of each leg.

The particle-carrying ability of a low-velocity system is confined to particles smaller than 10 μm. Because of the absence of larger particles, as well as the lower velocities involved, abrasion is low even at points of air direction change. This allows for the use of short radius or mitered elbows without fear of extreme wear.

From a design standpoint, dust control systems are engineered for expected pressure losses within the system. These losses are due to air friction and pressure losses across the dust collection unit. By reducing the air velocity, frictional losses in the ductwork and fittings are reduced, lowering overall power requirements.

Because of the behavior of low-velocity air, moderate orifice size changes can be made without completely upsetting the system. Even opening a branch completely will not drastically change the airflow in other branches because the pressure drop change is minimal. Should one branch of the network fail or change airflows dramatically, the system tends to stay in balance and other branches do not lose effectiveness. Therefore, the airflow can vary substantially on two identical pieces of equipment while maintaining excellent dust containment. Thus, overall losses through the dust collection system are lower.

The disadvantages of low-velocity systems are the higher initial cost and more complex design. In these systems, the ability to pass a certain volume of air (cfm) through the duct at a predetermined velocity (fpm) dictates the diameter of the duct, requiring larger diameter ductwork. Also, because ductwork cannot be run horizontally, a sawtooth design is used, which adds expense and complicates installation.

A common misconception when designing a low-velocity system is that an exhaust hood is not required due to the low-velocity ducting design. To the contrary, a properly designed exhaust hood on a low-velocity system is still a requirement for a good dust containment system. A properly designed hood will help maintain lower system pressure losses by minimizing the shock loss that occurs from the airflow entering the system. Hood design is essentially the same for both high- and low-velocity systems. However, all exhaust hoods should be located away from the point of material impact to lower the likelihood of material being drawn into the hood. The principle of low-velocity dust containment is to exhaust air from a well-enclosed area. This will impart a negative inward pressure, trapping fugitive dust within the enclosure. The shape of the exhaust hood is also important for two reasons:

1. A well-designed hood has a lower resistance to air movement.
2. A well-designed hood prevents material from being vacuumed into the dust containment system. Even with low-velocity ducting, coarse material can still be drawn into the system. It is therefore important to have a hood with a face velocity of 200 to 300 fpm or less in order to avoid entraining saleable product.

Finally, the design and layout for a low-velocity system requires more engineering. Because of the constraint that runs cannot be made horizontally, it is sometimes difficult to physically find room for the required sawtooth design of the ducting. However, a properly designed and balanced low-velocity system provides a virtually nonplugging and low-maintenance dust control system. More specific engineering principles for low-velocity systems are detailed below.

Airflow Target of 1,800 fpm

Because of the nature of a low-velocity system, it is not necessary to maintain a specific air velocity within the ductwork. However, the desired target of 1,800 fpm should move most of the respirable dust particles (minus 10 µm) through the system to the filter equipment while containing it within the confines of the individual pieces of dust-producing equipment/hood. The basic goal of low-velocity dust control is to minimize the amount of contaminants (respirable dust) that escape the system, thus becoming a hazard or nuisance. The 1,800 fpm target maintains a slightly negative pressure within the piece of equipment, while providing adequate airflow to discourage dust particles from escaping the unit. This ensures that any airflow at the equipment is moving inwards rather than outwards.

Avoidance of Horizontal Ductwork

As a result of the system's low-velocity airflow, it is imperative that ductwork is not run horizontally. This is because the heavier dust particles drawn into the airstream may drop out and attach themselves to the sides of these ducts. The net result is an eventual narrowing of the duct opening, leading to an increase in velocity and ductwork erosion. Therefore, ductwork should be designed for a minimum upflow angle of 45 degrees and a minimum downflow angle of 30 degrees (Figure 1.11). This allows any particles drawn into the airstream to slide back to their source. This design also requires that a dropout point be located at the low point of all duct runs to allow these particles to fall out, as shown in Figure 1.10.

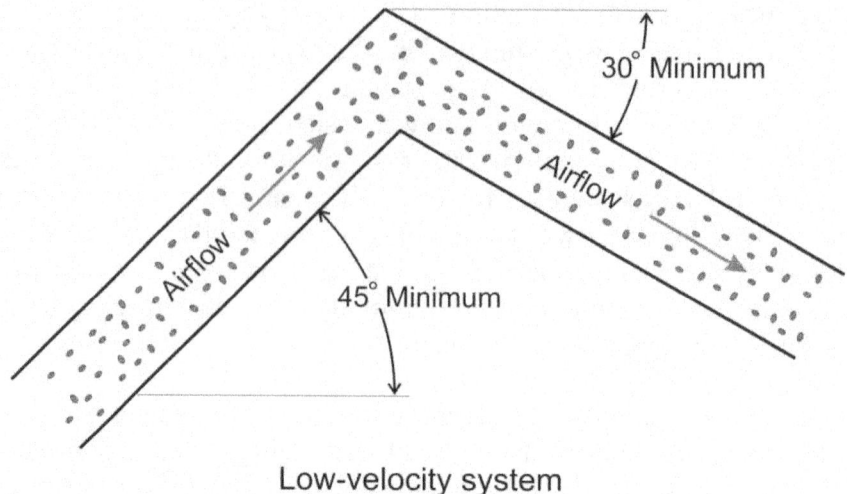

Figure 1.11. Duct design angles for low-velocity systems.

The Use of Main Duct Trunk Lines

From a design standpoint in low-velocity systems, consideration should be given to using one or more main duct trunk lines, running from the air filter to a centrally located proposed dust control point in the plant. This allows for smaller branch duct lines to emanate from the main duct trunk line and enables easy connection to individual pieces of equipment. These trunk lines can be either rectangular or round. In areas where space is a problem, rectangular duct cross sections that are shallow in depth and wide are easier to install. Many times these trunk lines can be run vertically up the side of an existing column or bucket elevator. As previously noted, a dropout point—possibly the bucket elevator—must be provided at the low point of all duct runs.

Minimizing Field Fits and Welds

New ductwork and fittings should be shop fabricated and be as complete as possible before being sent to the field for erection. Welding or cutting duct joints in the field are costly, and proper engineering design can minimize the welding required. Extra time spent at the design stage preplanning and prefabricating hoods, fittings, and duct segments, pays dividends later during installation.

Frequently, field fits must be made, and these can be anticipated before the material is sent to the field. It is best to make field cuts and welds at the flange point of the ductwork. In these instances the ductwork can intentionally be shop fabricated to lengths longer than necessary. Also, a rolled angle ring can be shop installed at the end of the duct, then tack welded. This keeps the ductwork from becoming distorted during handling and transit, and the tack weld can be cut in the field, the duct shortened, and the flange moved, rotated, and welded as needed.

Flanged connections are an efficient and easy means to install ductwork in the field. Rolled angle ring flanges are readily available, inexpensive, and easy to install. These flanges should normally be used on all shop fabricated fittings, hoods, and duct segments. These types of connections also help eliminate costly welding requirements and allow for ease of replacement sections should maintenance be required.

Avoiding Mitered Elbows Greater Than 90 Degrees

As with high-velocity systems, pressure losses increase rapidly in ductwork with abrupt directional changes greater than 90 degrees. Since duct pressure losses are a squared function of the velocity, the low-velocity system still results in lower pressure losses. However, a segmented elbow can be used in areas of unusually abrupt air directional change if maintaining low pressure losses in the system is critical.

Sizing and Locating Orifice Plates

In many complex ventilation systems, orifice plates are used to balance airflows by changing the pressure requirement in a selected section. It is recommended that an orifice plate, sized for a minimum pressure loss of 2 inches, be installed in each individual branch line. This allows for easy expansion of the system in the future. As additional branch lines are added to the main trunk line, the orifice plate can be resized for a larger, and therefore less restrictive, orifice opening.

Figure 1.12 demonstrates inappropriate locations for orifice openings. Orifice plates should be installed a minimum of 4 to 5 duct diameters upstream and 4 diameters downstream from air directional changes. This permits the airflow to become closer to laminar and to reestablish normal velocity. If an upstream orifice plate is placed too near a downstream air directional change, the higher velocities encountered through the orifice opening will produce a sandblasting effect on the far-side wall. Orifice plates should be installed in the upflow branch leg whenever possible. This helps avoid a buildup of heavier dust particles above the orifice plate, as could happen in the upflow branch line installation, where particles are unable to drop out through the higher velocities of the orifice plate.

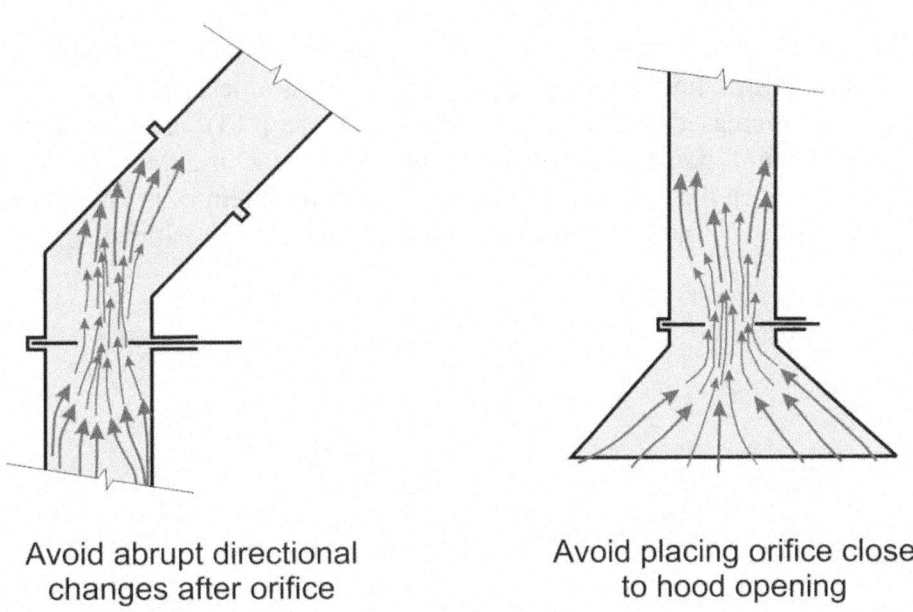

Avoid abrupt directional changes after orifice

Avoid placing orifice close to hood opening

Figure 1.12. Demonstration of poor orifice plate placement.

Orifice plates should also not be installed near hood inlet openings. The higher velocities near the orifice tend to capture unwanted material. Hood capture velocities below 500 fpm should be maintained. On fine-grind materials, a target capture velocity of 200 fpm is desirable.

Minimizing the Use of Flexible Hoses

Flexible hoses should not, as a rule, be used as part of the ductwork in low-velocity systems. A failure could compromise the dust control ability of the branch lines in the general area. Flexible hoses also increase the frictional losses in the system. If flexible hoses are used, they should be placed between the orifice plate and the inlet hood where their failure would have the least impact.

Modified Low-Velocity (MLV) System

The objective of the modified low-velocity (MLV) system is to combine the advantages of the high- and low-velocity systems, while avoiding the disadvantages of both. The key to the success of the MLV resides at the pickup hood. Here the first duct run must be vertical for at least 6 feet, and preferably 8 to 10 feet. The velocity in this first vertical run is extremely low, even compared to the standard low-velocity system (i.e. 1,000 to 1,200 fpm). As in other systems, dust is contained, but only the smaller micron particles are transported. The 6- to 10-foot vertical rise is necessary to reduce turbulence, to achieve a smooth laminar flow, and to optimize elutriation (separation of lighter and heavier particles) at the first bend.

Because there are minimal particles in the airstream after this first vertical rise, the need to maintain high velocities in subsequent horizontal runs is not necessary. Similarly, there is no need for an overall sawtooth design or mid-run discharge points. Velocity in the first horizontal run following a 1,200-fpm vertical duct should be about 2,000 fpm.

Another important design consideration with MLV systems is that as additional lateral lines are connected to the main trunk line, the exit velocity of the trunk line after the initial lateral connection should be increased by about 100 fpm (see Figure 1.13). Even in long runs with many lateral lines, it is not likely that the line velocity upon reaching the collector will exceed 2,500 to 3,000 fpm. Given the low pressure drops in such a system, it is not necessary to construct expensive elbows, mitered bends, or laterals. Instead, box elbows and fittings can be used.

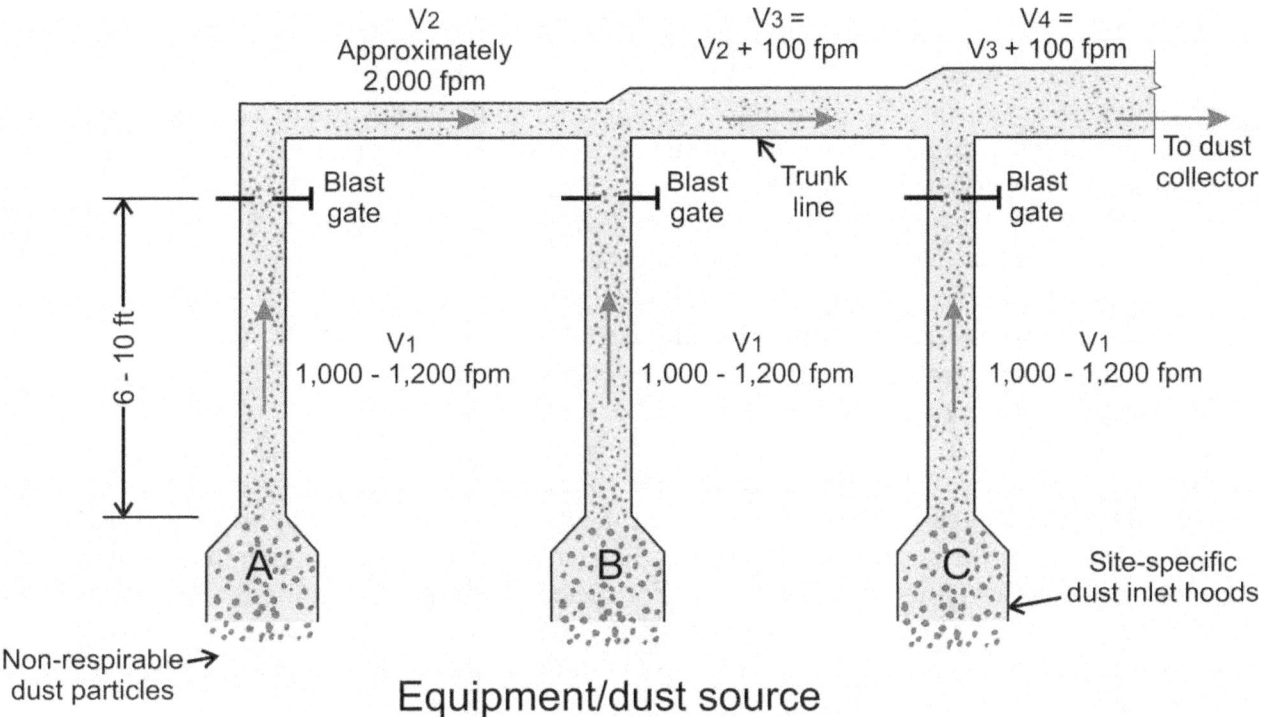

Figure 1.13. Depiction of the horizontal and vertical velocity relationship in a modified low-velocity system.

The MLV system contains all the advantages of the low-velocity system relative to low abrasion, reduced maintenance, reduced power, and stability of the system balance. At the same time, it does not require the space-consuming sawtooth installation. Duct size (diameter) is a compromise between the high- and low-velocity systems, since velocities in all trunk lines will be somewhere between the two. Another advantage of the MLV system is that it can readily be used in extensions to, or modifications of, existing high-velocity systems. Only under certain conditions can it be used in an extension of a low-velocity system. One would need to compare velocities at the connecting points, recognizing that it is not feasible to move from a higher to a lower velocity within a duct. A low-velocity extension can, however, be made to a MLV system.

The basic design tips for low-velocity systems as detailed previously can be applied to MLV systems, with a few exceptions:

1. The reference to avoiding horizontal ducts does not apply.
2. Orifice plates or blast gates should be installed at the top of the low-velocity vertical leg, not the bottom. Plates/gates should be as near to the higher-velocity leg as possible.
3. Vertical leg velocity should be 1,000 to 1,200 fpm.
4. Horizontal leg velocity should begin at a minimum of at least 2,000 fpm.

AIR CLEANING DEVICES

Air cleaning devices used within the industrial minerals mining industry are used to clean ventilation airstreams of harmful particulate matter. The choice of air cleaner for any particular installation will depend on the following:

- dust concentrations and dust characteristics,
- particle size,
- efficiency of particulate removal required,
- airstream temperature, airstream moisture content, and
- methods of disposal.

Distinguishing dust characteristics that affect the collection process include abrasive, explosive, sticky or tacky, and light or fluffy. The shape of the dust particle is also important because it factors into whether the particles are agglomerating (irregular) or nonagglomerating (spherical), which is important when using a filter cloth. For collection purposes, agglomerating particles are ideal as they allow dust cakes to build up easily on the filter cloth, allowing for more efficient collection at the dust collector. However, agglomerating particles may have a tendency to not release from the filter cloth very easily.

The types of dust control equipment used for air cleaning range from very crude gravity separators to more sophisticated electrostatic precipitators. The following is a list of the types of collectors used for particulate removal.

1. Gravity separators (drop-out boxes).
2. Centrifugal collectors or cyclones.
3. Baghouse collectors.
4. Cartridge collectors.
5. Wet scrubbers.
6. Electrostatic precipitators (ESPs).

A brief overview of each type of collector, along with its advantages and disadvantages, follows.

Gravity Separators (Drop-Out Boxes)

Gravity separators (also called drop-out boxes) are large chambers where the velocity of the airstream is drastically reduced in order to facilitate the vertical drop of particles. The separator works by not only slowing down the air but by changing its direction as well. Airflow enters horizontally and is immediately directed vertically downward by a target plate (Figure 1.14). As the air slows and moves downward, gravity takes over the large particles and drops them out of the airstream. Finer particles not affected by this will continue to flow in the airstream and exit the separator.

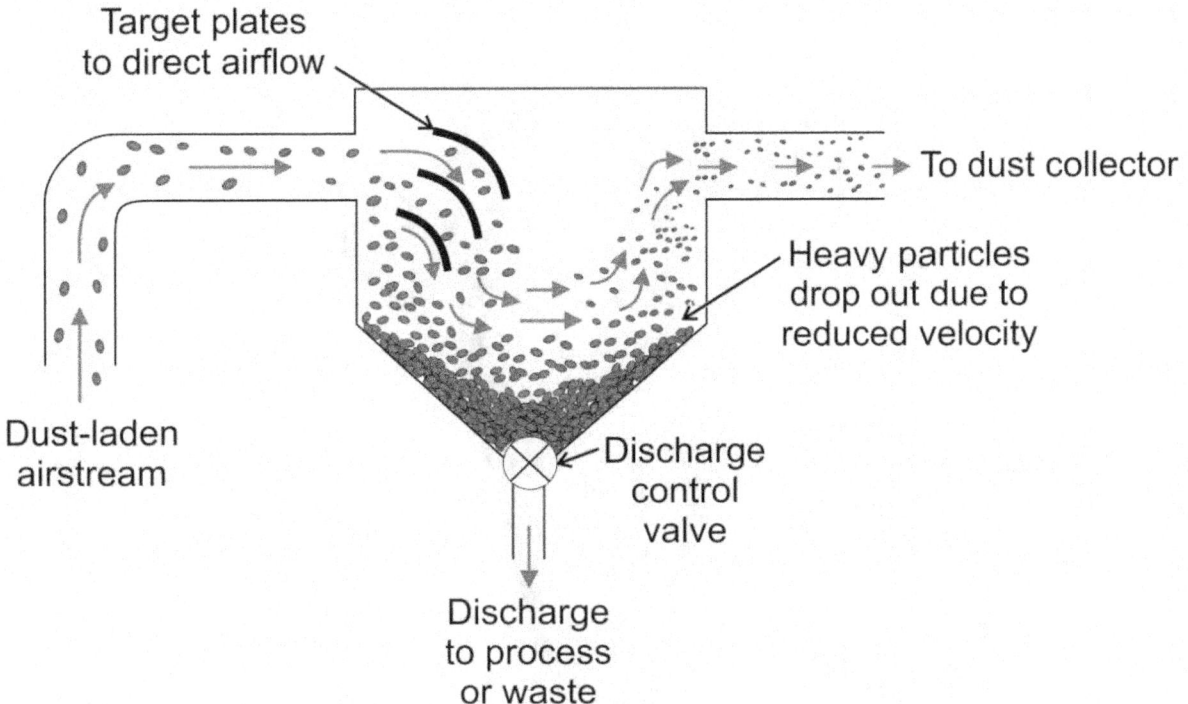

Figure 1.14. Typical design of gravity separator (drop-out box).

The benefits of using gravity separators are that they require little maintenance and they reduce the load on the primary dust collector. However, they also take up significant plant space and have a low collection efficiency.

Centrifugal Collectors or Cyclones

Cyclones are a dust collection device that separates particulate from the air by centrifugal force. The cyclone works by forcing the incoming airstream to spin in a vortex. As the airstream is forced to change direction, the inertia of the particulates causes them to continue in the original direction and to be separated from the airstream (Figure 1.15). Although, the cyclone is simple in appearance and operation, the interactions inside a cyclone are complex. A simple way to explain the action taking place inside a cyclone is that there are two vortices that are created during operation. The main vortex spirals downward and carries the coarser particles. An inner vortex, created near the bottom of the cyclone, spirals upward and carries finer dust particles.

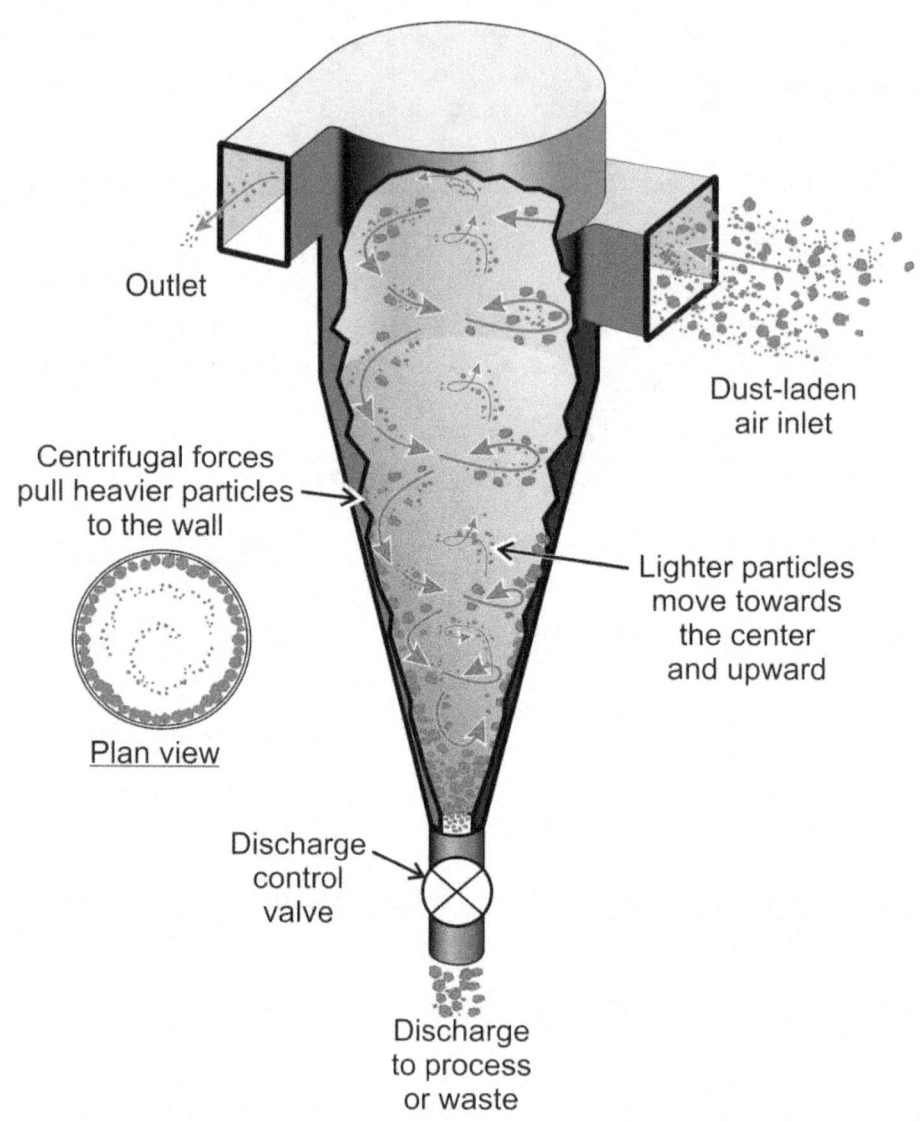

Figure 1.15. Typical design of a cyclone dust collector.

Cyclones are cost-effective and low-maintenance devices, and they can handle high temperatures. They also reduce loading on the primary collector and allow for the dry recovery of product. However, it is difficult to predict the performance of cyclones and they pose particular design challenges. Accurate inlet data are necessary and they require significant plant space.

Cyclones have low efficiencies in removing fine particulate. They are typically used as a precleaner to remove coarser particles that could otherwise damage the bags in fabric collectors or plug wet scrubbers. It should be noted that adding a cyclone to a ventilation system may not reduce the overall system resistance because the drop in resistance at the baghouse, due to lower dust loading, may be more than offset by the pressure drop of the inertial cyclone collector. Pressure drops range from 3 inches wg for low efficiency inertial cyclone collectors and up to 8 inches wg for higher efficiency models.

30 Fundamentals of Dust Collection Systems

Baghouse Collectors

Baghouse dust collectors capture the particulate in an airstream by forcing the airflow through filter bags. A baghouse works by taking the inlet dust-laden air and initially reducing the velocity to drop out larger particles, then filtering the remainder of the particles by passing the air through a fabric bag (Figure 1.16). Separation occurs by the particles colliding and attaching to the filter fabric and subsequently building upon themselves, creating a dust cake. Since the dust has been deposited on the outside of the bag, when the dust cake is removed from the bag or cleaned, it falls by gravity into the collection hopper located below the bag section. Collected dust is then removed from the collector through a hopper valve.

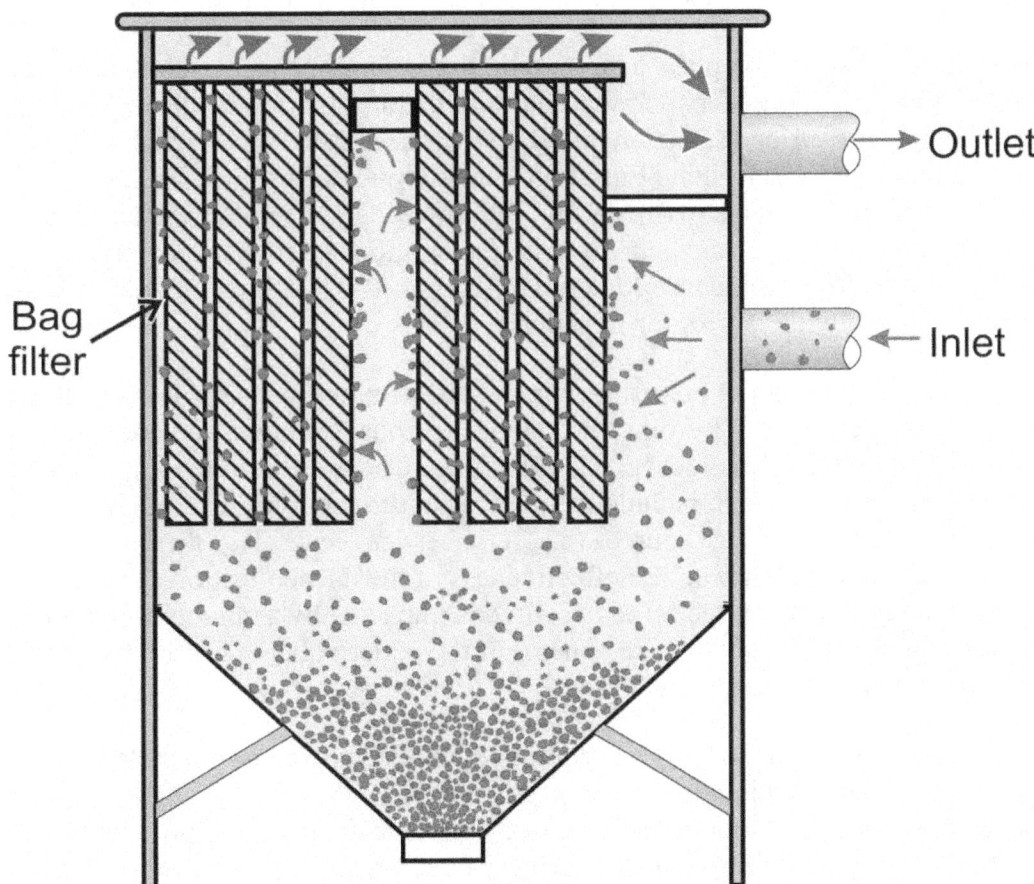

Figure 1.16. Basic design of a baghouse dust collector.

Baghouse collectors are generally designed and sized to operate with a differential pressure between 4 and 6 inches wg. These collectors can achieve air cleaning efficiencies of more than 99.97 percent (high-efficiency particulate air, or HEPA) for fine particles. The fabric bags can be made from cotton, synthetic materials, or glass fiber. The type of bag used depends on the type of fabric collector and application. For most applications involving ambient temperature, a cotton bag is the most economical. However, in a corrosive or high-temperature environment a bag material other than cotton should be employed. Since bags must be changed periodically, fabric collector designs that facilitate bag changes should be purchased. Designs where the bags can be changed from outside the collector are preferred.

Baghouse systems can also be designed for economic optimization. For a given emission control problem, factors such as the overall pressure drop, filtration cleaning cycle, and total filtration surface area can be addressed simultaneously. The article "Baghouse System Design Based on Economic Optimization" [Caputo and Pacifico 2000] provides a useful model, in particular for operations in the preliminary design phase.

Bulk density of the material requires special engineering attention. The effect that upward velocity (interstitial velocity) can have on the operation of a dust collector can be enormous. Materials with low bulk density (<30 pounds per cubic feet) must have specialized designs. In these cases, collector designs must be modified to accommodate lower interstitial velocities. Typical modifications include wider bag to bag spacing, shorter length bags, or high side inlets.

Finally, particle size distribution plays a key role in determining the air to cloth ratio and filter bag selection. It is generally understood that the finer the dust the lower the air to cloth ratio needed. Proper bag or cartridge selection based on the material to be collected is fundamental to a successful system. The article "Fine Filtration Fabric Options Designed for Better Dust Control and to Meet $PM_{2.5}$ Standards" [Martin 1999] provides a useful fabric characteristics and capabilities chart, matching fabric type to operating conditions. Another recommended resource is the article "Pick the Right Baghouse Material" [Mycock 1999], which includes a chart detailing properties of textile fabrics for filtration.

Inlet loading refers to the amount of dust arriving at the inlet of the dust collector. It is typically expressed in pounds per minute (lbs/min) or pounds per hour (lbs/hour) and converted into grain loading expressed in grains/cubic foot (gr/cf) of airflow. The grain loading within an airstream is dependent on many factors, which include the number of dust sources serviced by the dust collection system, the types of dust sources (e.g., crushers, screens, etc.), the dust emissions from these individual sources, and the capture effectiveness of the dust collection system at each source. The amount of dust emitted by each source is impacted by a number of parameters, including the particle size distribution (dustiness) of the material being handled in the process, the moisture content, and the throughput rate.

The Environmental Protection Agency has compiled data on dust emission factors for a number of processes that are involved in mineral processing [EPA 1995]. Table 1.2 lists some of these factors, which are averages of dust emissions from available data and should be viewed as a general starting point when attempting to determine inlet loading. These emission factors would have to be multiplied by the processing rate at the specific source to calculate inlet loading in lbs/min. Recommendations based on experience from dust collector manufacturers and from filter cloth manufacturers should also be utilized in efforts to effectively quantify inlet loading. This inlet loading or grain loading helps determine the air to cloth ratio, filter media, type of collector, type of inlet to be used, and how the filter cleaning system will be configured.

Table 1.2. Emission factors for crushed stone processing operations (lb/ton)*

Source	Total PM-10**
Tertiary crushing	0.0024
Screening	0.0087
Conveyor transfer point	0.0011
Wet drilling—unfragmented stone	0.00008
Truck unloading—fragmented stone	0.000016
Truck loading—conveyor, crushed stone	0.0001

*Uncontrolled emissions in lb/ton of material throughput.
**PM-10 is particulate matter less than 10 micrometers (μm), in size.

The following formula can be used for converting lbs/min into grains/ft^3:

$$\frac{7{,}000 \text{ grains(lbs/min)}}{\text{cfm}} = \text{gr/ft}^3 \qquad (1.7)$$

where lb/min = loading rate of the material, and cfm = airflow of the dust collector.

The following expressions define various inlet loading relationships:

- low concentration = <1 gr/ft^3;
- typical concentration = 1–5 gr/ft^3;
- high concentration = 5–10 gr/ft^3;
- very high concentration = >10 gr/ft^3.

The air to cloth ratio is a measure of the volume of gas or air per minute per unit area of bag (ft^3 air per minute/ft^2 of bag area), and can be expressed mathematically as follows:

$$\text{air to cloth ratio} = Q/A$$

where Q = quantity of gas (air) in cfm and, A = area of the filter cloth or total number of bags in ft^2.

The porosity of the filter media (cloth) will determine the amount of air that can be drawn through before the static pressure becomes too high for the collector housing or the fan capacity. By estimating the inlet loading, one can also determine the grains of dust striking each square foot of filter media per unit time.

Air to cloth ratio is also a measure of the average velocity of the air moving towards the bags. Inlet loading directly affects the air to cloth ratio. The heavier the inlet loading the lower the air to cloth ratio should be. High inlet loading results in more dust being retained on the filter media and causing higher pressure drops. Air to cloth ratios below 4:1 are considered low, from 4:1 to 7:1 are moderate, and above 7:1 are considered high. By lowering the air to cloth ratio, the filter has more filter area to distribute the dust, thereby helping to keep the pressure drop lower on the residual dust cake.

There are basically two methods to reduce the air to cloth ratio—lowering the cfm of air or increasing the filter cloth area. However, in overall dust collection system design, changing the airflow volume can be impractical. Therefore, increasing the filter cloth area is more commonly used.

In addition to air to cloth ratio, inlet loading can also affect the method used for cleaning the bags. Since virtually all filter materials work better with a consistent filter cake to optimize collection efficiency, varying the duration of the cleaning cycle, in conjunction with the rest period between cycles, will allow the operator to maintain a consistent filter cake. This is manifested by a low variation in pressure drop across the filter media. This approach is called "on-demand" cleaning. Bag cleaning is initiated at a predetermined high pressure drop and stops when the pressure drop reaches a predetermined low set point. This method ensures that the bags always have a sufficient amount of dust cake. Experience and manufacturers' recommendations are the best means of determining the optimum cleaning cycle for each system.

There are three techniques used with baghouse collectors to clean the dust cake from the filter media. These techniques are accomplished by mechanical shaker collectors, reverse air collectors, and reverse jet collectors.

Mechanical Shaker Collectors

Mechanical shakers use a mechanical rapping device to remove the excess dust cake; however, the airflow through the collector must be temporarily stopped to clean the bags. These shakers involve low maintenance and low operating costs, but they require large amounts of space, have limited design flexibility, and have low air to cloth ratios (2:1).

Shaker collectors employ tubular filter bags fastened on the bottom and suspended from a shaker mechanism at the top. Dust-laden air enters the collector and is deposited on the outside of the tubular bags. Continuous processes use compartmentalized collectors where airflow can be diverted to other compartments during the bag cleaning cycle (Figure 1.17). Bag materials must be made from woven fabrics such as cotton to withstand shaking.

Some shaker collectors have been converted to pulse jet collectors. However, this is an expensive operation and should not be necessary if the collector is properly maintained.

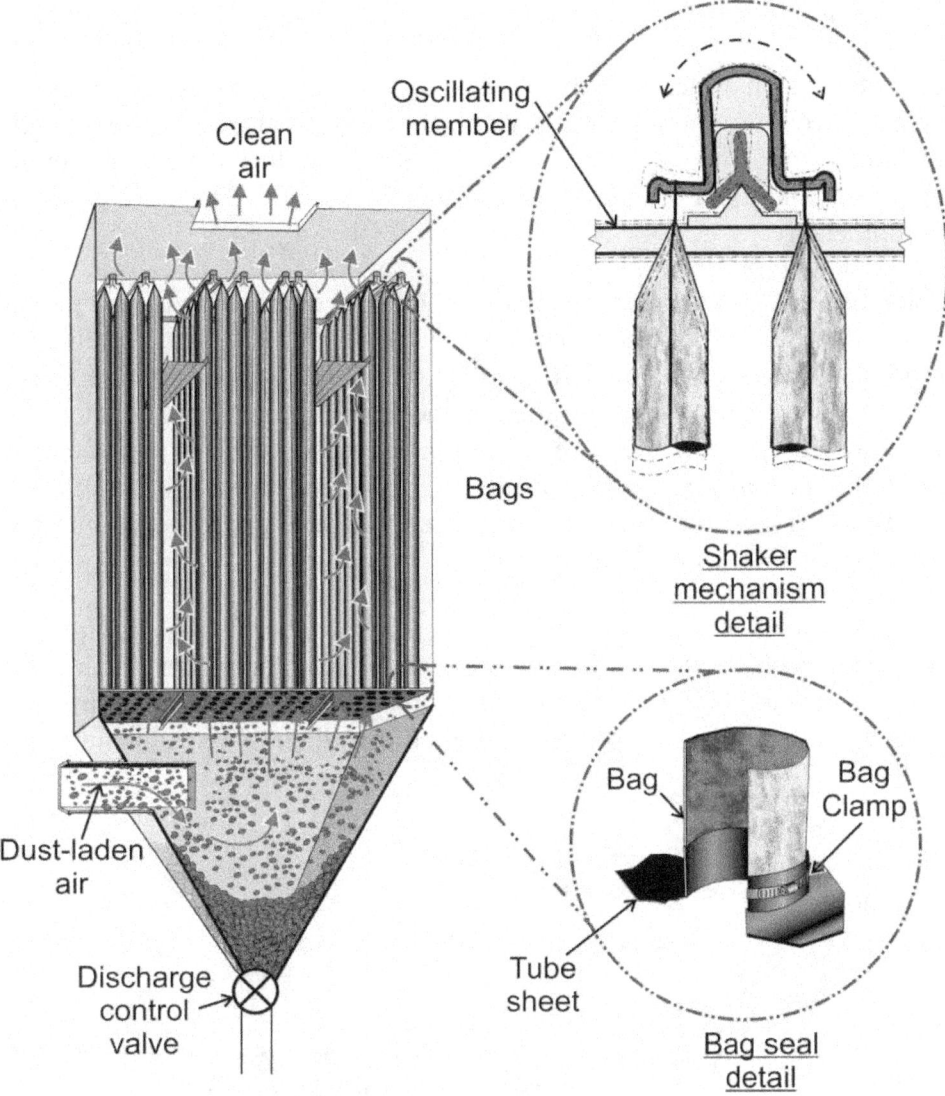

Figure 1.17. Typical design of a mechanical shaker dust collector.

When using mechanical shakers, a number of recommendations should be followed:

- Magnehelic gauges should be installed on each compartment to monitor differential pressures during the cleaning cycles.
- Differential pressures should be as close as possible to 0.0 inches wg to ensure that the dust cake breaks and is released from the bags. Differential pressures above 1/4-inch wg during the cleaning cycle will dramatically interfere with bag reconditioning, which can lead to reduced airflow, high static pressures, and low bag life.
- Bags should only be shaken when differential pressures across a section of bags have increased by 1/2-inch wg.
- Experiments to determine the optimum time interval between the shaking of bags should be performed so that bags are not shaken excessively. This also lessens wear on the bags as well as the mechanical parts.

Reverse Air Collectors

Reverse air collectors use a traveling manifold to distribute low-pressure cleaning air (3–7 psig) to the filter bags for reconditioning. They have no compressed air requirements and involve no freezing of air lines. On the negative side, they have larger footprints than conventional collectors and maintaining their cleaning mechanisms is difficult. They also perform poorly in corrosive environments.

Reverse air collectors employ tubular bags fastened onto a cell plate on the bottom and suspended from the top of the collector. The collectors must be compartmentalized for continuous service. Dust-laden air enters the collector and deposits dust on the outside of the bags. Before a cleaning cycle begins, filtration in the compartment to be cleaned is suspended. Bags are cleaned by blowing low-pressure air into the compartment in the reverse direction to normal airflow. This causes the bags to partially collapse and release the dust cake. The bags have rings located at various intervals to prevent total collapse so that the dust cake can escape and fall into the hopper (Figure 1.18).

Reverse air collectors were originally developed primarily with fragile glass cloth for use in high-temperature operations. These collectors have declined in popularity with the use of new materials that can withstand high temperatures and greater physical action. Air to cloth ratios for reverse air collectors are similar to those for shakers due to the low bag cleaning efficiency of the reverse air.

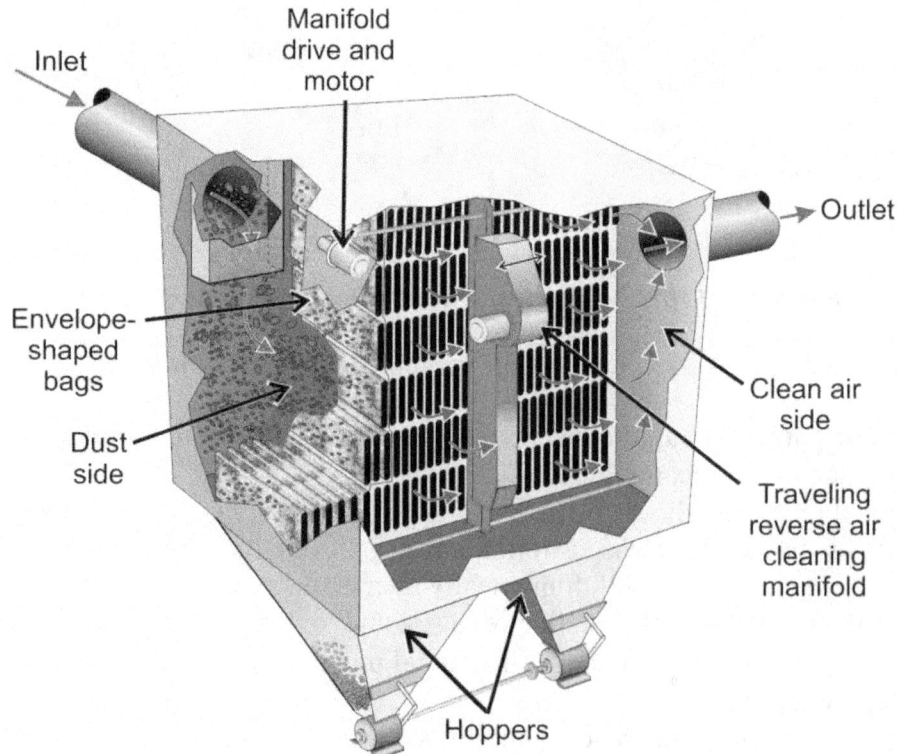

Figure 1.18. Typical design of a reverse air dust collector.

36 Fundamentals of Dust Collection Systems

Reverse Jet (Pulse Jet) Collectors

Reverse jet (also called pulse jet) collectors use bags supported from a metal cage fastened onto a tube sheet at the top of the collector (Figure 1.19). Dust-laden air enters the collector and flows from outside to inside the bag. The dust cake deposits on the outside of the bag and is cleaned by short bursts of compressed air injected inside the bag. The burst of air causes the bags to flex, breaking and releasing the dust cake. The compressed air must be clean and dry or moisture can build up on the bags, hindering the bag cleaning efficiency. Pulse jet collectors are not compartmentalized, allowing bags to be reconditioned without removing a section from service.

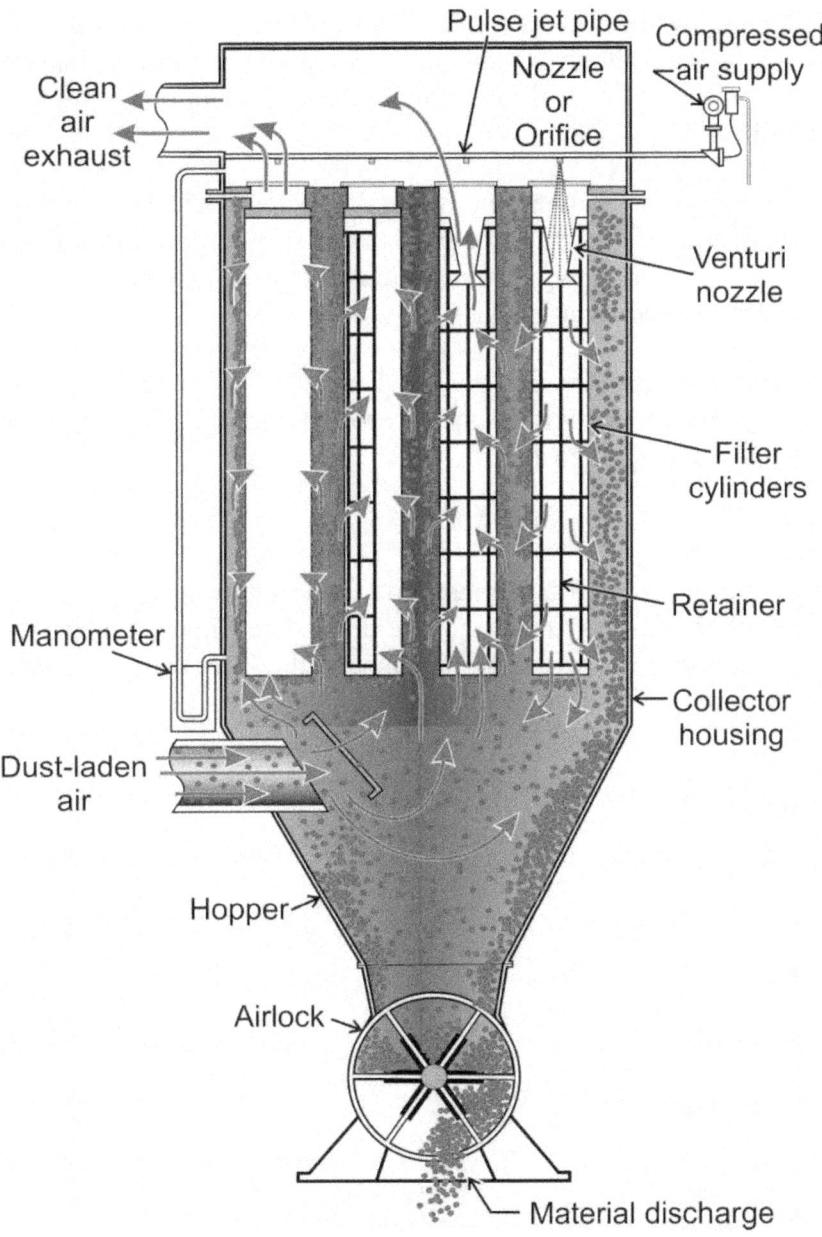

Figure 1.19. Typical design of reverse jet (pulse jet) dust collector.

Reverse jet collectors use timed blasts of compressed air (60–90 psig) for filter bag reconditioning. This is the most common type of baghouse collector and has been in use since the late 1950s. Reconditioning or cleaning of the filter bags allows the dust collection system to maintain pressure drops and operate at designed airflows. The advantage of using reverse jet collectors is high product recovery and high collection efficiency. They also enjoy high flexibility of application with many inlet design options. Their limitation is that their performance varies with temperature and moisture.

Due to more frequent cleaning intervals, these collectors provide more complete bag cleaning than previous styles of collector already discussed. Thus, the air to cloth ratios can be higher—typically, suppliers specify air to cloth ratios of 6:1 or higher. However, ratios of 4:1 should be used for applications involving abrasive minerals. High air to cloth ratios can cause high air velocity impingement on the bags, resulting in dust re-entrainment on neighboring bags after cleaning and low bag life.

Bags are typically made from nonwoven felt material. Woven materials are not used in pulse jet collectors because they require a buildup of a permanent dust layer to provide efficient air cleaning. Since pulse jet collectors aggressively clean the bags, excessive penetration of dust particles through the woven fabric can occur.

Pulse jet collectors are more cost-effective than earlier styles of collector such as mechanical shaker collectors. They can operate with higher air to cloth ratios, have no moving parts to maintain, and involve lower capital costs.

Cartridge Collectors

Cartridge collectors capture particulate from an airstream by forcing the air through filter canisters in which the filter media is fabricated in a pleated configuration. There are two basic configurations of cartridge collectors: those that suspend the filter canisters vertically and those that are mounted horizontally (Figure 1.20).

Figure 1.20. Cartridge collector with dust filter canisters.

Cartridge collectors are the latest generation of fabric collector. Unlike other fabric collectors in which the filtering media are woven or felt bags, this type of collector employs cartridges that contain a pleated filtering media. The pleated cartridges can be made from a variety of media including polyester or synthetic material. Due to the pleated design, the total filtering surface area is greater than with conventional bags of the same diameter. However, the pleated design cartridges produce very high approach velocities and thus greater re-entrainment onto the cartridge can occur. Filtration velocities are therefore limited to air to cloth ratios of less than 2:1.

The cartridge filter works like the pulse jet baghouse collector in that dust is filtered on the outside of the elements. Cleaning air (compressed) is blown into the center of the elements, discharging the excess collected material. Cartridge collectors involve lower capital costs than a baghouse because the cartridges have more area than bags, therefore fewer are required. Cartridge collectors have lower headroom requirements, because the cartridges are inserted from the side of the unit as opposed to the top. Cartridges are shorter than bags, and there are fewer of them. Media can also be changed rapidly, facilitating a higher level of maintenance. Limitations to cartridge collectors include a lack of flexibility based on extreme temperature and moisture applications, since there are not as many varieties of filter fabrics available as there are for baghouses. Because of the high filter area per unit and the pleated design, media replacement costs are high on a per-cartridge basis.

The main advantages of cartridge collectors are the compact design and ease of cartridge changing, resulting in reduced dust exposure to workers. New cartridges are packaged in a cardboard box, and to change cartridges the worker simply removes the new cartridge from the box, extracts the used cartridge and places it in the box, then installs the new cartridge, with minimal dust exposure to the worker. Within the collector, the cartridges are cleaned by a conventional pulse jet cleaning system. In practice, the cartridges can have a two-year life in abrasive applications. However, purchasing this type of collector may require the user to purchase cartridges from the equipment supplier, reducing a competitive pricing advantage.

It should be noted that cartridge collectors do not work well with moist or sticky materials, and are generally restricted to applications with temperatures less than 180°F. The inlet loading is typically lower than with bags, because with the pleated design the cartridges do not clean as well. In addition, the horizontal alignment of the cartridges allows dust from the cartridges above to fall on the cartridges below during cleaning.

Wet Scrubbers

Wet scrubbers accomplish particulate collection by employing water or another liquid as the collection media. There are many scrubber designs; most particulate scrubbers work by creating a wetted target for particle collection. This wetted target plate can be a bed of water or a zone where the particle and water droplet collide (Figure 1.21). Benefits to wet scrubbers are their ability to perform in various moisture and temperature conditions, their resistance to chemical corrosion, and their low maintenance needs. However, wet scrubbers require a significant amount of water, which must be disposed of along with the collected particulate, all of which lower the collection efficiency and increase the energy costs. Settling ponds are one common method of collected particulate disposal in the mineral processing industry, and the water from these ponds can often be reused within the operation.

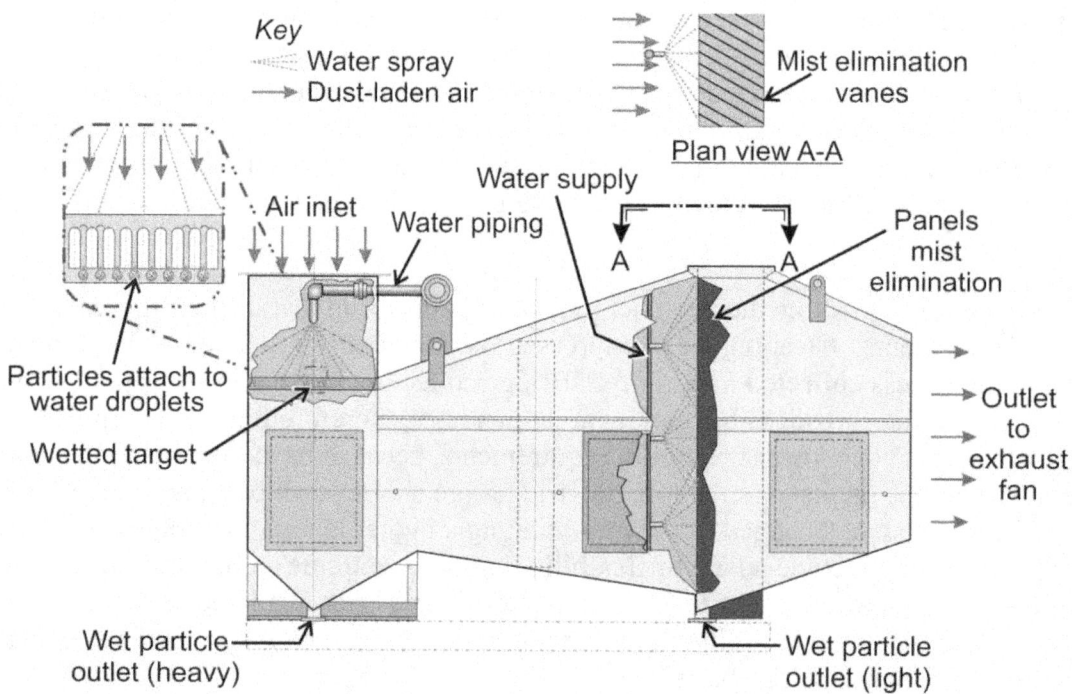

Figure 1.21. Typical design of a wet scrubber dust collector.

The air cleaning efficiency of this type of collector primarily depends on its pressure drop. Scrubbers with high differential pressures have higher air cleaning efficiencies than scrubbers with lower differential pressures. These pressures range from 1 to over 15 inches wg. The style of scrubber chosen for a particular application depends on the air cleaning required, the dust loading, and the particle sizes involved.

Wet scrubbers are particularly advantageous when handling moist hot gases. Problems such as bag blinding and condensation can arise when fabric collectors are used to clean moist hot gases. These problems are eliminated when using scrubbers. However, wet scrubbers discharge contaminated water, which requires further treatment in a settling pond or sewage system.

Venturi Scrubbers

Venturi scrubbers are a type of wet scrubber and consist of a venturi-shaped inlet and a separator (Figure 1.22). Dust-laden air is accelerated to velocities between 12,000 and 36,000 feet per minute in the throat of the venturi. These high velocities atomize the coarse water spray and create pressure drops ranging from 5 to over 15 inches wg. The venturi can be ceramic-lined for abrasive applications. The extreme turbulence promotes collision between water droplets and dust particles in the throat of the venturi. An inertial separator then removes these agglomerates.

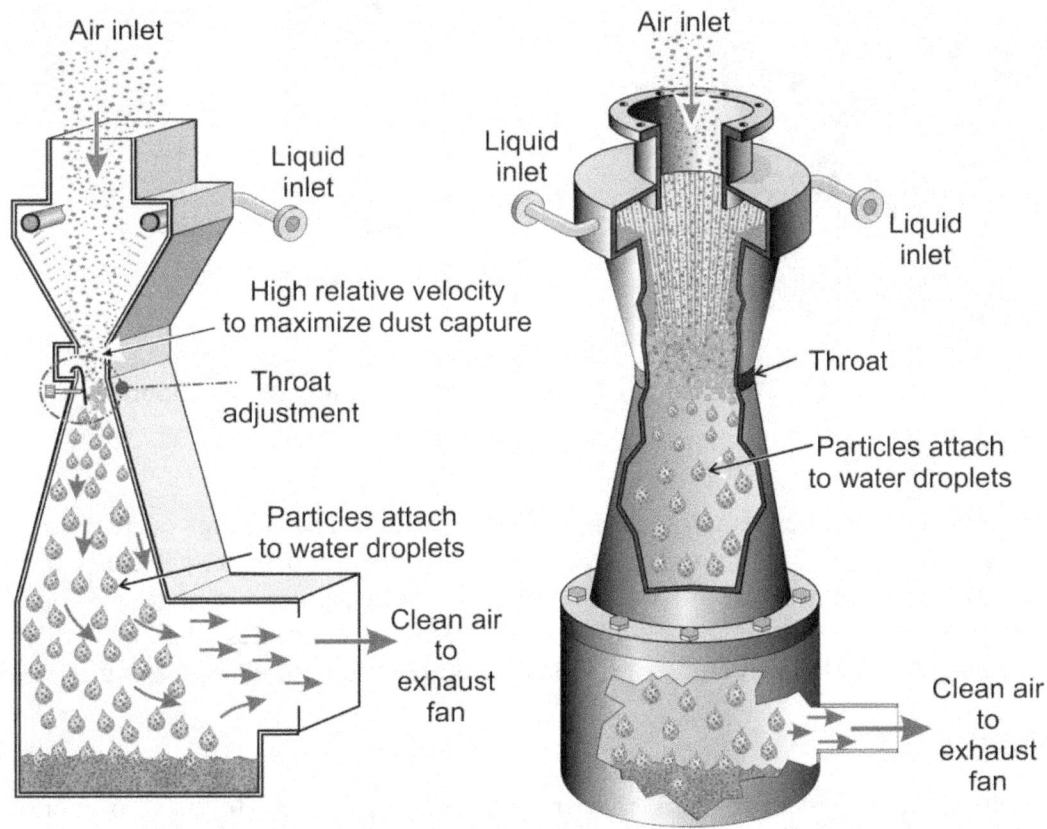

Figure 1.22. Typical designs of a venturi scrubber.

Impingement Plate Scrubbers

In another type of wet scrubber—an impingement plate scrubber (see Figure 1.23)—the exhaust passes upward through openings in perforated plates, which hold a layer of water. An impingement baffle is located above each hole, resulting in the formation of small droplets. Intimate gas/liquid contact results in efficient particle collection. A pressure drop of 4 inches wg is typical.

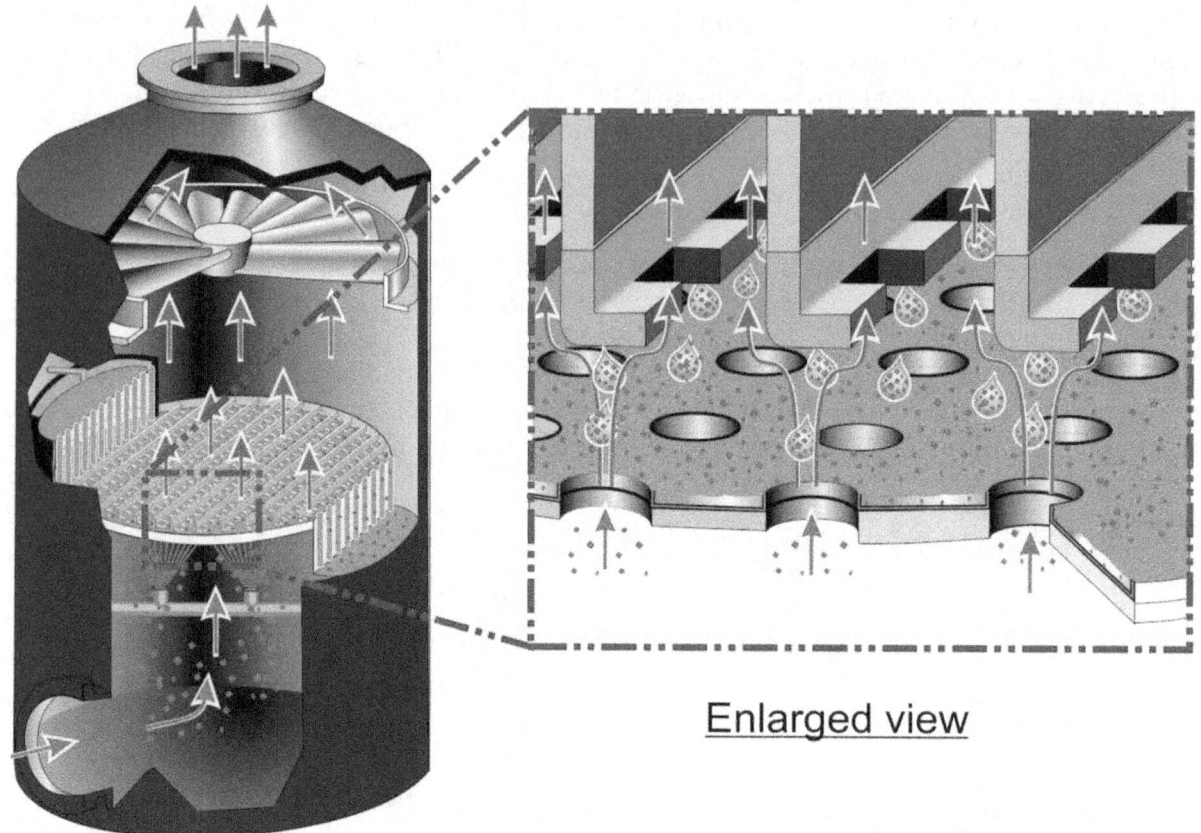

Figure 1.23. Typical design of an impingement plate scrubber.

Spray Tower Scrubbers

Gravity spray tower scrubbers employ atomized water, which falls counter-current through a rising dust-laden airstream, to remove dust particles. These scrubbers are generally of lower efficiency and operate at pressures of 1 to 2 inches wg. They are approximately 70 percent efficient on 10-μm particles and above with poor efficiency on particles smaller than 10 μm. However, they are capable of treating high dust concentrations without becoming plugged.

Wet Cyclone Scrubbers

Similar to a dry cyclone, wet cyclone scrubbers use centrifugal forces to throw particles on the collector's wetted walls. Water is introduced from the top of the scrubber to wet the walls and wash the particles away. Pressure drops for these collectors range from 2 to 8 inches wg with a good efficiency for removing particles 5 μm and larger.

Electrostatic Precipitators (ESPs)

Electrostatic Precipitators (ESPs) are particulate control devices that use electrical forces to move particles from the airstream to collection plates. Particles passing through the precipitator are given a negative electrical charge by being forced to pass through a region, called a corona, in which gas ions flow. Once the particle has been negatively charged, it is forced to the positively charged plate. Particles are removed from the plate by knocking action.

Electrostatic precipitators normally have a higher initial cost than local exhaust ventilation systems, but a number of advantages that make them worth considering. Once installed, ESPs require very little maintenance because there are no moving parts. The installation time and the operating costs are also lower than for a local exhaust ventilation system. One last advantage is that the product is easily recovered and recycled right back into the process.

There are four basic types of ESPs: plate and wire (dry), flat plate (dry), wet, and two-stage. ESPs provide a large air volume, operate favorably in various temperatures, and require little maintenance. Their limitations include their physical size, operation expenses, and inconsistent collection efficiencies. A more thorough discussion of electrostatic precipitators, including distinctions between single-stage and two-stage types, is available in the ACGIH handbook, *Industrial Ventilation: A Manual of Recommended Practice for Design* [ACGIH 2010].

COLLECTOR DISCHARGE DEVICES

The dust collected in a baghouse falls into a hopper below the bags. This hopper must have a discharge device that not only releases the accumulated dust, but ensures a seal between the baghouse and outside air to maintain negative pressure within the unit. Common hopper waste discharge devices available for baghouses are the rotary airlock valve, double dump valve (flap valve), tilt valve, and vacuum valve (dribble valve). Each of these is readily available in various designs from several manufacturers.

Rotary Airlock Valves

Rotary airlock valves are normally used in applications where both an active airlock and material metering are required. Although they are used in many baghouse installations, such common use may not be the best application for this valve. The initial cost is high, and maintenance costs may also be high in abrasive applications.

Double Dump Valves

Double dump valves are dual flapper valves mounted in tandem off the discharge of the baghouse and can be either automated or weight-based. For automated double dump valves, an automated rotating cam briefly opens the normally closed spring-loaded door to each valve. The dual valves are opened individually in a sequence that allows the top door to open, dump, and close before the lower door follows the same sequence. This allows a negative pressure to be maintained in the discharge hopper. The weight-based valve is very similar except that the valve opens based upon the weight of the product and not on an automated basis (Figure 1.24). Both types of double dump valves require sufficient vertical room below the baghouse hopper. Some wear can be expected at the actuator cam lever of each door.

Figure 1.24. Double dump weight-based valves.

Tilt Valves

Tilt valves are similar to double dump valves and are normally used in pairs. These valves commonly have a near horizontal and counterweighted flap door that is adjustable for closed tension. The function of the valve is to allow material to build up on the near horizontal flap gate until it overcomes the pull of the negative pressure in the baghouse. The material then dumps to the next valve as the gate shuts behind it. Typically a divider chute 24–30 inches long is used between the two valves to allow one valve to close completely before the other opens.

Tilt valves are less expensive than double dump and rotary valves, but can be hard to adjust. They have been known to stick open after being in service for a period of time, and they require 4 feet 0 inches clearance below the baghouse for installation.

Vacuum or Dribble Valves

A vacuum valve basically consists of a soft gum rubber tube mounted in a steel housing, which attaches to the bottom of a baghouse hopper discharge. Manufacturers have various designs available for purchase, some using a "fishtail" dribble bladder mounted inside a flanged pipe. The valve works using the negative pressure of the hopper to collapse the rubber tube and seal the discharge until sufficient material builds up and "dribbles" through. This valve normally has low initial cost, high reliability, and low maintenance, and can fit in a short discharge space.

FILTER FABRICS

The majority of development work that has been done in the industry over the last ten years has not been in the area of the collector itself but in filter fabrics. Some of these fabrics have been designed to handle the more challenging applications. Problems that particular fabrics address may be related to specific applications or the fabrics may simply provide a more efficient way of optimizing the collector's potential.[4] A wider variety of filter fabrics are now available, giving the design engineer more options to address the problem.

One of the most critical aspects to any dust collector device is the filter fabric's ability to remove the dust particles from the airstream as the dust-laden air passes through the filter fabric material in the collector. The term "filter fabric" can be used interchangeably with the term "filter media." For use in dust collector devices, there are numerous types of filter fabrics with a wide variety of properties and characteristics produced by different manufacturers. Regardless of this variability, filter fabric effectiveness is fundamentally dependent on the ability to capture or remove dust particles from the airstream. A filter fabric's ability to let air pass through the media (i.e., its permeability) is usually defined as the volume of air that can pass through one square foot of filter fabric each minute, at a pressure drop of 0.5 inches wg.

Along with permeability, another measure of a filter fabric's effectiveness is its efficiency at removing dust particles. For instance, a HEPA quality filter is one that is 99.97 percent efficient at removing 0.3-μm particles. A practice that is becoming more prevalent in the industry is to refer to a fabric's dust removal efficiency by its minimum efficiency reporting value (MERV) rating (see Table 1.3). As Table 1.3 illustrates, the higher the MERV rating, the more efficient the filter is at capturing the various size ranges of dust particles.

Table 1.3. Minimum Efficiency Reporting Values (MERV) according to ASHRAE Standard 52

Group	MERV Rating	Average Particle Size Efficiency (PSE) 0.3–1.0 μm	Average Particle Size Efficiency (PSE) 1.0–3.0 μm	Average Particle Size Efficiency (PSE) 3.0–10.0 μm
1	1			< 20%
	2			< 20%
	3			< 20%
	4			< 20%
2	5			20–34.9%
	6			35–49.9%
	7			50–69.9%
	8			70–84.9%
3	9		< 50%	≥ 85%
	10		50–64.9%	≥ 85%
	11		65–79.9%	≥ 85%
	12		80–89.9%	≥ 90%
4	13	< 75%	≥ 90%	≥ 90%
	14	75–84.9%	≥ 90%	≥ 90%
	15	85–94.9%	≥ 90%	≥ 90%
	16	≥ 95%	≥ 95%	≥ 95%

---

[4] For instance, many applications that involve temperature and corrosive dust or gases had been previously handled by using wet scrubbers. However, as emission requirements become more stringent, the use of scrubbers must diminish.

A filter fabric becomes more efficient, or achieves a higher MERV rating, by its ability to capture particles by *interception, impaction, diffusion,* and *electrostatic charge*. Distinctions among these capture processes are as follows:

- *Interception* occurs with smaller-sized particles that follow the airstream flow lines and come within one particle radius of a fabric fiber, then adhere to the fabric.
- *Impaction* occurs with larger particles that are not able to stay on airstream flow contours, travelling through the filter media and then embedding into one of the fibers directly. The amount of impaction increases with higher airflow velocities and with decreasing distances between the fibers.
- *Diffusion* occurs when the smallest dust particles collide with gas molecules, especially those smaller than 0.1 μm, and then alter their flow path so that they are captured by either interception or impaction.
- With *electrostatic collection*, the filter fabric is made from a fabric material or media that is able to sustain an electrical charge. Because dust particles also have an electrical charge, as they pass through the filter fabric, they are attracted to and adhere to the fabric material.

Filter fabrics are constructed of either natural or man-made fibrous material and are either woven or nonwoven. Woven fabrics are normally identified by a weight per unit area as well as a thread count. Nonwoven fabrics are also identified by the weight per unit area, but typically include a filter thickness classification. Nonwoven material is usually more efficient than woven fabric of the same thickness because the open areas or pores are smaller. It is important to note that, from a dust collection standpoint, any type of filter fabric can be made more efficient or be given a higher MERV rating, simply by the use of smaller fiber diameters, the inclusion of a greater weight of fibers per unit area, or by the fibers being packed more tightly.

With any filter fabric, the efficiency or MERV rating improves as the filter forms a filter dust cake. After the initial collection by one of the methods described above, as particle collection continues, a layer of dust particles forms on the fabric material and this layer then becomes the principal collection medium. As more dust particles continue to impinge on the previously collected particles, this dust filter cake continues to grow. Over a period of time, this dust filter cake will grow to such an extent that it will become increasingly difficult for the air molecules to pass through the fabric material, and this is evidenced by an increasing pressure differential across the filter. Once this pressure reaches a certain level, the dust filter cake needs to be removed from the filter fabric by some type of mechanical process (such as by a shaker, reverse air, or reverse jet process, described earlier).

Particulate Collection of Fabrics

There are two methods for collecting particulate on fabrics. The first is *depth loading*. This means that the fabric depends on a filter cake to optimize collection efficiencies. The second is *surface loading,* meaning that the fabric does not depend on a filter cake to obtain maximum collection efficiencies.

Felted fabrics are depth loaded. Polyesters, polypropylene, aramids, and similar fabrics rely on the ability of the dust collection system to develop and maintain a dust cake or filter cake at all times. When considering the depth of loading onto felted fabrics, the filter bag is actually acting as a support for the dust cake. Dust is allowed to penetrate into the fabric to form a layer of dust on the bag. Thus, the dust creates a "porous" cake with much smaller air passages, allowing the dust to filter itself. Without this cake, the filter would bleed through and never obtain acceptable outlet emissions. Therefore, in the design and operation of a baghouse using felted media, it is important to use appropriate air to cloth ratios, cleaning cycles, and cleaning pressures. The optimum collector operation maintains a constant pressure drop of 3–6 inches wg and minimizes cleaning cycles and compressed air usage. This does not mean that a system running higher than a 6-inch pressure drop is not operating satisfactorily. Some systems can operate effectively at higher pressure drops. The system is operating correctly as long as the pressure drop is constant and the required air volume is being achieved.

Surface loaded media do not require a dust cake to achieve optimum performance. As the name implies, the dust is collected on the very surface of the fabric. Some surface loaded media use membranes composed of a very thin layer (1 mil) of PTFE (polytetrafluoroethylene) material that is laminated to a substrate. This substrate material is a conventional felt. The laminate is applied in a special process that controls the size of the pore openings; thereby the collection efficiency is predictable. These pore openings are so small that they will only allow extremely small submicron particles to pass, making these fabrics extremely efficient.

Dust collectors using felted media typically run higher pressure drops than those operating with surface-loaded media because the filter or dust cake adds resistance to the system. As previously mentioned, the pressure drop tends to range from 3–6 inches wg, while surface-loaded media operating without a filter cake typically run in the 2–4 inches wg range.

The standard fabrics used in shaker collectors are woven cotton and polyester sateens and lightweight felts. Pulse jet and reverse air collectors use felts ranging in weight from 14 to as heavy as 22 ounces. Woven fabrics are not used in the last two types of collectors because the cleaning system is much more aggressive than shaking. Pulsing woven fabrics opens up the weave and causes leakage and undesirable emission levels. Cartridge filters typically use cellulose/synthetic blends and spun-bonded polyester media.

When matching fabric choice to dust collection needs, the basic criteria are temperature, inlet loading, particle size distribution, gas composition, abrasion, static charge, release, and efficiency requirements. The following are some basic guidelines for each criterion.

1. *Temperature.* Inlet gas temperature should be matched with a fabric capable of continuous operation at or above the maximum temperature the system might experience.
2. *Inlet loading.* Inlet grain loading (normally expressed in grains/dry standard cubic feet) is especially important when the loading is very light. Light loadings can make it difficult for the media to operate with sufficient dust cake.
3. *Particle size distribution.* Finer particle size distributions often require using higher efficiency media.
4. *Gas composition.* The designer must make sure that if corrosive compounds or hydrocarbons are present in the airstream, then the media selected will be able to withstand that environment.

5. *Abrasion.* When the dust is abrasive, the designer should consider using heavier felt or adding wear cuffs to the bottom of the filter bags on hopper entry inlets.
6. *Static charge.* Explosive dusts and gases or dust that can develop a static charge in the duct system require some type of static grounding. This can come in the form of a braided grounding wire sewn into the filter bag and attached to the tubesheet. Grounding can also be accomplished through the bag material by way of epitropic fibers (using carbon fibers, graphite fibers, or 3–4 percent stainless steel fibers throughout the bag). In these cases, it is important to make sure the bag collar (in top-loaded bags) is fabricated out of the same material so that there is conductivity to ground.
7. *Release.* A properly designed dust collector has two functions: the first is to collect the dust; the second is to release the dust off of the filter bag after it is collected. In the case where the dust is oily or sticky, the fabric may require a treatment to the filtering surface in order to aid in the release. This can take the form of modifying the surface with treatments such as singeing or glazing. The filtering surface can also be treated with a coating such as a PTFE bath or even a membrane. There are also specialty fabrics that treat the fiber before the felting process, providing a longer lasting surface than typical coatings.
8. *Efficiency.* Expressed as a mass loading (grains per standard cubic feet, or gr/dscf), the required collection efficiency is determined by the environmental engineer based on EPA, state, or regional regulations that correspond to the particular plant's allowable discharges to the atmosphere.

DESIGNING MINE/PLANT DUST COLLECTION SYSTEMS

Applying the above principles presented in this chapter for simple systems provides the foundation for designing more complex exhaust ventilation systems. In reality, a complex exhaust ventilation system is just a combination of a number of simple systems combined and pieced together (Figure 1.25). When designing a complex system, the following basic approach should be taken:

1. Consider the layout of the building, equipment, supports, etc.
2. Begin the design at the hood farthest away from the fan.
3. Create a line sketch of the proposed duct system layout (including plan and elevation dimensions), fan location, collector location, and equipment locations, with each branch and section of main on the line sketch numbered or lettered for convenience.
4. Select from an existing design or design an exhaust hood tailored to suit the operation and determine its airflow rate specifications.
5. Create a rough sketch design of the desired hood for each piece of equipment, including orientation and elevation of the outlet.

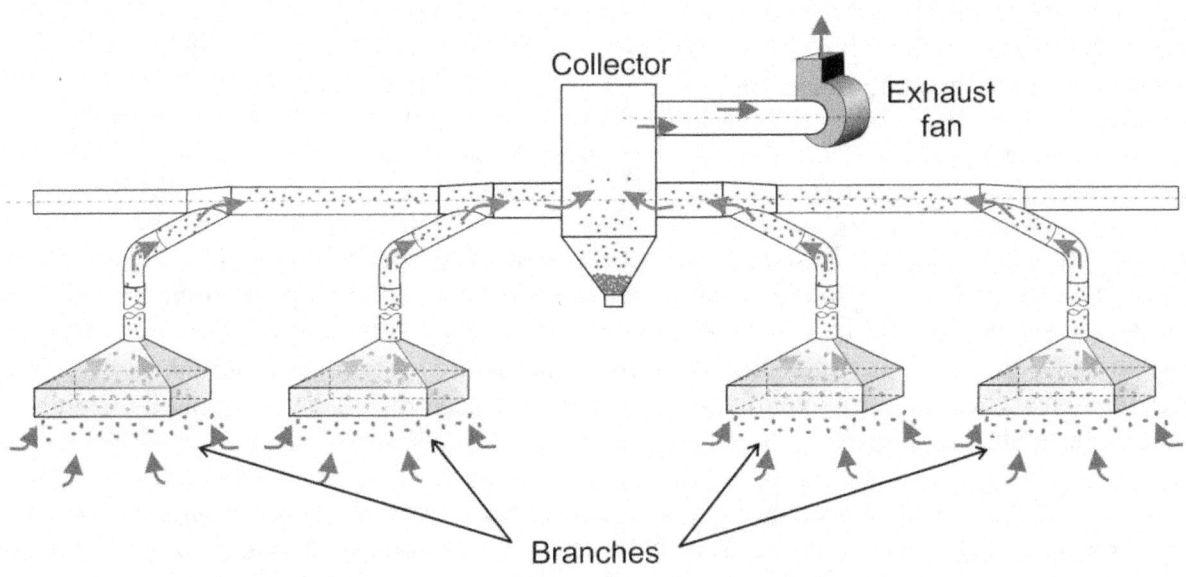

Figure 1.25. Demonstration of how a complex exhaust system is a combination of branches linking simple exhaust systems [adapted from ACGIH 2010].

To begin designing a dust control system, the following are some basic preliminary considerations:

- The amount of dust emissions to be collected by a system may be the most important design consideration. Different collector systems possess different capabilities at removing particulate.
- The overall system pressure (total head) helps to determine the type of collector to use. Most fugitive dust applications will have inlet static pressures below 20 inches wg. Standard baghouses and cartridge filters are capable of handling this pressure.
- Some applications require higher system pressures (in some cases, pressures exceed 40 inches wg), and therefore the equipment must be reinforced.
- In most LEVs, square or rectangular dust collector housing designs are adequate; however, in high-pressure systems (>40 inches wg), cylindrical housings, which are inherently stronger, would be used.

Once the entire system is laid out including all the hoods, ductwork, and dust collector system(s), this information is then used to determine the required fan capacity for the system. Many times, fan manufacturers will provide information and assistance with determining the correct fan and settings.

Clay Application

Particular caution should be exercised in designing a system for any product that contains clay dust. Typically, clay products are dried to 10 percent moisture compared to silica products, which are dried to 0.5 percent moisture or less. The higher moisture content of the clay becomes a significant factor if the conveying air temperature approaches dew point temperature. This will usually occur at the wall of the ductwork or baghouse. The surface of clay particles will become sticky and adhere to any surface contacted. Over a period of time, plugging or blinding will occur. Therefore, additional measures may be required to prevent material buildup, such as insulation of additional equipment, duct heaters, location of equipment indoors, etc.

FANS

Fans are a critical feature in the design of ventilation systems for dust control. They are used to move the air through the ventilation system, whether to create an exhausting or blowing ventilation system. In an exhausting system the fan is located at the end or discharge of the ventilation system and is used to "pull" air through the entire system. In a blowing system, the fan is located at the inlet of the ventilation system and is used to "push" air through the entire system [Hartman et al. 1997].

There are different types of fans used in ventilation systems, with their selection being dependent upon their operating characteristics. Several basics of fan operation need to be understood in order to properly select a fan for a ventilation system.

Fan Operating Characteristics

The operating characteristics of the fan are provided by the manufacturer. This information can be in the format of fan performance tables and/or fan performance curves. Fan performance tables provide only the minimum information, i.e., static pressure, airflow, and brake horsepower (BHP) required for selecting a fan [Greenheck Fan Corporation 1999]. Static pressure is based upon the amount of pressure required to overcome the friction loss of the entire ventilation system. Airflow is the amount of air required for the ventilation system. BHP is the horsepower required of the motor to operate the fan at the desired static pressure and airflow.

Fan curves or performance curves are graphs which also describe the performance characteristics of the fan. The fan performance curve is provided for a particular model of fan at a given revolutions per minute (RPM). There can be a series of fan performance curves to cover the performance of a selected fan with each graph representing a different RPM. A typical fan performance curve is in the format shown in Figure 1.26, with static pressure on the y-axis and airflow on the x-axis. Additionally, BHP curves for the fan can be included on the graph with the BHP scale on a separate scaled y-axis. Information about mechanical efficiency and noise can sometimes be included, if the manufacturer provides this information [Hartman et al. 1997].

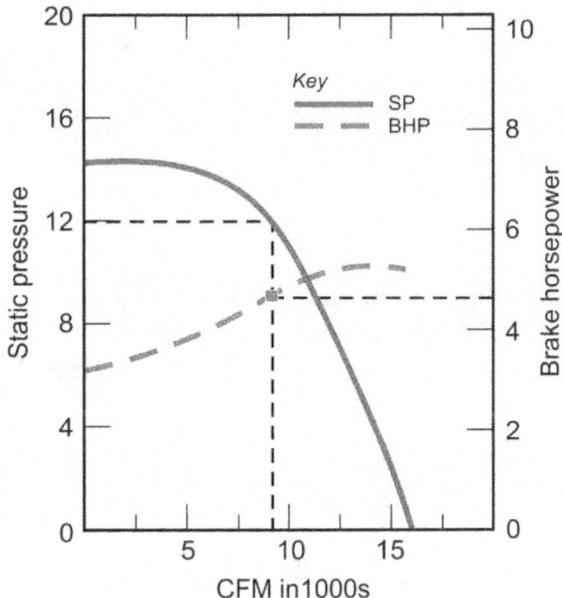

Figure 1.26. A typical fan performance curve. Each fan curve is associated with a certain fan model at a selected RPM.

Fans are selected based upon the required static pressure and airflow of the ventilation system. It is important that the static pressure and airflow be located in the operating range of the fan curve—*not* in the stalling range of the curve. Figure 1.27 shows a graph presenting the operating and stalling ranges for the fan curve.

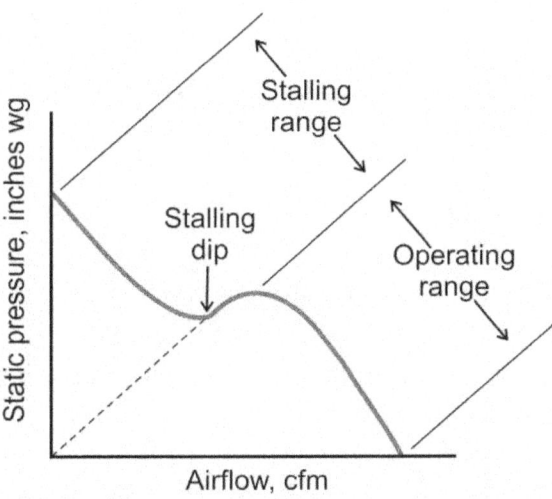

Figure 1.27. Fan performance curve showing stalling and operating ranges.

When the fan operates in the operating range, the airflow passes through the blades smoothly and quietly without any eddies. Once the static pressure and airflow are located in the operating range of the fan curve, the BHP of the motor required to operate the fan at this condition can be determined from the graph. If the static pressure and airflow are located in the stalling range of the fan, it will not operate correctly, operating in a condition known as aerodynamic stall. Stall occurs when the airflow and resulting air velocity becomes so low that the airflow is not able to follow the designed contours of the fan blade. The blades throw the approaching inlet air outward to produce high static pressures with low airflows, which creates high turbulence and eddies. Thus, the fan will operate noisily and with low efficiency due to the air turbulence and eddies caused by this condition [Bleier 1998].

Fan Laws

Several equations can be used to determine the effects to airflow, static pressure, and BHP from changes in RPM. These simplified equations are the general fan laws and are applicable as long as fan diameter and air density are constant [Bleier 1998]. The laws are represented in the following way.

$$\frac{CFM_2}{CFM_1} = \frac{RPM_2}{RPM_1} \tag{1.8}$$

$$\frac{SP_2}{SP_1} = \left(\frac{RPM_2}{RPM_1}\right)^2 \tag{1.9}$$

$$\frac{BHP_2}{BHP_1} = \left(\frac{RPM_2}{RPM_1}\right)^3 \tag{1.10}$$

$$N_2 - N_1 = 50\log_{10}\frac{BHP_2}{BHP_1} \tag{1.11}$$

where RPM = speed of fan, revolutions per minute;
CFM = airflow, cfm;
SP = static pressure (this can be for total, velocity, or static pressure as long as the pressures used are consistent—i.e., all total, all velocity, or all static pressure);
BHP = horsepower (this can be for air or total horsepower or brake horsepower as long as the horsepower used is consistent); and
N = noise level, decibels.

The subscripts represent the different fan characteristics—i.e., 1 for the original fan characteristics and 2 for the new/modified characteristics. There are also equations that can account for changes in fan diameter and air density, but these are beyond the scope of this chapter [Hartman et al. 1997]. Further information on changes in fan diameter and air density can be found by reviewing the references Bleier [1998] and Hartman et al. [1997].

Fan Types

There are two basic types of fans: axial-flow fans and centrifugal fans. There are also other fan designs that use or combine the concepts of axial or centrifugal flow; these are axial-centrifugal fans and roof ventilators. Again, the selection of the fan type is based upon the requirements of the ventilation system design.

Axial-Flow Fans

The axial-flow fan category includes propeller fans, tubeaxial fans, vaneaxial fans, and two-stage axial-flow fans. Axial flow fans move the air in a direction that is "axial" or parallel to the axis of rotation of the fan. Propeller fans are the most common type. They are generally mounted in the wall of the building near a heat source to exhaust the hot air into the outside atmosphere (Figure 1.28). They can have different drive configurations, being either direct drive or belt drive. The wall-mounted fans can also be constructed with shutters that close when the fan is turned off. Propeller fans are designed to move large volumes of air at low static pressures [Bleier 1998].

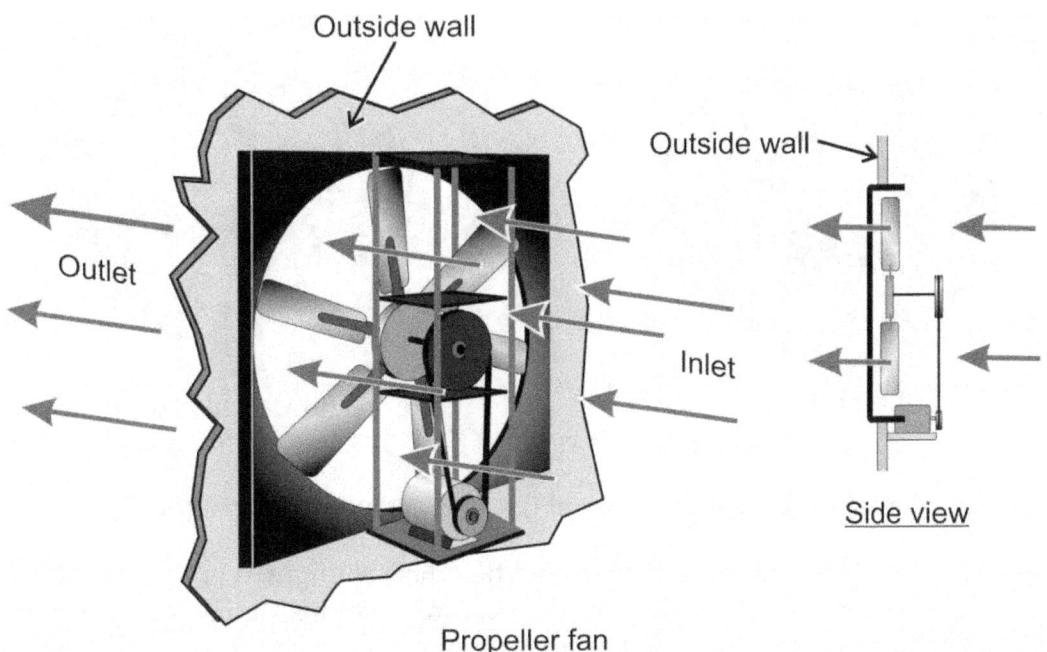

Figure 1.28. Wall-mounted propeller fan.

Tubeaxial fans are used for exhausting air from an inlet duct (Figure 1.29, left). They consist of a fan with many blades in a cylindrical housing. The blades are generally shaped as an airfoil to help with air movement. The hub diameter can be 30 to 50 percent of the blade outside diameter [Bleier 1998]. The housing is connected to the inlet duct and also contains the motor support. These fans are used in conditions that require moderate static pressures (higher than those required for propeller fans).

Vaneaxial fans are similar to tubeaxial fans. The vaneaxial fan has a housing that contains guide vanes that are oriented parallel to the airflow (Figure 1.29, right). These vanes are used to recover the tangential airflow velocity from the fan blade to convert it into static pressure. This tangential airflow velocity is not recovered as static pressure in propeller or tubeaxial fans (this component is lost energy). The hub diameter of a vaneaxial fan is much larger being 50 to 80 percent of the blade outside diameter [Bleier 1998]. The vaneaxial fan is used for high static pressure conditions.

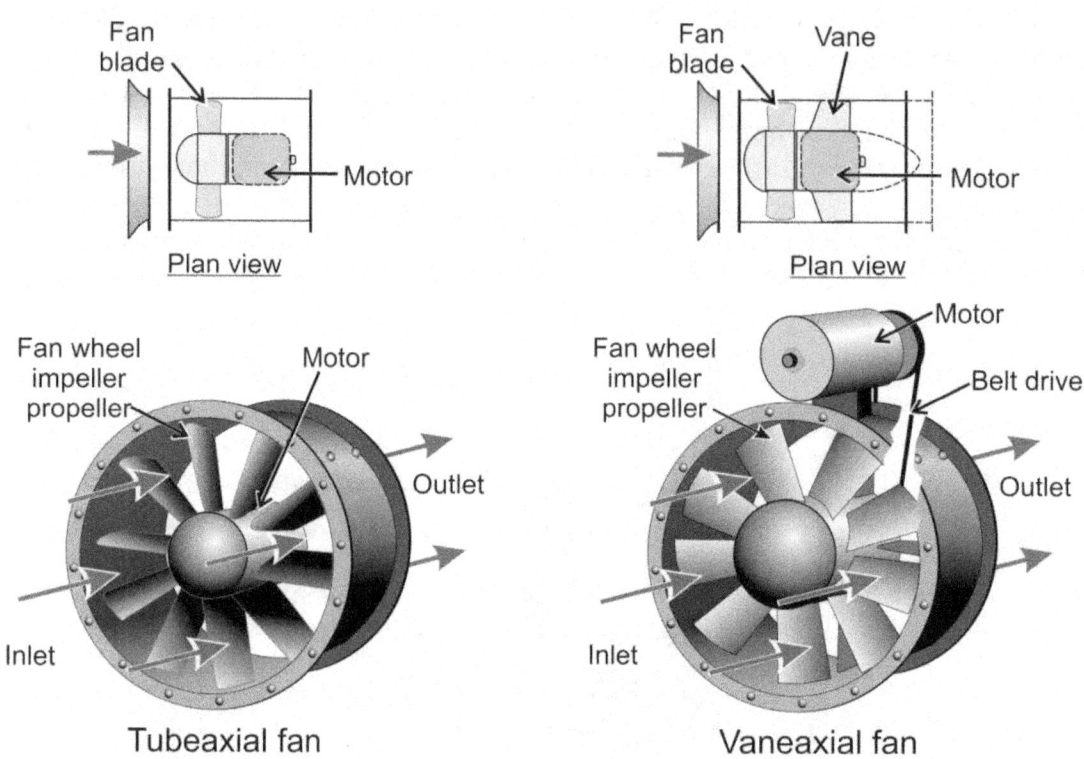

Figure 1.29. Typical tubeaxial fan (left) and vaneaxial fan (right).

Two-stage axial-flow fans are basically two axial flow fans configured in series. This configuration allows for operation in high static pressure conditions as each fan's operating pressure is added together when in series. This fan can be designed to have each fan rotate in the same direction with guide vanes located between each of the fans, or the fans may counter-rotate.

Centrifugal Fans

The airflow for a centrifugal fan is different from that of axial flow fans. For a centrifugal fan the airflow is drawn into a rotating impeller and discharged radially from the fan blade into a housing. The resulting flow of air is perpendicular to the axial rotation or parallel to blade motion [Hartman 1997] and the housing is used to direct the airflow to the desired location (Figure 1.30).

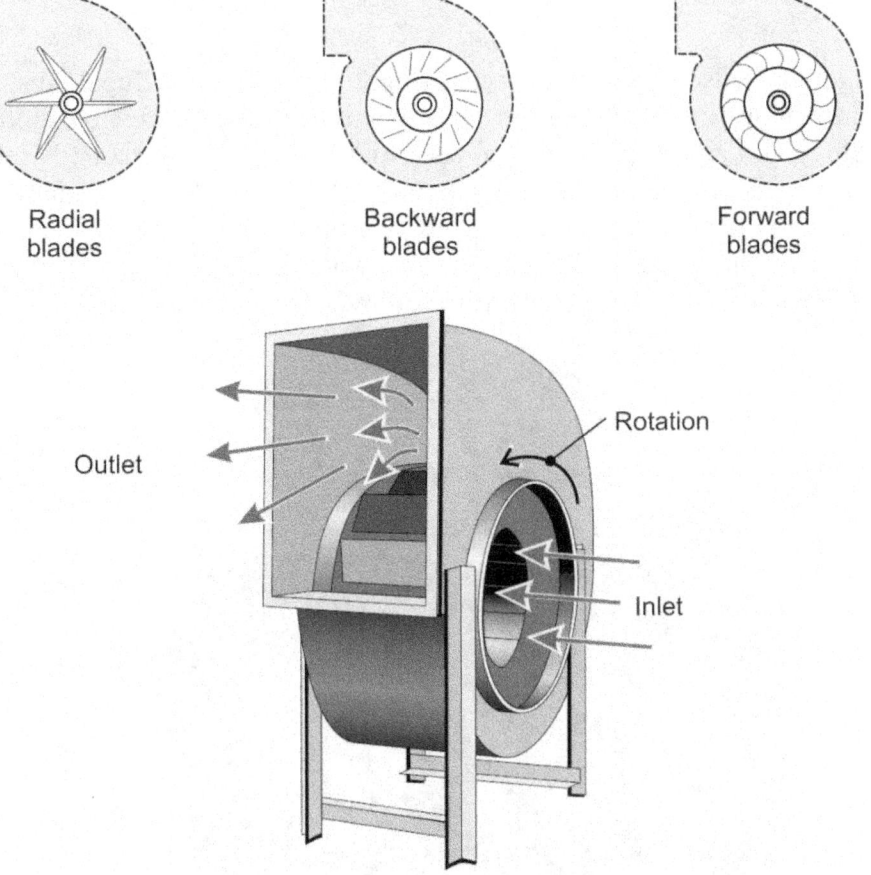

Figure 1.30. Typical centrifugal fan.

There are numerous types of centrifugal fans. The flow through the fan is basically the same for all types, the difference being in the configuration of the blades. Each blade type has its advantages for different applications, as detailed below.

- *Airfoil type blades* have the best mechanical efficiency and lowest noise level.
- *Backward curved blades* have slightly lower efficiencies compared to airfoil blades. These blades are better suited to handle contaminated air because they are single thickness and can be made of heavier material that can resist the effects to fan blades by the contaminated air.
- *Backward inclined blades* have lower structural strength and efficiencies. They are easier to produce due to the elimination of the blade curvature.
- *Radial tip blades* are curved at the tips. These types are used mainly in large diameters (30 to 60 inches) under severe conditions of high temperatures with minimal air contamination [Bleier 1998].
- *Forward curved blades* produce airflow rates higher than other centrifugal fans of the same size and speed. This allows for the fan to be more compact than other types of centrifugal fans. These fans are often used in furnaces, air conditioners, and electronic equipment cooling.
- *Radial blades* are rugged and self-cleaning, but have low efficiencies. They are suited for airflows containing corrosive fumes and abrasive material from grinding operations.

Other Fan Types

Axial-centrifugal fans, also called tubular centrifugal or in-line centrifugal fans, use a centrifugal fan to move air in an in-line configuration. To accomplish this, the air flows into the inlet and makes a 90-degree turn at the fan blade, travels radially along the blade, and at the tip makes another 90-degree turn in order to flow out the outlet (Figure 1.31). These fans are easily installed in-line with the ductwork, and they produce more static pressure than vaneaxial fans of the same fan diameter and speed. However, their mechanical efficiency is lower than for vaneaxial fans due to the two 90-degree turns required of the airflow.

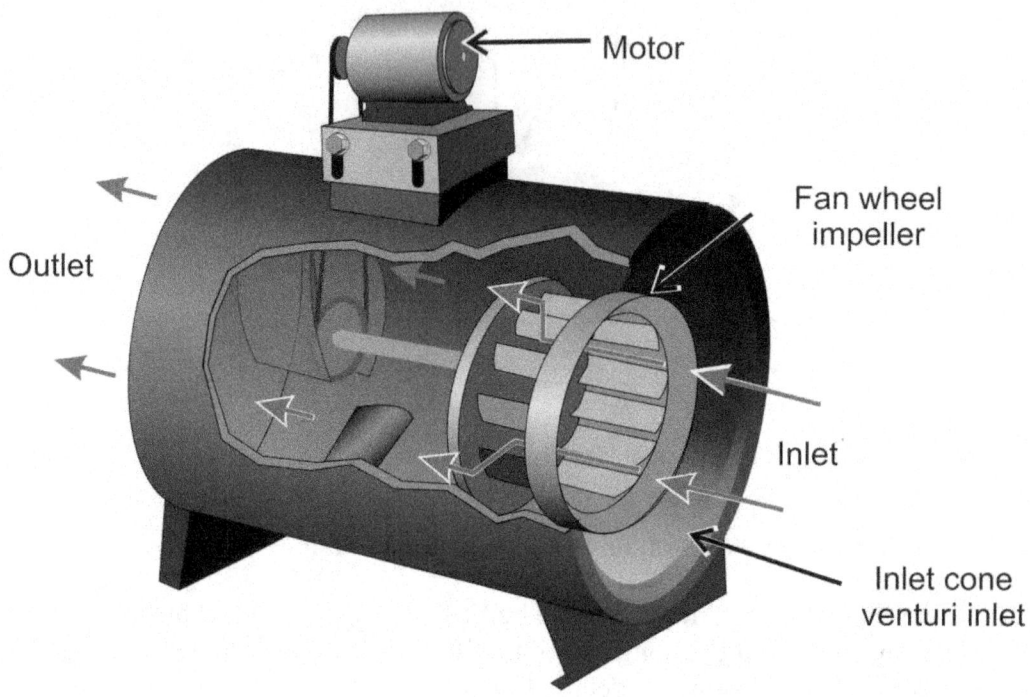

Figure 1.31. Axial centrifugal fan showing airflow [adapted from Bleier 1998].

Roof Ventilators

Roof ventilators use either axial or centrifugal fans in their configuration (Figure 1.32). They are generally installed in the roof of the mill building and are an integral part of total structure ventilation design as described in Chapter 8—Controls for Secondary Sources. Most roof ventilators are used for exhausting air from the building, creating an updraft throughout the building. However, they can also be used for supplying clean air by blowing it down into the building by creating a downdraft if required by the ventilation design. The exhaust discharge can be radial or upblast, with upblast used mainly for air laden with oil/grease or dust. Roof ventilators using axial fans generally do not require ductwork. However, those using centrifugal fans in their units may require some ductwork.

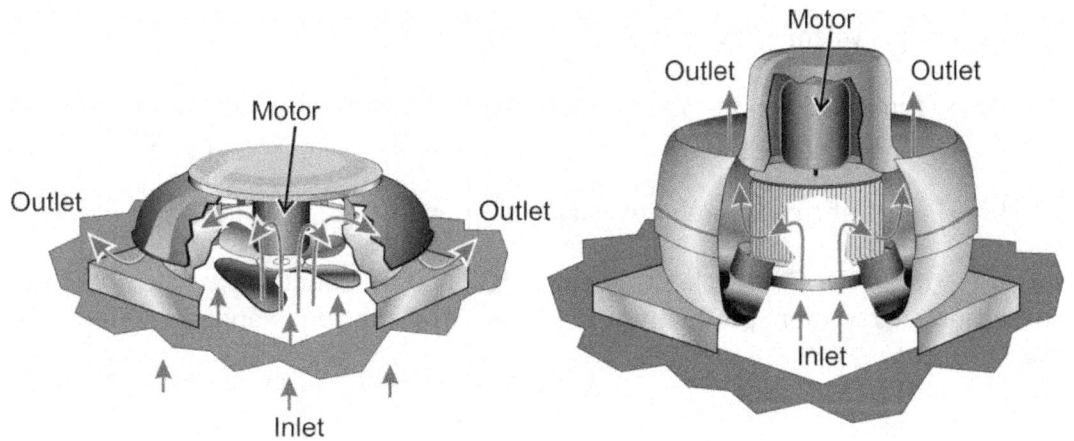

Figure 1.32. Typical roof ventilators using axial fans (left) and centrifugal fans (right).

REFERENCES

ACGIH [2010]. Industrial ventilation: a manual of recommended practice for design. 27th ed. Cincinnati, OH: American Conference of Governmental Industrial Hygienists.

Bleier FP [1998]. Fan handbook: selection, application, and design. New York: McGraw-Hill.

Caputo AC, Pacifico MP [2000]. Baghouse system design based on economic optimization. Env Prog *19*(4):238–245.

DallaValle JM [1932]. Velocity characteristics of hoods under suction. ASHVE Transactions *38*:387.

EPA [1995]. AP 42, Fifth Edition, Compilation of air pollutant emission factors, volume 1: stationary point and area sources. Environmental Protection Agency. [http://www.epa.gov/ttn/chief/ap42].

Fletcher B [1977]. Centreline velocity characteristics of rectangular unflanged hoods and slots under suction. Ann Occup Hyg *20*:141–146.

Greenheck Fan Corporation [1999]. The basics of fan performance tables, fan curves, system resistance curves, and fan laws. Schofield, WI: Greenheck Fan Corp. [http://www.greenheck.com/library/articles/10].

Hartman HL, Mutmansky JM, Ramani RV, Wang YJ [1997]. Mine ventilation and air conditioning. 3rd ed. New York: John Wiley & Sons, Inc.

Hinds WC [1999]. Aerosol technology, properties, behavior, and measurement of airborne particles. 2nd ed. New York: John Wiley & Sons, Inc.

Johnson GQ [2005]. Dust control system design: knowing your exhaust airflow limitations and keeping dust out of the system. Powder and Bulk Eng *19*(4):51–59.

Martin CT [1999]. Fine filtration fabric options designed for better dust control and to meet $PM_{2.5}$ standards. Cement Industry Technical Conference IEEE IAS/PCA. pp. 385–393.

Mycock JC [1999]. Pick the right baghouse material. Power Eng *103*(7):43–46.

Swinderman RT, Marti AD, Goldbeck LJ, Strebel MG [2009]. Foundations: the practical resource for cleaner, safer, more productive dust & material control. Neponset, Illinois: Martin Engineering Company.

CHAPTER 2:
WET SPRAY SYSTEMS

CHAPTER 2: WET SPRAY SYSTEMS

Probably the oldest and most often used method of dust control at mineral processing operations is the use of wet spray systems. In essence, as the fines are wetted each dust particle's weight increases, thus decreasing its ability to become airborne. As groups of particles become heavier, it becomes more difficult for the surrounding air to carry them off. The keys to effective wet spray dust control are proper application of moisture, careful nozzle location, controlling droplet size, choosing the best spray pattern and spray nozzle type, and proper maintenance of equipment.

In the vast majority of cases for mineral processing operations, the wet spray system used is a water spray system. Although the use of water sprays is a very simple technique, there are a number of factors that should be evaluated to determine the most effective design for a particular application. The following two methods are used to control dust using wet sprays at mineral processing operations.

- Airborne dust prevention, achieved by direct spraying of the ore to prevent dust from becoming airborne.
- Airborne dust suppression, which involves knocking down dust already airborne by spraying the dust cloud and causing the particles to collide, agglomerate, and fall out from the air.

Most operations will use a combination of both of these methods in the overall dust control plan.

PRINCIPLES OF WET SPRAY SYSTEMS

To use wet sprays effectively, it must be remembered that each ore type and application point is a unique situation and needs to be evaluated separately to achieve the optimal design. For example, wet sprays cannot be used with all ores, especially those that have higher concentrations of clay or shale. These minerals tend to cause screens to blind and chutes to clog, even at low moisture percentages. Also, water cannot be used at all times throughout the year in various climates where low temperatures may cause freezing.

Water Application

When water is used to control dust, it has only a limited residual effect due to evaporation, and will need to be reapplied at various points throughout the process to remain effective. Over-application in the amount/volume of moisture can be a problem in all operations and may impact the equipment as well as the total process and transportability of the final product if shipped in bulk. In most cases, a properly designed spray system using finely atomized water sprays will not exceed 0.1 percent moisture application [USBM 1987]; however, systems that address prevention over larger areas with larger droplet sprays may add up to 1.5 percent moisture to the process.

The vast majority of dust particles created during crushing are not released into the air, but stay attached to the surface of the broken material. Therefore, adequate wetting is extremely important because it ensures that the dust particles stay attached to the broken material. Uniformity of wetting is also an important issue for an effective system. By far the best dust reductions can be achieved by spraying the ore with water and then mechanically mixing the ore and water together to achieve a uniformity of wetting.

Ideally, the spray system should be automated so that sprays are only activated when ore is actually being processed. For dust knockdown, or suppression, a delay timer may be incorporated into some applications to allow the spray system to operate for a short time period after a dust-producing event.

Nozzle Location

Due to the unique characteristics of each application, there are no hard and fast rules for specifically locating spray nozzles in dust control applications; however, the following guidelines will contribute to spray system efficiency.

- For wet dust prevention systems, nozzles should be located upstream of the transfer point where dust emissions, in most cases, are being created [Blazek 2003].
- Care should be taken to locate nozzles for the best mixing of material and water [Blazek 2003].
- For airborne dust prevention, the nozzles should be located at an optimum target distance from the material—far enough to provide the coverage required but close enough so that air currents do not carry the droplets away from their intended target [Blazek 2003]. Droplet size also needs to be considered when setting the correct target distance.
- For airborne dust suppression, nozzles should be located to provide maximum time for the water droplets to interact with the airborne dust.

Figure 2.1 illustrates a common dust control application at a conveyor dump point into a bin. In this dust suppression application, spray nozzles are positioned in a manner which allows the spray patterns of the individual nozzles to properly interact with the dust particles and at a distance where the droplets will not be carried off by air currents.

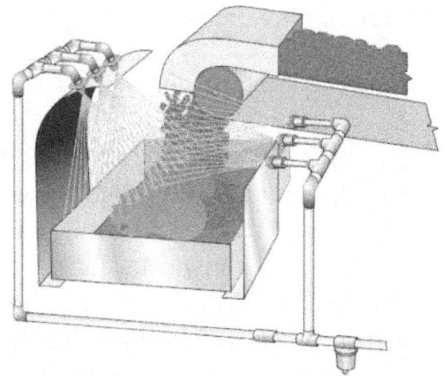

Figure 2.1. Common dust control application illustrating nozzle positioning [SSCO 2003].

Controlling Droplet Size

When using sprays, one of the primary considerations is the droplet size. If the droplet diameter is much greater than the diameter of the dust particle, the dust particle simply follows the air stream lines around the droplet. If the water droplet is of a size comparable to that of the dust particle, contact occurs as the dust particle follows the stream lines and collides with the droplet (Figure 2.2). For optimal agglomeration, the particle and water droplet sizes should be roughly equivalent. The probability of impaction also increases as the size of the water spray droplets decreases, because as the size of the droplets decreases, the number of droplets increases [Rocha 2005a].

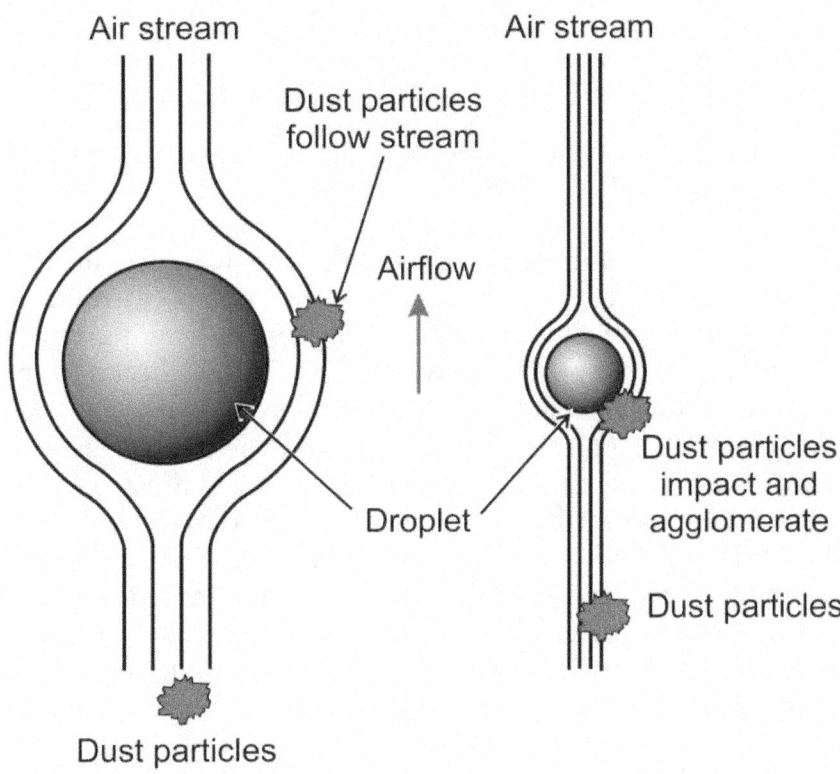

Figure 2.2. Effect of droplet size on dust particle impingement.

When wetting the bulk ore to achieve airborne dust prevention, droplet sizes above 100 micrometers (μm) (preferably 200 to 500 μm) should be used. In contrast, for airborne dust suppression, where the goal is to knock down existing dust in the air, the water droplets should be in similar size ranges to the dust particles. The intent is to have the droplets collide and attach themselves (agglomerate) to the dust particles, causing them to fall from the air. To achieve this goal, droplets in the range of 10 to 150 μm have been shown to be most effective [Rocha 2005b]. For context, Table 2.1 relates particle/droplet size to common precipitation classification.

Table 2.1. Particle/droplet size in comparison to common precipitation classification [Bartell and Jett 2005]

Particle size in micrometers (µm)	Reference	Seconds for particle to fall 10 feet
5,000 to 2,000	Heavy rain	0.85 to 0.90
2,000 to 1,000	Intense rain	0.9 to 1.1
1,000 to 500	Moderate rain	1.1 to 1.6
500 to 100	Light rain	1.6 to 11
100 to 50	Mist	11 to 40
50 to 10	Thick fog	40 to 1,020
10 to 2	Thin fog	1,020 to 25,400

Methods of Atomization

Atomization is the process of generating droplets by forcing liquid through a nozzle, which is accomplished by one of two methods.

- *Hydraulic* or *airless atomization* controls droplet size by forcing the liquid through a known orifice diameter at a specific pressure. This method utilizes high liquid pressures and produces relatively small- to medium-sized droplets in uniformly distributed fan, full cone, or hollow cone spray patterns. Hydraulic fine spray nozzles are preferred in most areas because operating costs are lower since compressed air is not required.
- *Air atomizing* controls droplet size by forcing the liquid through an orifice at lower pressures than the hydraulic atomizing method, using compressed air to break the liquid into small droplets. This method produces very small droplets and uniform distribution in a variety of spray patterns. However, it is more complex and expensive because it requires compressed air. In most cases, air atomizing nozzles are effective in locations where dust particles are extremely small and the nozzles can be located in close proximity to the dust source, although some applications will require large capacity air atomizing nozzles to throw their sprays long distances to reach the dust.

Chemical Additives to Control Droplets

Surfactants are sometimes used in wet spray applications because they lower the surface tension of the water solution, which has the following effects:

- reduced droplet diameter;
- an increase in the number of droplets for a given volume of water; and
- a decrease in the contact angle [Blazek 2003], defined as the angle at which a liquid meets a solid surface (θ as shown in Figure 2.3).

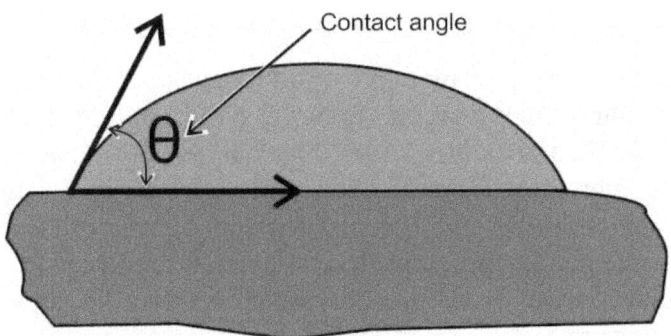

Figure 2.3. Contact angle resulting from a liquid meeting a solid surface [NDT 2009].

The use of surfactants increases the rate at which the droplets are able to wet or coat dust particles; thus less moisture is used to produce the same effects as a typical water application. Small amounts of surfactants can be injected into the spray water, typically in a dilution range of 1:700 to 1:1500, to improve the wetting and subsequent control of dust particles [Swinderman et al. 2002]. Despite the effectiveness of chemical additives, it must be noted that they are not often used in the metal/nonmetal mining industry based upon several limitations.

- Surfactants are significantly more expensive than a typical water application.
- They can alter the properties of the mineral or material being processed.
- They can damage some equipment such as conveyor belts and seals.
- Surfactant systems require more upkeep and maintenance than typical water systems.
- Surfactants have limited usefulness in the metal/nonmetal mining industry, as opposed to in the coal industry, since ore or stone are much easier to wet than is coal due to its hydrophobic nature [NIOSH 2003].

The effectiveness of chemical additives depends on:

- the type of wetting agent;
- hydrophobic nature of the mineral particles;
- dust particle size;
- dust concentration;
- water pH; and
- minerals present in the water used [Rocha 2005c].

NOZZLE TYPES AND SPRAY PATTERNS

Spray nozzles are categorized by the type of atomization method used and by the spray patterns they produce. The most commonly used spray nozzles produce full cone, hollow cone, round, or flat fan patterns. Air atomizing nozzles (which utilize compressed air) are typically used to produce round or flat fan spray patterns, while hydraulically atomizing nozzles are typically used to produce full or hollow cone spray patterns; however, some hydraulically atomizing nozzles can also produce flat fan spray patterns.

Air Atomizing Nozzles

Air atomizing nozzles are sometimes called two-fluid nozzles because they inject compressed air into the liquid stream to achieve atomization. Figure 2.4 depicts two styles of air atomizing nozzles known as internal mix and external mix. Internal mix nozzles use an air cap that mixes the liquid and air streams internally to produce a completely atomized spray and external mix nozzles use an air cap that mixes the liquid and air streams outside of the nozzle. With an internal mix nozzle, the atomization air pressure acts against the liquid pressure to provide additional liquid flow rate control. With an external mix nozzle, the liquid pressure is unaffected by the atomization air pressure.

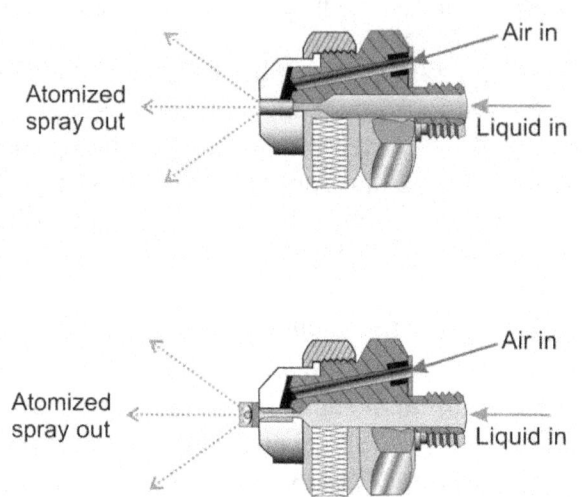

Figure 2.4. Typical internal mix nozzle (top) and external mix nozzle (bottom).

Internal mix nozzles can produce either round or flat spray patterns and external mix nozzles produce flat spray patterns as illustrated in Figure 2.5.

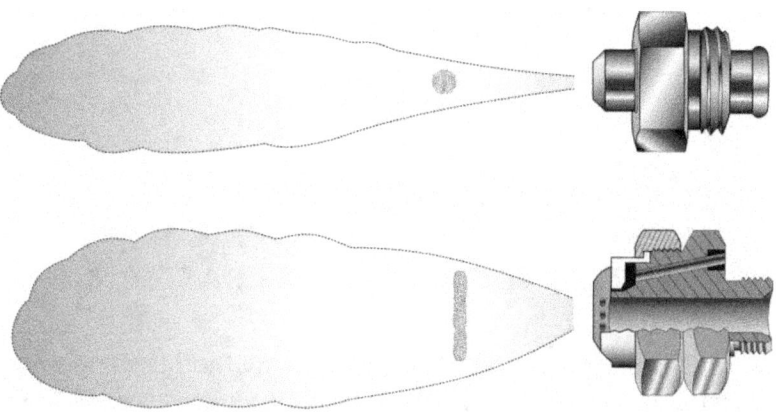

Figure 2.5. Typical air atomizing nozzle round spray pattern (top) and fan spray pattern (bottom).

Hydraulically atomizing nozzles force liquid through a fixed orifice at high pressures to achieve atomization into droplets. The orifice geometry produces various spray patterns and droplet sizes.

Hydraulic Full Cone Nozzles

Hydraulic full cone nozzles produce a solid cone-shaped spray pattern with a round impact area that provides high velocity over a distance (Figure 2.6). They produce medium to large droplet sizes over a wide range of pressures and flows. They are normally used when the sprays need to be located further away from the dust source [Bartell and Jett 2005].

Figure 2.6. Typical full cone nozzle and spray pattern.

Hydraulic Hollow Cone Nozzles

Hydraulic hollow cone nozzles produce a circular ring spray pattern and typically produce smaller drops than other hydraulic nozzle types of the same flow rate (Figure 2.7). They also have larger orifices which results in reduced nozzle clogging. Hollow cone nozzles are normally useful for operations where airborne dust is widely dispersed and are available in two different designs: whirl chamber and spiral sprays. In most whirl chambers, the spray pattern is at a right angle to the liquid inlet; however, in-line designs are also available. Both produce a more uniform pattern with smaller droplets (Figure 2.7). Spiral sprays are used when greater water flow is needed, resulting in less clogging due to the large orifices. There is also less pattern uniformity and larger droplets are created (Figure 2.8) [Bartell and Jett 2005].

Figure 2.7. Typical hollow cone whirl chamber nozzle and spray pattern. Right angle design shown.

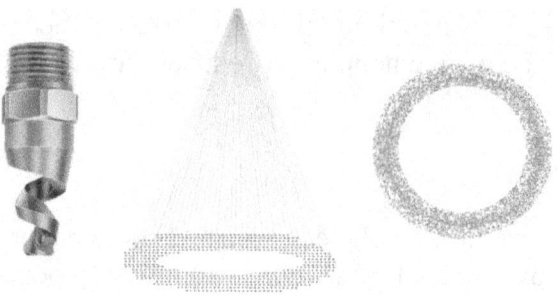

Figure 2.8. Typical hollow cone spiral nozzle and spray pattern.

Hydraulic Flat Fan Nozzles

Hydraulic flat fan nozzles produce relatively large droplets over a wide range of flows and spray angles and are normally located in narrow enclosed spaces (Figure 2.9). These nozzles are useful for wet dust prevention systems. Flat fan nozzles are available in three different designs: tapered, even, and deflected [Bartell and Jett 2005].

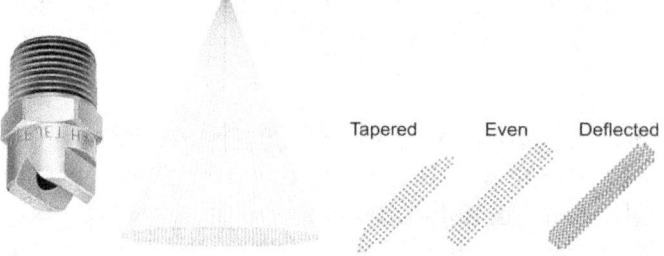

Figure 2.9. Typical flat fan nozzle and spray patterns.

Table 2.2 lists some common dust control application areas and the type of spray nozzle typically used for that application.

Table 2.2. Typical applications by spray nozzle type

	Air Atomizing	Hydraulic Fine Spray	Hydraulic Full Cone	Hydraulic Hollow Cone	Hydraulic Flat Fan
Typical Liquid Pressures	10–60 psi	30–1,000 psi	10–300 psi	5–100 psi	10–500 psi
Typical Air Pressures	10–70 psi	N/A	N/A	N/A	N/A
Airborne Dust Suppression					
Jaw crushers	•	•		•	
Loading terminals	•	•			
Primary dump hoppers	•	•			
Transfer points	•	•			
Dust Prevention					
Stackers, reclaimers			•	•	
Stockpiles			•		•
Transfer points			•	•	•
Transport areas/roads				•	

Figure 2.10 illustrates a typical dump application with both dust suppression and prevention spray nozzles. In this case, dumping action produces airborne dust in the upper level of the enclosure and the impact of the falling material with the material already in the enclosure causes additional dust to become airborne. The dust suppression spray nozzles in the headers at the top of the enclosure would be either air atomizing or hydraulic fine spray hollow cone nozzles because of the smaller droplet sizes they produce. The dust prevention spray nozzles at the lower level of the enclosure would be hydraulic atomizing full cone nozzles because of their larger droplet sizes and full cone spray pattern for large surface area coverage.

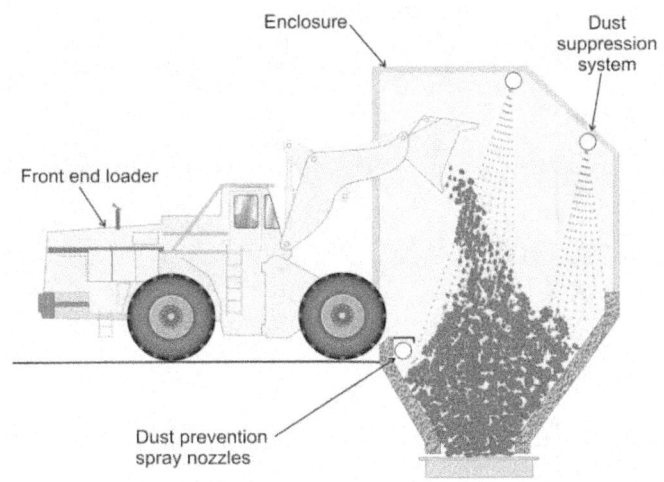

Figure 2.10. Typical loader dump dust control application.

Figure 2.11 illustrates a typical conveyor application with both dust suppression and prevention spray nozzles. In this case, dust prevention spray nozzles are positioned above the moving conveyor to prevent dust on the material from becoming airborne. Dust suppression nozzles are positioned at the discharge end transfer point to control airborne dust.

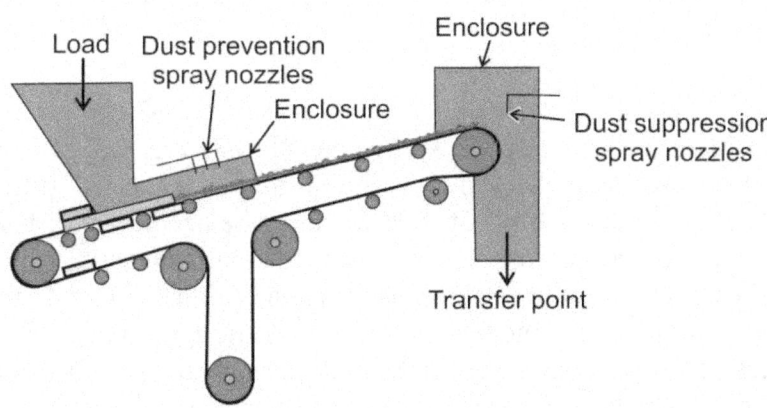

Figure 2.11. Typical conveyor dust control application.

Figure 2.12 shows the airborne suppression performance of the different spray nozzles performing at different operating pressures. As shown, atomizing sprays are the most efficient for dust knockdown or suppression, followed by the hollow cone sprays. Hollow cone sprays are a good choice for many applications in mineral processing operations because significant coverage or wetting of the ore occurs, even at low moisture percentages. Full cone sprays would be most applicable in the early stages of the process where the quantity of moisture added is not as critical. Flat fan sprays are most appropriate for spraying into a narrow rectangular space because less water is wasted by spraying against an adjacent rock or metal surface.

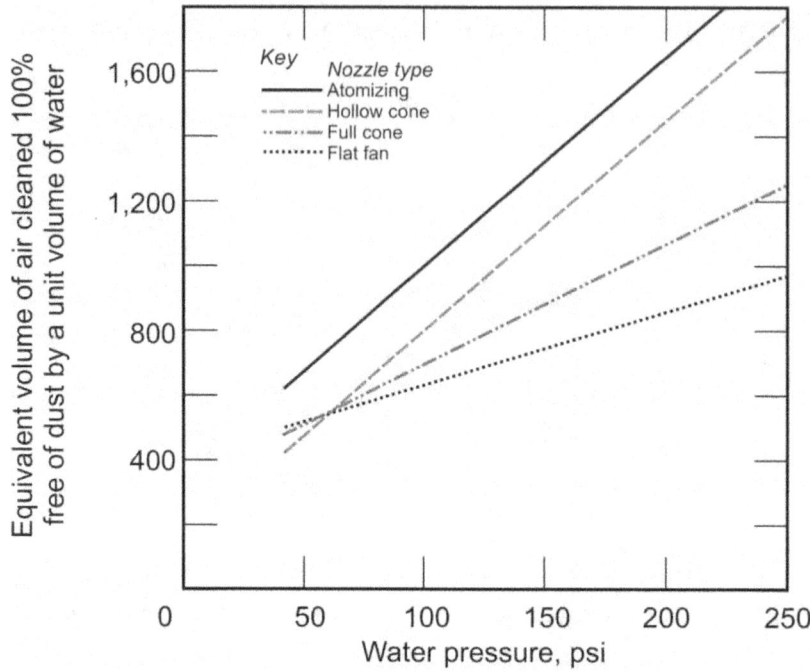

Figure 2.12. Airborne suppression performance of four types of spray nozzles [NIOSH 2003]. Atomizing nozzle is air atomizing, and hollow cone, full cone, and flat fan spray nozzles are hydraulically atomizing.

MAINTENANCE ISSUES WITH WET SPRAY SYSTEMS

Water Quality

A fundamental consideration with any wet spray control system is the quality of the water being sprayed, with water hardness and cleanliness being of greatest concern. If the available water supply contains a high level of minerals, nozzle wear can be accelerated. The user should consider nozzles fabricated in stainless steel if this condition is present. A high level of minerals in the supply water can also introduce caking, which is a buildup of carbonates on the nozzle, and can increase the amount of maintenance required. Although it may not be possible to modify the hardness of the water being sprayed, a properly designed and applied maintenance program can help to minimize the adverse effects of high levels of minerals in the available water supply.

If spray nozzles become plugged with sediment or debris, they render the spray system ineffective. Since the water to be used for spray systems at most mineral processing operations is drawn from settling ponds, water purity is a concern. It is recommended to use a water filtering system to eliminate the possibility of sediment and debris clogging the water sprays. Most spraying companies offer filtering systems for this purpose.

An effective method of supplying water to the system is to use a self-contained water delivery system specifically designed for the application. Such a system includes a water pump selected to provide water at the specific flow and pressure required for the application. Also included are a manual inlet valve, an inlet strainer, a flow switch to prevent the pump from operating without any water supply, pressure indicators, a bypass circuit to allow manual adjustment of water flow, a manual outlet valve, and a manual or timer-based control panel. A typical water delivery system is illustrated in Figure 2.13.

Water supply systems can be as simple as one which delivers a fixed water flow rate or one which can adjust the water flow rate in reaction to the needs of an application. Today's electronic technology offers the ability to sense factors such as the dust loading and automatically adjust the liquid flow rate and pressure to optimize the dust control system performance.

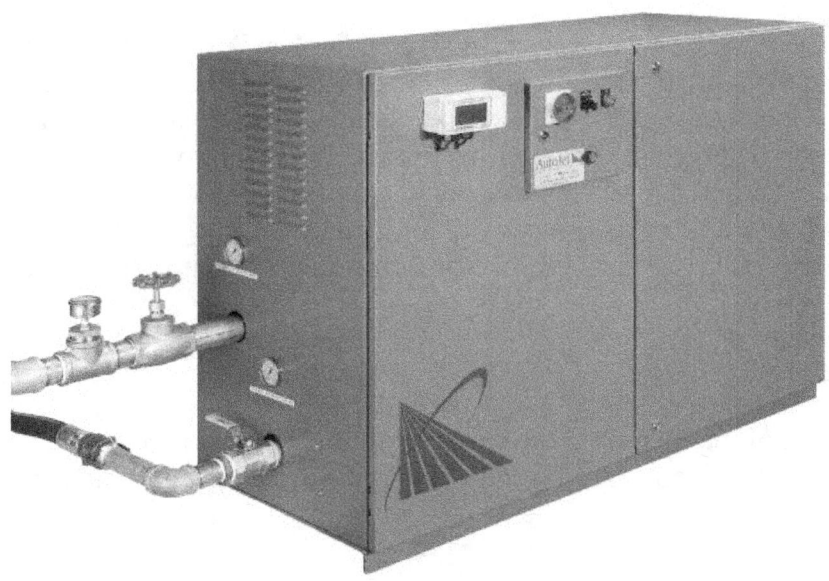

Figure 2.13. Typical self-contained water delivery system [SSCO 2006].

In applications where water quality is poor and where the water contains excessive amounts of particulate, such as water drawn from settling ponds, additional filtration should be provided. In these cases, it is recommended that a duplex basket strainer be used prior to the water delivery system inlet. These units provide two basket strainers and a manual valve, which allows the operator to switch from one strainer to the other to allow the inactive strainer to be removed and cleaned. The strainer mesh should be selected based on the particulate size present in the water supply. The function of the strainer is to stop particles that could ultimately clog the spray nozzle orifice. For this reason the strainer mesh should be such that it will stop solid particles that are larger diameter than the orifice of the spray nozzles in the system. A typical duplex basket strainer is shown in Figure 2.14.

Figure 2.14. Typical duplex basket strainer [SSCO Catalog 70].

Nozzle Maintenance

Nozzles require maintenance, regular inspection, cleaning, and even replacement to preserve final production quality and to maintain production processes on a cost-efficient basis. The type and frequency of the maintenance schedule depends on the particular application. In some operations, nozzles can spray usefully after hundreds of hours of operation, while in others, nozzles require daily attention. Most nozzle applications fall between these extremes. At a minimum, nozzles should be visually surveyed for damage on a regular basis. Additional maintenance will depend on application specifications, the water quality used, and nozzle material.

Erosion and Wear

The gradual removal of material from the surfaces of the nozzle orifice and internal flow passages causes them to become larger and/or distorted (Figure 2.15), which can affect flow, pressure, and spray pattern.

Figure 2.15. Nozzle erosion—new versus used [Schick 2006].

Corrosion

The chemical action of sprayed material or the environment causes corrosion breakdown of the nozzle material (Figure 2.16).

72 Wet Spray Systems

Figure 2.16. Nozzle corrosion—new versus used [Schick 2006].

Clogging

Unwanted dirt or other contaminants blocking the inside of the orifice can restrict the flow and disturb spray pattern uniformity.

Caking

Overspraying, misting, or chemical buildup of material on the inside or outer edges of the orifice from evaporation of liquid can leave a layer of dried solids and obstruct the orifice or internal flow passages (Figure 2.17).

Figure 2.17. Nozzle caking—new versus used [Schick 2006].

Temperature Damage

Heat may have an adverse effect on nozzle materials not intended for high-temperature applications (Figure 2.18).

Figure 2.18. Temperature damage—new versus used [SSCO 2004].

Improper Reassembly

Misaligned gaskets, over-tightening, or other repositioning problems can result in leakage as well as poor spray performance.

Accidental Damage

Inadvertent harm to an orifice can be caused by scratching through the use of improper tools during installation or cleaning (Figure 2.19).

Figure 2.19. Nozzle damage—new versus used [SSCO 2003].

Checking Spray Nozzle Performance

Visual nozzle inspections alone will not always indicate whether the nozzle is performing to specifications. The spray tips pictured in Figures 2.18 and 2.19 illustrate that factors such as wear that affect performance are not always obvious.

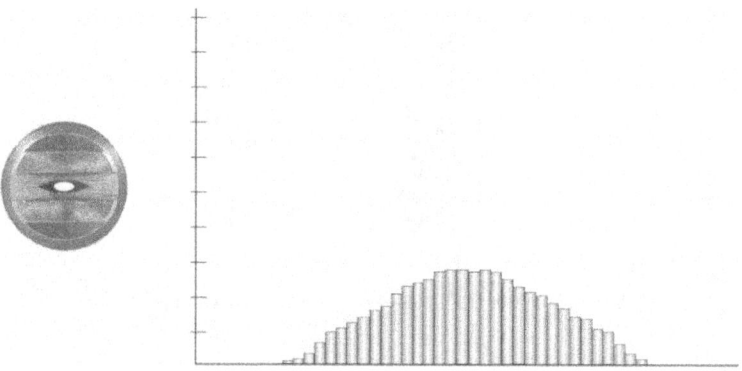

Figure 2.20. Good spray tip showing pattern and distribution graph [Schick 2006]. Height of bar indicates distribution of water over pattern width and indicates relatively uniform flow over the width of the pattern.

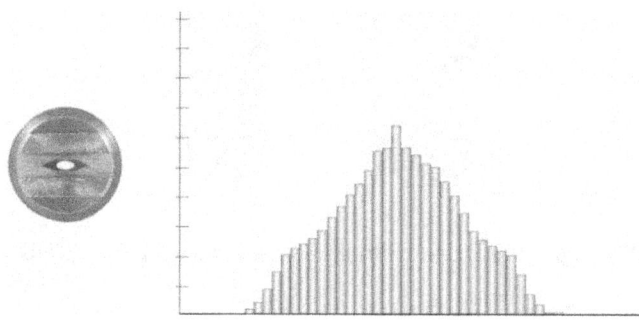

Figure 2.21. Worn spray tip showing pattern and distribution graph [Schick 2006]. Height of bar indicates distribution of water over pattern width and indicates an increase of and excessive flow in the center of the pattern due to orifice wear.

Performance Testing of Equipment

Due to the number of possible maintenance issues, performance testing should be done to monitor the condition of spray equipment on a regular basis, considering numerous factors.

- *Application-specific analysis or measurements.* The dust content of the air should be surveyed and checked either visually, to determine the effectiveness of the spray at removing dust, or by using dust monitoring equipment that measures dust content in air. Visual inspection consists of simply observing the dust-producing area to determine if the spray system is effectively controlling the dust condition. Because this method of monitoring does not quantify the dust loading and is subject to operator interpretation, it does not specifically measure the efficiency of the spray system. Dust particle monitoring equipment is available that will quantify the dust loading in the problem area and provide a more accurate assessment of the system efficiency.
- *Flow rate.* The eye cannot necessarily detect increased flow, so the flow rate of each nozzle should be checked periodically by reading the flow meter or collecting the spray in a container. The results should be compared to specifications or to the performance of new nozzles.
- *Spray pressure.* Pressure in the nozzle manifold should be checked using a properly calibrated pressure gauge.
- *Spray pattern.* In many instances visual inspection is adequate for monitoring pattern uniformity. Changes caused by orifice damage, clogging, or caking are usually noticeable. However, to detect gradual orifice wear, special measuring equipment will be required. A flow meter may be used to compare the flow of a new nozzle at a specific liquid pressure to the flow of a used nozzle at the same pressure. Nozzle wear causes the flow to increase so the flow meter reading is a quantifiable indicator of the degree of orifice wear.
- *Nozzle alignment.* To provide uniform coverage, nozzles should be oriented correctly in relation to one another so that all like patterns are parallel.

REFERENCES

Bartell W, Jett B [2005]. The technology of spraying for dust suppression. Cement Americas, May/June pp. 32–37.

Blazek CF [2003]. The role of chemicals in controlling coal dust emissions. Presented at the American Coal Council PRB Coal Use: Risk Management Strategies and Tactics Course. Dearborn, Michigan, June 2003.

NIOSH [2003]. Handbook for dust control in mining. By Kissell FN. Pittsburgh, PA: U.S. Department of Health and Human Services, Centers for Disease Control and Prevention, National Institute for Occupational Safety and Health, NIOSH Information Circular 9465, DHHS, (NIOSH) Publication No. 2003–147.

NDT Educational Resource Center, Iowa State University. [http://www.ndt-ed.org/EducationResources/CommunityCollege/PenetrantTest/PTMaterials/surfaceenergy.html]. Date accessed: February 13, 2009.

Rocha E [2005a]. PowerPoint presentation slide # 30 presented by E. Rocha, General Manager, Spraying Systems do Brasil Ltda. Spray Technology Workshop for Pollution Control at Spraying Systems do Brasil Ltda, São Bernardo do Campo, Brazil.

Rocha E [2005b]. PowerPoint presentation slide # 43 presented by E. Rocha, General Manager, Spraying Systems do Brasil Ltda. Spray Technology Workshop for Pollution Control at Spraying Systems do Brasil Ltda, São Bernardo do Campo, Brazil.

Rocha E [2005c]. PowerPoint presentation slide # 75 presented by E. Rocha, General Manager, Spraying Systems do Brasil Ltda. Spray Technology Workshop for Pollution Control at Spraying Systems do Brasil Ltda, São Bernardo do Campo, Brazil.

Schick R [2006]. Spray technology reference guide—understanding drop size. [http://www.spray.com].

SSCO (Spraying Systems Co. Catalog 70). Industrial spray products. [http://www.spray.com].

SSCO (Spraying Systems Co.) 2006. A guide to safe and effective tank cleaning. [http://www.spray.com].

SSCO (Spraying Systems Co.) 2003. Spray optimization handbook TM–410. [http://www.spray.com].

Swinderman RT, Goldbeck LJ, Marti AD [2002]. Foundations 3: the practical resource for total dust & material control. Neponset, Illinois: Martin Engineering.

USBM [1987]. Dust control handbook for mineral processing. U.S. Department of the Interior, Bureau of Mines. Contract No. J0235005. NTIS No. PB88–159108. [http://www.osha.gov/SLTC/silicacrystalline/dust/dust_control_handbook.html]. Date accessed: December 13, 2008.

CHAPTER 3: DRILLING AND BLASTING

CHAPTER 3: DRILLING AND BLASTING

The drilling process is used in both surface and underground mines for blasting operations which are conducted to fragment the rock. While the drilling process is similar for both types of mines, the different operating environments require specialized techniques to accomplish the same task. Drilling operations are notorious sources of respirable dust, which can lead to high exposure levels for the drill operator, drill helper, and other personnel in the local vicinity during operation. Therefore, dust controls on drills are necessary and involve both wet and dry methods.

Operator cabs are increasingly becoming an acceptable method for protecting the drill operator from respirable dust generated by the drilling operation, and the use of cabs is fully discussed in Chapter 9—Operator Booths, Control Rooms, and Enclosed Cabs. Enclosing the operator in an environmentally enclosed cab is extremely effective. However, the protection provided, is only available to the personnel inside the cab. Other personnel working in the vicinity of the drilling operations, the drill helper, shotfirer, mechanics, etc., cannot be protected in this manner. They can attempt to maintain a work location upwind of the drill to avoid respirable dust, but this is not always practical. Therefore, methods for dust control on drilling operations are still required.

There are two basic methods for controlling dust on drills: either a wet suppression system or a dry cyclone/filter type collector. Wet systems operate by spraying water into the bailing air as it enters the drill stem. Dust particles are conglomerated as the drill cuttings are bailed out of the hole. Dry collectors operate by withdrawing air from a shroud or enclosure surrounding the area where the drill stem enters the ground. The air is filtered and exhausted to the atmosphere. When dust controls are implemented effectively at drilling operations, both wet suppression and dry collection systems can achieve good dust control efficiency.

This chapter reviews methods for efficient and effective dust control for both underground and surface drilling operations. Much of the research for dust control has been conducted throughout the past century. The timeline for underground dust control research was during the 1920–1950 time period, when there was a significant program undertaken by the U.S. Bureau of Mines to prevent silicosis in underground miners. While this research could seem outdated, the principles still apply today and are confirmed to be still effective by practices in the field. Additionally, many of the results from underground research form the basis for dust controls developed for surface mine drilling. Surface mine dust control research began around 1980, when it was recognized that surface drillers were also susceptible to silicosis. Research continues today to discover new methods for dust control for both surface and underground drilling.

SURFACE DRILLING DUST CONTROL

Surface mine drilling is accomplished using both rotary and percussion drilling methods. Rotary drilling achieves penetration through rock by a combination of rotation and high down pressures on a column of drill pipe with a roller drill bit attached to its end. Percussion drilling also achieves rock penetration through rotation and down pressure, but with a pneumatic drill, which contains a piston which delivers hammer blows to the drill column or the drill bit, depending upon the location of the drill (top hammer or down-the-hole hammer), eliminating the need for the high down-pressures required in rotary drilling.

Typical holes can be any size up to 15 inches in diameter, with the larger hole diameters commonly produced using rotary drill bits. Generally, these holes are oriented vertically, although some operations do use angled holes in their blast design, and the holes are drilled in a pattern where they are aligned in rows. The type of drilling equipment can range from small surface crawler rigs to truck-mounted drills to large track-mounted drill rigs, as seen in Figure 3.1.

Dust control methods for surface drilling use wet drilling or dry drilling with dust collection systems. There are variations to these methods due to the operating environment of the surface drilling operation and the type of equipment used, but generally the principles of dust control presented in this chapter are applicable to all types of surface drilling; including small crawler, truck-mounted, and large track-mounted drills.

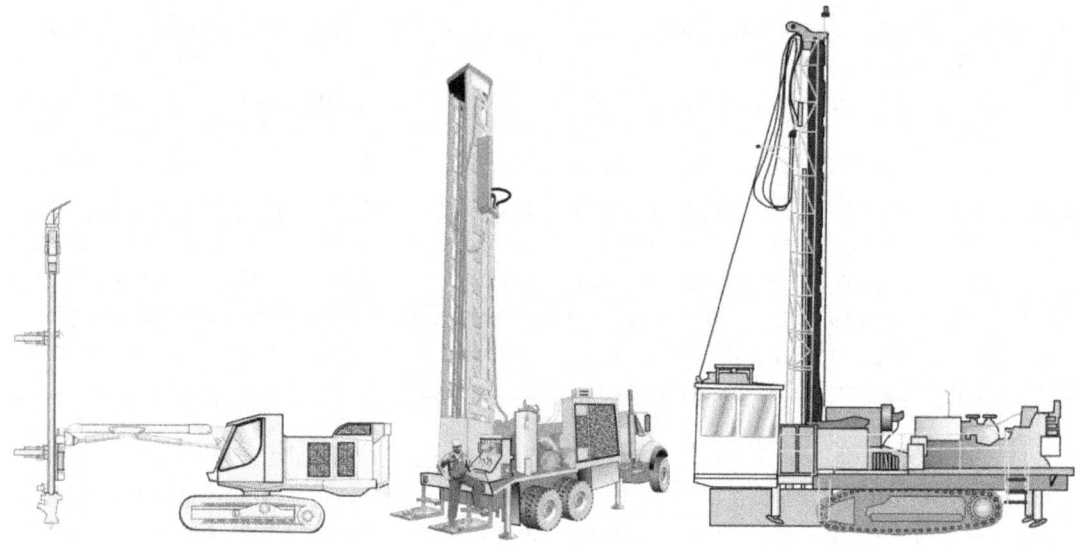

Figure 3.1. Illustrations depicting a small surface crawler drill rig, a truck-mounted drill rig, and a large track-mounted drill rig, respectively.

Wet Drilling

The best method for dust control in surface drilling is to use wet drilling techniques. Wet drilling injects water along with the air to flush the cuttings out of the hole as shown in Figure 3.2. Testing has shown that this technique can provide dust control efficiencies up to 96 percent [USBM 1987].

Water injection requires monitoring by the drill operator to efficiently control dust. The amount of water required for dust control is not large. Typical water flow rates in wet drilling systems generally range from 0.1 to 2.0 gpm, but this varies based on the drill type, geology, and moisture level of the material being drilled. For example, testing at a surface mine site showed that dust control efficiencies greatly increased from 0.2 to 0.6 gallons per minute (gpm), then leveled off above this flow rate. However, once the flow rate approached 1.0 gpm at this site, operational problems were encountered such as the drill bit plugging and the drill steel rotation binding due to the water causing the drill cuttings to become too heavy to be removed by the bailing air [USBM 1987]. Therefore, too little water reduces dust control efficiency while too much water creates operational problems. The amount of water needed would be dependent upon the surface drill type and the material being drilled.

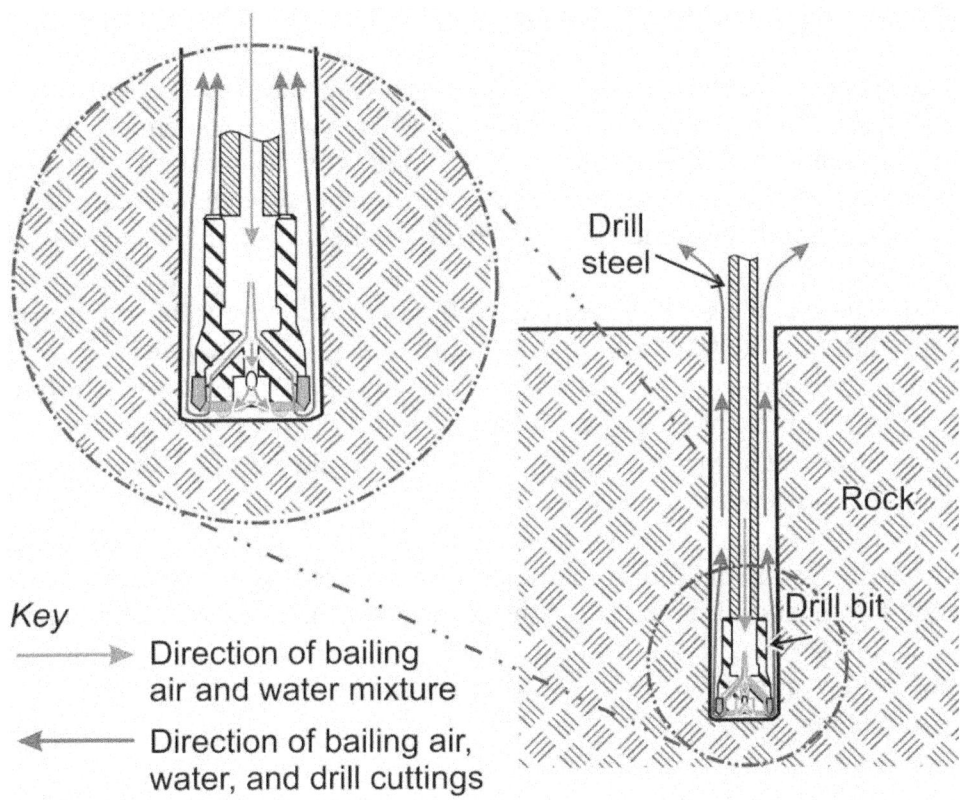

Figure 3.2. The air and water flow during drilling operation to demonstrate water flushing of the drill cuttings. The water flows through the center of the drill steel and out the end of the drill bit to remove the cuttings from the drill hole.

To provide for wet drilling, a water tank mounted on the drill is used to pump water into the bailing air, and the water droplets trap dust particles as they migrate up the annular space of the drilled hole (the annulus is the open area of the drill hole between the drill steel and the wall of the drill hole), thus controlling the dust as the air bails the cuttings. The drill operator controls the water flow from the cab, and some cabs are equipped with a meter to indicate water flow rate.

Advantages to Wet Drilling

A potential advantage to the use of water in percussion drilling operations, besides better dust control, is that the use of water may improve the penetration rate of percussion drilling for surface mining. There is very little information on wet drilling effects on penetration rate, but sources have shown that the use of water while percussion drilling in underground mining can produce an increase in the penetration rate of the drilling operation [Hustrulid 1982; USBM 1995]. Since this is applicable for underground mining, it may also correlate to increased penetration rates in surface mining situations. Further testing would be required to definitively assert that wet drilling would increase penetration rates of surface percussion drilling operations.

Recommendations for Proper Wet Drilling

Based on testing results and observational best practices, the following recommendations apply to surface drilling using wet drilling techniques.

- In order to operate at close to the optimum water flow rate, the operator should slowly increase the amount of water just to the point where visible dust emissions are abated. Addition of more water beyond this point will not provide any significant improvement in dust control, but will most likely create operational problems such as bit degradation (when using tri-cone bits) and possible seizing of the drill stem. Using less water will give poor dust control.
- It is important that the water be increased slowly to account for the lag time as the air/water/dust mixture travels from the bottom to the top of the hole.
- Continuous monitoring of the water flow during drilling is necessary to provide optimum dust control and prevent drill steel binding.
- The water used in the wet drilling system should be filtered to prevent debris from plugging the drill's wet drilling system.
- When using a wet suppression system in outside temperatures below freezing, the system must be heated when the drill is in operation, and during downtimes the system must be drained. For most drilling machines the proximity of the water tank and lines to the engine and hydraulic lines is sufficient to prevent freezing during operation (except in extreme cold weather situations). The water tank and lines must be drained when the drill is not in operation.

Disadvantages to Wet Drilling

While wet drilling may be an advantage when percussion drilling, there is a disadvantage to wet drilling when rotary drilling. The use of water degrades the tri-cone roller drill bits and shortens their lives by 50 percent or more [Page 1991]. This is due to rapid bearing material degradation through hydrogen embrittlement and accelerated bit wear. The bit wear is a result of operating in the abrasive rock dust-water slurry environment.

There is a solution to the disadvantage of short bit lives when wet drilling. In order to obtain acceptable drill bit life, the water must not reach the drill bit. This can be achieved through effective water optimization and water separation. Data gathered by the Bureau of Mines (BOM) from a mine where drilling occurred in monzonite, sandstone, limestone, and iron ore over a 14-year period showed that drill bit life averaged 1,938 ft/bit when wet drilling without water separation. With water separation, the drill bit life increased over 450 percent to an average of approximately 9,000 ft/bit [USBM 1988].

Water Separator Sub

Water separation can be achieved through the use of a water separator sub. A water separator sub uses inertia to remove the injected water from the bailing air. A sub is a short length of a drill collar assembly which is placed between the drill bit and the drill steel. This assembly has male threads on one end to accommodate the drill steel and female threads on the other for the drill bit. It generally incorporates stabilizer wear bars affixed on its circumference which are used to center the drill bit in the drill hole.

Water separation is accomplished by forcing the bailing air stream to make a sharp turn or turns within its course, just above the bit (see Figure 3.3). Basically, the air/water mixture flows down through the center of the drill steel, down past the standpipe, and makes a u-turn at the water reservoir. The airflow then flows up the outside of the stand pipe, through the slots, at which point the flow turns down the inside center of the standpipe, exiting through the nozzles in the drill bit. Due to the much higher inertia of the water, it cannot negotiate the turns and thus separates from the air. Since the drill stem interior is under positive air pressure, the accumulated water ejects through weep holes. These holes are in the separator perimeter above the bit and the water ejects into the drill stem annulus. All drill cuttings cleared by the airflow travel through this annulus and are wetted. This prevents water from reaching the drill bit and inhibits the formation of slurry at the bit/rock interface.

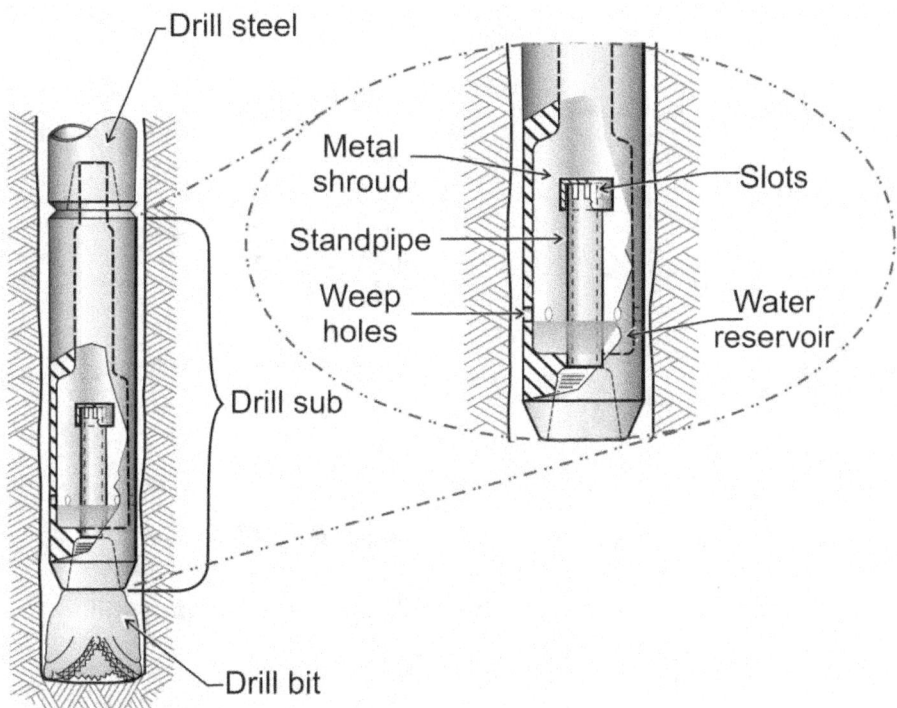

Figure 3.3. Internal workings of a water separator sub.

Testing has demonstrated that dust control efficiencies of up to 98 percent can be obtained using the water separator sub while dust control efficiencies of wet drilling without the water separator sub were 96 percent [Page 1991]. Most importantly, the use of the water separator sub increased bit life.

Dry Drilling

Dry drilling is accomplished without the use of water for dust control. Dust control is accomplished using a dust collection system mounted on the drill. These systems have the ability to operate in various climates, i.e., they are not subject to freezing at lower temperatures as with the use of water, and they can be up to 99 percent efficient if properly maintained [USBM 1987]. There are different types of dust collector configurations used which are dependent upon the size of the drill.

Medium- to Large-Diameter Drill Dust Collection Systems

Figure 3.4 shows a typical dry dust collection system on a medium- to large-diameter drill. Drill dust is generated by the bailing air, which is compressed air that is forced through the drill steel out the end of the drill bit and used to flush the cuttings from the hole (Figure 3.2 depicts a similar scenario, but includes water). In a properly operating collection system, these cuttings are contained by the drill deck and shroud located over the drilling area. This dusty air from underneath the shrouded drill deck is removed by the dust collector system. The collector is composed of an exhaust fan and filters which filter the air from underneath the shroud area. It is generally self-cleaning, using compressed air to pulse through the filters at timed intervals to clean them and prevent clogging. The filtered fine-sized material then drops out the bottom of the collector to the collector dump.

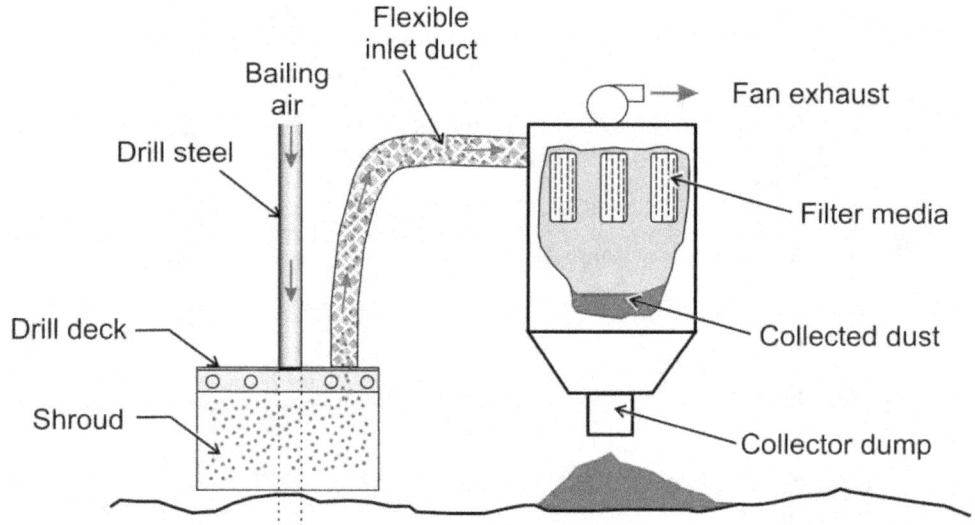

Figure 3.4. A basic dry dust collection system on a drill.

Sources of Dust Emissions from Collection Systems

Damaged, nonfunctional, or missing dust collector filters are generally the root cause of dusty exhaust air from the collector. Other emission sources in the collection system can be caused by damaged or worn collector system enclosure components or operational conditions impeding the enclosure components from containing and capturing dust.

The integrity of the drill deck shroud and its sealing capability to the ground are critical factors in the effectiveness of a dry collection system for medium- to large-diameter drills. Over half of the dry dust collector emissions are from shroud and drill stem bushing leakage [USBM 1985]. Openings or gaps greater than six inches between the drill deck shroud and ground can significantly diminish the dust collector's inlet capture effectiveness [USBM 1986; Zimmer et al. 1987]. Three areas of vulnerability have been identified related to dry dust collection:

- deck shroud leakage,
- collector dump discharge, and
- drill stem/deck leakage.

Deck Shroud Leakage Solutions

With the exception of air track drills, most rock drills have the same basic configuration and collector airflow volumes. The only parameter that varies is the shroud leakage area, and such variances obviously change the capture efficiency. Some of the most common shroud enclosure openings are between the bottom of the shroud and ground, caused by uneven or sloping drill bench surfaces and at the corners of the drill deck.

<u>Rectangular Shroud</u>

Some shrouds have gaps at the corners of the front side of the shroud, which can be hydraulically lifted up so that drill cuttings are not dragged over the hole when the drill moves with its boom up. Because most deck shrouds are rectangular and constructed from four separate pieces of rubber belting attached to the deck, leakage can also occur as the seams separate from one another. To combat this problem, corner flaps can be added, attached with brackets, to help preserve the integrity of the shroud (see Figure 3.5), or the shroud can be constructed with a single piece of material (rubber belting) encompassing the drill deck perimeter.

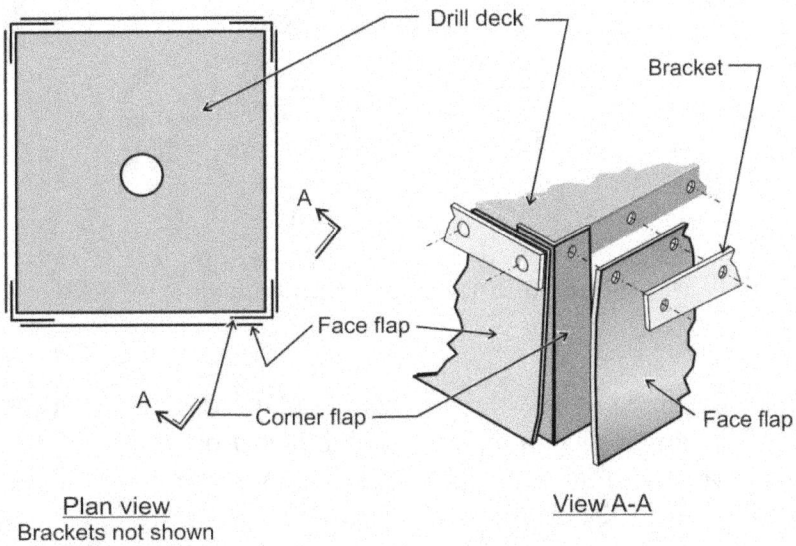

Figure 3.5. Reinforcing flaps added to each corner of a rectangular deck shroud to reduce leakage.

<u>Circular Shroud</u>

In contrast to the use of a typical rectangular shroud, a specialized circular deck shroud can be used to effectively accomplish dust capture during dry collection. A circular deck shroud is slightly conical in design with no open seams, and steel banding is used to attach the shroud to the bottom of the drill deck. The shroud can be hydraulically raised to the drill deck and lowered to make contact with the ground. A steel band is also attached to the bottom of the shroud to help maintain its shape and provide weight for lowering it to the ground, and guide wires are attached to the bottom of the steel band and a hydraulic cylinder (see Figure 3.6). A thin sheet rubber material is used on the shroud to give it flexibility, and the shroud has a small trap door, operated manually so that the cuttings can be shoveled from inside the shroud without compromising capture efficiency.

During drilling, a large amount of cuttings may be generated and it is frequently necessary to raise the circular drill shroud to prevent the cuttings from falling back into the hole. As a result, there are times when it is unavoidable to have a broken seal between the shroud and the ground or cuttings. Therefore, it is important that the driller make a conscientious effort to keep the leakage area to a minimum. This may involve making frequent inspections underneath the shroud by raising the drill shroud to visually assess the circumstances.

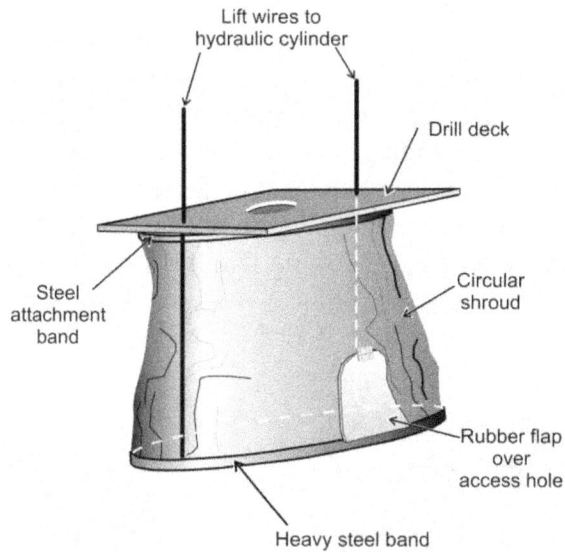

Figure 3.6. Circular shroud design.

<u>Collector to Bailing Airflow Ratio</u>

Dry collection can be performed most efficiently by maintaining an appropriate dust collector to bailing airflow ratio. The dust collector airflow to bailing airflow ratio, which can be represented by Q_C/Q_B, has been found to be an important indicator of dust collection performance. Dust collector airflow (Q_C) in standard cubic feet per minute (scfm) is the quantity of airflow generated by the dust collector fan, which pulls the air from underneath the drill shroud. Bailing airflow (Q_B) in scfm is the quantity of compressed air blown down the drill stem through the bit in order to flush the cuttings out of the hole.

Bailing airflow (Q_B) can be obtained from the drill manufacturer as the quantity airflow rating of the drill's compressor, which has units in scfm, while measurements of the collector airflow (Q_C) can be reasonably made by using a hot wire anemometer, vane anemometer, or pitot tube at the collector exhaust. More accurate measurements can be obtained by attaching a short (4-foot) duct extension to the collector exhaust and inserting the hotwire or pitot tube, to measure the airflow in cubic feet per minute (cfm), into a hole made at the halfway point in the duct. This extension can be simply made from cardboard and fitted to the outside of the collector exhaust duct. The collector airflow can be converted to scfm, but this is not necessary as the corrections from the ambient conditions generally have a small effect on the measurement. Also, due to the changes in the pressure drop across the filters caused by dust buildup, the quantity of air from the dust collector can vary by an amount that is much more significant than changes from standard atmospheric conditions.

Common collector to bailing airflow ratios on operating drilling rigs were found to be as high as 3:1 in the field. However, 2:1 ratios were more typical of operating collectors with normal filter loading [Page and Organiscak 2004]. Poorly operating collectors were found to operate at ratios of 1:1 or lower. Testing has shown a notable reduction in dust concentrations escaping through the bottom of the deck shroud gap by increasing the dust collector to bailing airflow ratio [NIOSH 2006; Organiscak and Page 2005]. Figure 3.7 shows the results of changing the dust collector to bailing air ratio at various drill deck shroud gap heights from the ground. As shown, the largest decrease in dust levels is observed when the dust collector to bailing airflow ratio is increased from 2:1 to 3:1, with further decreases in dust levels as the ratio is increased to 4:1.

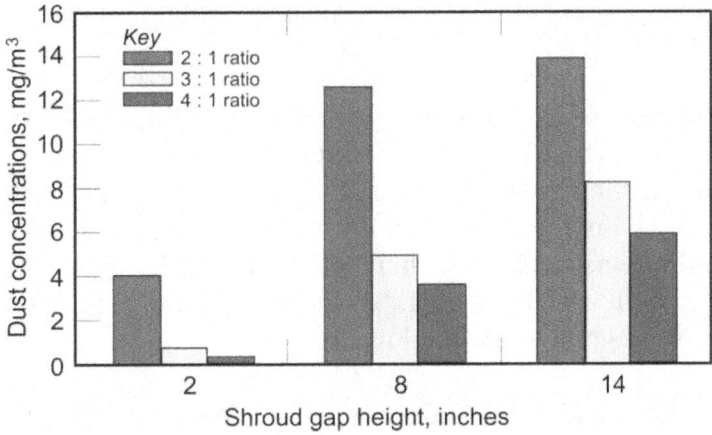

Figure 3.7. Results of changing dust collector to bailing air ratio at various shroud gap heights.

Airflow Maintenance for Dust Control

The collector and bailing airflows need to be maintained to sustain the high collector to bailing airflow ratios that prevent high dust emissions. The bailing airflow from the drill's compressor is the maximum rated airflow and generally does not vary much over the life of the drill. Therefore, its impact on the collector to bailing airflow ratio is minor compared to the impact of the collector airflow. Performing the normal maintenance on the compressor as required by the manufacturer will keep it in proper operating condition.

The collector airflow has a more substantial impact on the ratio due to the many components that make up the collector. The dust collector airflow can decrease over time through damaged components and neglected maintenance. Therefore, proper maintenance of the collector to maintain maximum airflow is required. Collector maintenance involves five actions by the operator:

- ensuring that filter backflushing is operating properly and within specifications;
- verifying that the intake duct and collector housing are tightly sealed and free of holes;
- changing filters at recommended intervals or when damaged;
- ensuring that the tubing and inlet to the collector are free of obstructions; and
- ensuring that the collector fan is operating properly and within its specified speed.

<u>Use of Collector to Bailing Airflow Ratios for Dust Control</u>

The collector to bailing airflow ratios are useful in that they form the basis of a model to predict the relative severity of the drill dust emissions and to estimate how much of a reduction is possible by measuring several basic parameters of the drilling operation. The parameters needed are: dust collector airflow (Q_C), bailing airflow (Q_B), drill deck shroud cross-sectional area (A_S), and shroud leakage area (A_L) or an approximate estimate for the leakage area [Page et al. 2008a,b].

The model depicted as a graph in Figure 3.8 (top) shows the relative reductions possible by reducing the leakage area [Page et al. 2008a,b]. To demonstrate how the graph can be used, the following example is given.

> An operator has a drill rig with drill deck dimensions of 4 feet by 5 feet. The rated compressor Q_B is 260 scfm and Q_C was measured at 530 scfm which represents a collector to bailing airflow ratio Q_C/Q_B of approximately 2. The area of the shroud is calculated by multiplying the width by the length resulting in $A_S = 20$ ft^2. A_L is calculated by multiplying the leakage height (LH) in feet by the perimeter of the shroud which results in $A_L = $ LH x 18 ft. Therefore, the ratio $A_S/A_L = 20$ ft^2/(LH x 18 ft) and can be calculated by estimating LH.
>
> Since $Q_C/Q_B = 2$, the top graph in Figure 3.8 can be used to show how reducing the leakage gap between the shroud and the ground will reduce the severity of the dust concentrations. A gap of 14 inches corresponds to $A_S/A_L = 0.95$, showing a relative airborne respirable dust concentration of approximately 16 mg/m^3, while a gap of 2 inches corresponds to $A_S/A_L = 6.7$, resulting in a relative airborne respirable dust concentration of approximately 5 mg/m^3. This example demonstrates that reducing the leakage height of the drilling operation will result in a substantial improvement in dust reductions.

It should be noted that any leakage area due to vertical shroud seam gaps should also be included by estimating the area for vertical leakage and adding this value to the shroud leakage area (A_L), but many times vertical leakage may not be significant. Graphs for Q_C/Q_B greater than 2 are similar with the difference being that airborne respirable dust values at any value of A_S/A_L will become smaller as Q_C/Q_B increases [Page et al. 2008a], as demonstrated in the comparison of the two graphs in Figure 3.8.

It is important to keep in mind that the calculated value of airborne respirable dust is a relative value only and is not as important as the estimated value of A_S/A_L. The key considerations are where on the curve the drill operates, as determined by A_S/A_L, and which curve is applicable (i.e., what value is Q_C). These determinations will indicate the long-term average improvement that can be expected from either increasing the collector airflow (installing clean filters or a larger collector) or reducing the amount of shroud leakage.

To demonstrate, a drill currently operating on the left side of the curves in Figure 3.8 can readily make significant airborne respirable dust reductions. A drill operating on the right side of the curves indicates that only minimal reductions are achievable. However, operation on the right side of the curves also usually indicates a drill that has good operating dust controls [Page et al. 2008b]. As stated previously, typical values of Q_C/Q_B in actual operation with dirty filters are on the order of 2. Therefore, on surface blasthole drill rigs it is important to maintain collector to bailing airflow ratios greater than 2, with ratios of 3 or more being more desirable.

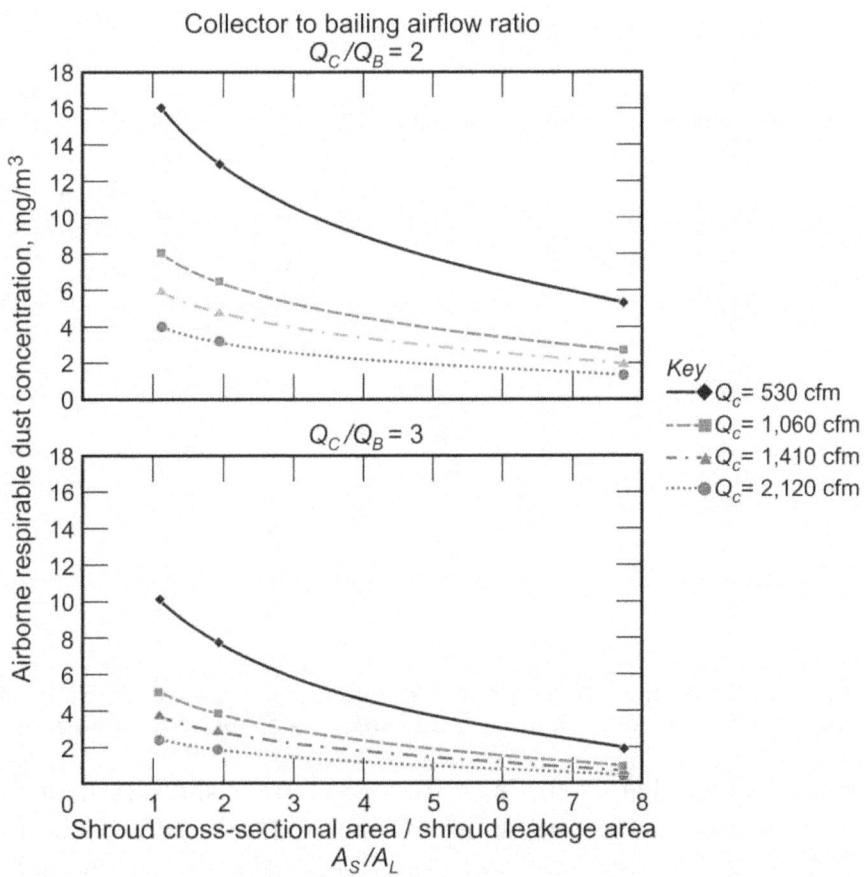

Figure 3.8. Graphs of the model that represents the severity of dust emissions based upon collector to bailing airflow ratios, showing that the respirable dust concentrations become lower as the collector to bailing airflow ratio increases.

Air-Blocking Shelf

The air-blocking shelf is a dust control that has been found to be effective for medium- to large-sized track-mounted blasthole drills. However, it would be effective for any drill with a large-sized drill shroud with approximate minimum dimensions of 4 feet by 4 feet. A 6-inch-wide shelf is placed underneath the drill shroud along the inside perimeter of the shroud. Its purpose is to reduce dust emissions from the drill shroud while the drill is operating. It was developed by observing the airflow patterns underneath the drill shroud at a drill shroud test facility [Potts and Reed 2008; Reed and Potts 2009].

Normally during drilling, the airflow pattern occurs as shown in Figure 3.9 (left), and consists of the bailing air exiting the drill hole along with an influence from the dust collector. The bailing air travels from the drill hole through the middle of the shrouded area, maintaining its course along the drill steel to the underneath side of the drill table, where it exhibits a coanda effect (a moving fluid's tendency to be attracted to a nearby surface) as it fans out across the bottom of the drill table and continues down the sides of the shroud. All of this occurs at a high velocity. Dust emissions at the ground surface occur when the air strikes the ground and fans out from underneath the shroud enclosure.

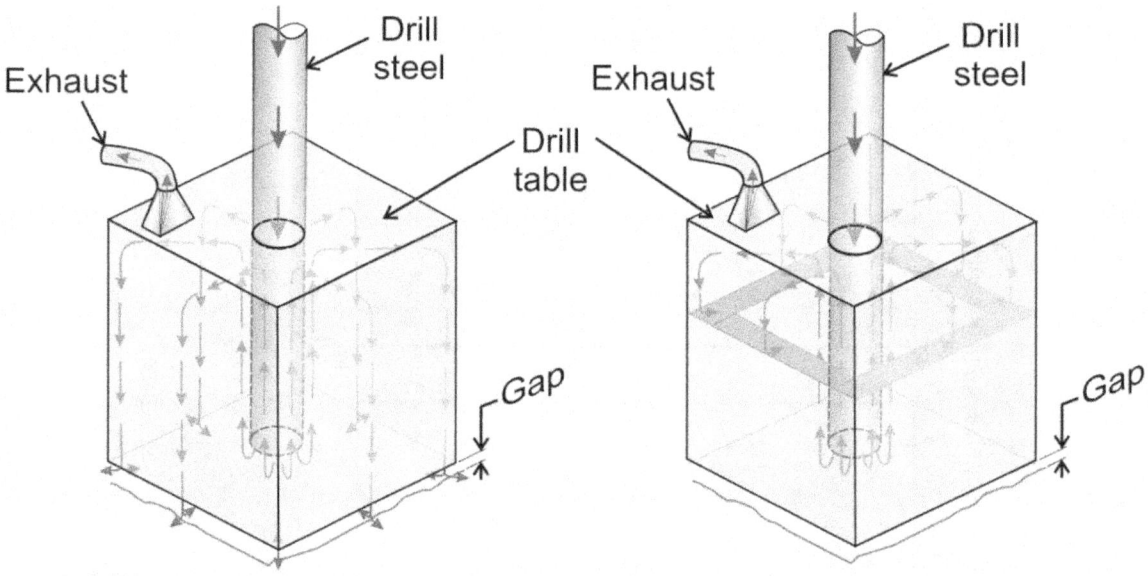

Figure 3.9. Qualitative models of airflow patterns underneath the shroud without the air-blocking shelf (left) and with the air-blocking shelf (right).

The 6-inch-wide shelf is placed along the inside perimeter of the shroud where it disrupts the airflow and redirects it toward the center of the enclosure to prevent the air from striking the ground (Figure 3.9, right). This redirection of air reduces dust leakage from underneath the shroud [Potts and Reed 2008; Reed and Potts 2009].

Figure 3.10 shows a shelf installed on the blasthole drill during testing. The shelf was constructed of 6-inch conveyor belting material that was bolted to 2-inch light-gauge angle iron. The angle iron was then bolted to the inside perimeter of the shroud. A section of shelf (not pictured) was added to the door flap of the shroud to ensure complete coverage of the parametric cross-section. The shelf was installed at a location that is approximately halfway between the top of the shroud and the ground surface. Laboratory testing demonstrated that the individual shelf pieces, which were located on each side of the shroud, did not have to be installed on the same horizontal plane in order to be effective. Rather, coverage of the parametric cross section of the shroud seemed to be the important design criterion (i.e., no gaps should be allowed at the corners or along the length of the shelf). Installation was simple with two persons able to assemble the shelf in less than an hour [Potts and Reed 2008].

The air-blocking shelf requires no maintenance once installed, unless it is damaged during tramming from hole to hole. Field testing of the air-blocking shelf on blasthole drills has shown dust reductions from the shroud ranging from 66–81 percent [Potts and Reed, 2010] with the differences being attributable to differing wind directions and rock type drilled. However, it should be noted that when the dust levels from drilling operations are very low (<0.5 mg/m^3), the effectiveness of the air-blocking shelf is reduced. This is due to the effect of the many other dust sources (dust collector, table bushing, other mining operations, etc.) surrounding the drill which can impact the dust levels [Potts and Reed 2010].

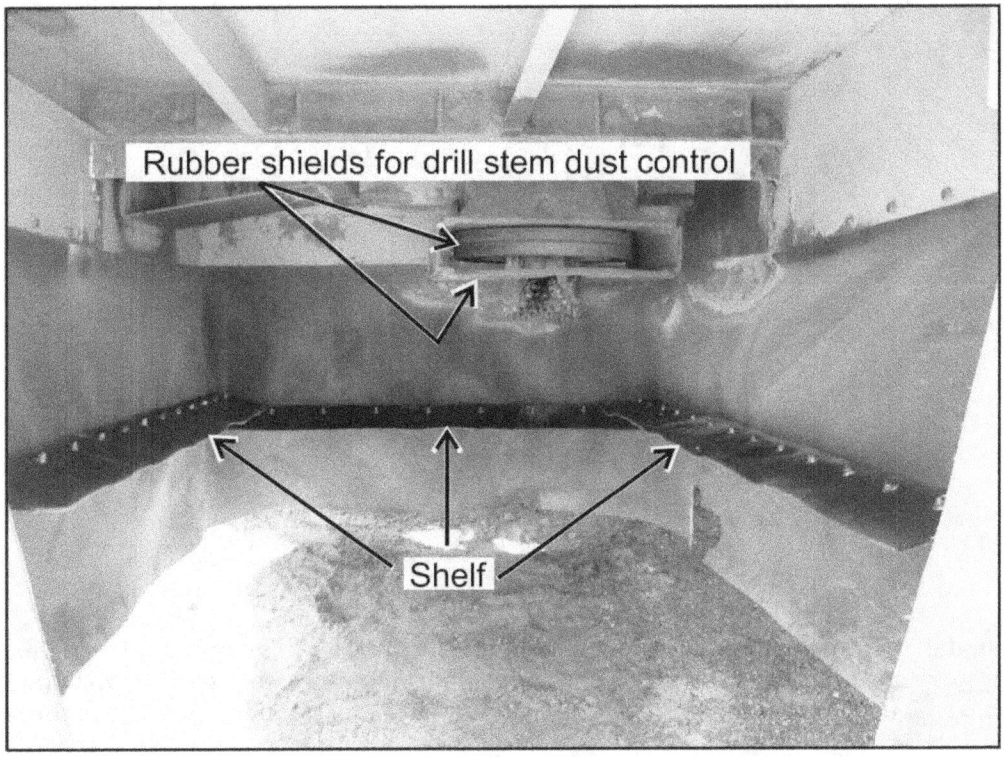

Figure 3.10. Air-blocking shelf installed on the inside perimeter of the drill shroud of a blasthole drill. Note the overlap in the rear corners to eliminate any gaps in the shelf perimeter.

A disadvantage that was found with the shelf is that material buildup from the drill cuttings can occur on the air-blocking shelf (Figure 3.11, left). This can cause significantly higher dust levels when lowering the mast due to the additional built-up material falling off the air-blocking shelf (Figure 3.11, right). Dust emissions from material falling from the shroud during drill mast lowering normally occur without the shelf. Therefore, the shelf does not create a new problem of dust emissions from mast lowering, but it does have the potential to exacerbate this existing problem. This disadvantage can be overcome by installing the shelf at a 45 degree angle from the horizontal and constructing it from different material (Figure 3.12).

Figure 3.11. Buildup of material on the air-blocking shelf (left) and the resulting dust emissions that are created during lowering of the drill mast (right).

Figure 3.12. Air-blocking shelf with modifications incorporated into installation. This shelf was installed in short sections in an attempt to prevent potential damage that could occur during tramming with the drill mast up. Note that each of the sections overlap, eliminating any gaps in the shelf perimeter.

When installing the shelf at a 45-degree angle, it is important that the shelf width be increased from 6 inches to 8 inches to maintain an equivalent 6-inch horizontal width. The shelf should be constructed of 1/4-inch high-density polyethylene (HDPE). The HDPE has a slippery surface which minimizes adhesion of the dust to its surface. It can be purchased in 4 foot by 8 foot sheets and cut to the required dimensions. These modifications eliminate the buildup of material on the air-blocking shelf. Additional modifications include shortening the shelf sections to prevent possible damage during tramming, and the addition of chains to support the shelf to prevent it from sagging. It is important that when using shortened shelf sections that they overlap to eliminate any gaps that may occur in the air-blocking shelf perimeter. These modifications should not diminish the effectiveness of the air-blocking shelf.

Collector Dump Discharge

The collector dump cycle can account for as much as 40 percent of the respirable dust emissions [USBM 1985]. The dust collector dump cycle operates in one of two modes: trickle mode or batch mode. In the first mode (trickle mode), the collector filter cleaning mechanism operates at preset time intervals, e.g., every 1 minute. This allows the fine material from the collector to trickle out onto the ground. In the second mode (batch mode), the collector cleaning mechanism operates when the bailing airflow stops. This condition occurs during intervals when adding or removing drill steels, or when removing the drill bit from the hole. The batch mode creates the highest dust levels because of the longer accumulation time of fine material in the collector before dumping. During the batch mode, the dust is usually emitted into the ambient air when the fine material is dropped, which creates significant dispersion. Also, this fine material strikes the ground, producing more dust dispersion. Finally, the drill itself and service vehicles frequently drive through the dust pile formed by dumping, stirring up clouds of dust. Avoiding disturbing the collector dump piles will prevent dust from being entrained into the air.

Preventing Collector Dump Dust Entrainment

The dust collector dump point is generally anywhere from 24 to 36 inches above the drill bench. Dumping the fine material from this height causes entrainment of the respirable sized fraction of this material into the air. To reduce this entrainment and lower the respirable dust concentrations at the collector dump point, a piece of brattice cloth can be attached to the dust collector dump using a large hose clamp. Figures 3.13 A, B, and C show the installation of the dust collector dust shroud. This dust shroud was installed over the existing rubber boot attached to the dust collector dump.

Figure 3.13. Dust collector dump point prior to shroud installation (A). Two men installing the shroud onto the dust collector dump point (B). The dust collector dump point after installation of the shroud (C).

This simple procedure of creating a dust collector dump shroud has been shown to be very effective in reducing the respirable dust. Respirable dust concentrations measured after installation of the dust collector shroud ranged from 0.16 to 0.24 mg/m^3. Figure 3.14 shows the reduction of the respirable dust concentrations near the collector dump point with and without the dust collector shroud. It can be seen that the respirable dust generated by the dust collector dump point can be reduced by between 63 and 88 percent using the shroud [Reed et al. 2004]. It should be noted that this reduction is highly dependent upon wind direction and wind speed. Advantages to this method of respirable dust reduction are that the material is inexpensive and requires almost no maintenance. If the shroud becomes damaged, it can easily be replaced in 10–15 minutes requiring little, if any, downtime for the drill.

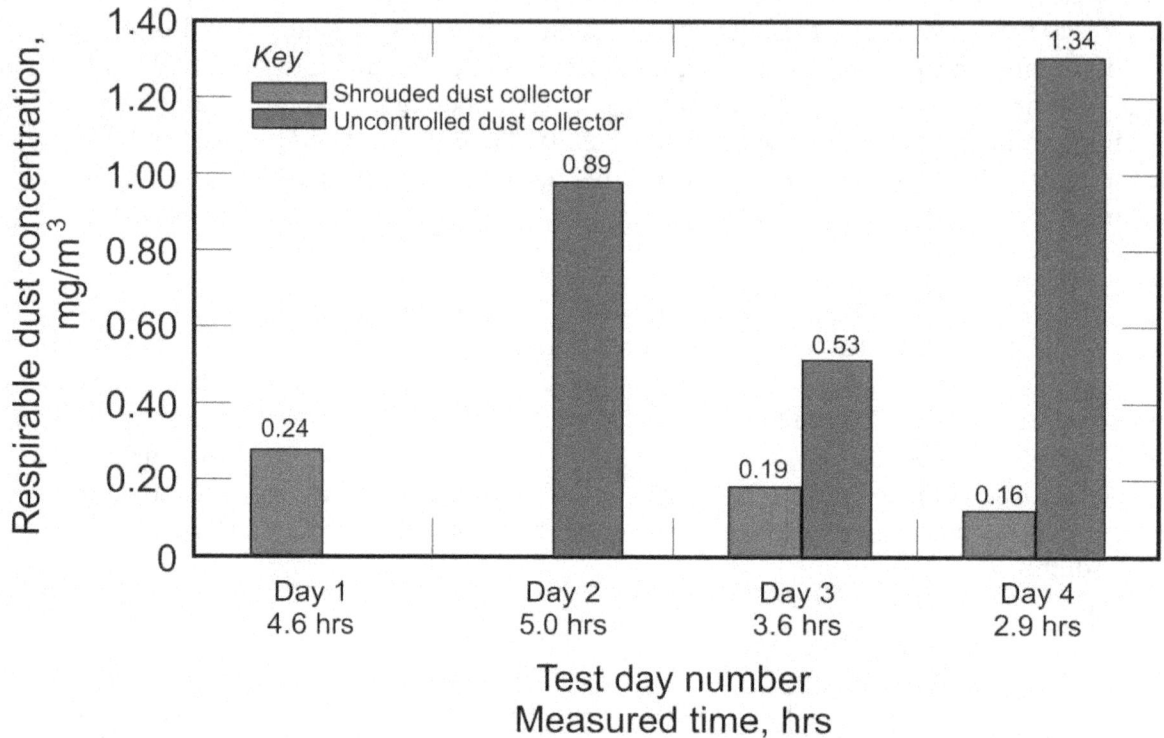

Figure 3.14. Comparison of dust concentrations of uncontrolled dust collector dump to shrouded dust collector dump.

To install the shroud, the length of brattice cloth (or similar material) should be sufficient to allow it to extend from the dust collector dump point to the ground. It should be cut so that it is only long enough to just touch the ground when the drill is lowered. When wrapping the cloth around the dust collector dump, the overlap should be placed so that it is on the outside of the dust collector dump (i.e., it should be visible as the installer looks directly at the dust collector dump, as shown in Figure 3.13 C). This overlap allows the cloth to expand as fine material is dumped to the ground, while containing the entrained respirable fraction within its confines. Placement of the overlap on the outside also keeps the fine material off the drill tracks, which otherwise could cause re-entrainment of the respirable size fraction of the material when the drill starts in motion.

Drill Stem/Deck Leakage

Drill cuttings and dust leaking through the drill table bushing can be a large source of respirable dust exposure to the drill operator and surrounding personnel. The table bushing, when new, generally prevents drill cuttings from coming through the drill deck. However, the table bushing is susceptible to wear during drilling operations, creating gaps between the bushing and the drill steel and deck surfaces which can allow the drill cuttings and dust to escape from underneath the shroud.

<u>Preventing Drill Stem/Deck Leakage</u>

In an attempt to minimize this source of dust, many drill operations install a rubber shield (generally a piece of used conveyor belting) underneath the drill deck covering the area of the table bushing. A hole is placed in this shield to allow the drill bit to pass through. This shield is also susceptible to wear and damage, and over time dust emissions through the drill table bushing soon occur. Figure 3.10 shows this rubber shield, which is located at the drill bit underneath the drill deck.

An innovation to help keep dust from escaping around the drill stem/deck is the Air Ring Seal (AIRRS). The AIRRS is a nonmechanical, virtually maintenance-free system consisting of a donut-shaped compressed air header with closely spaced holes (see Figure 3.15) along its inside perimeter [Page 1991]. When compressed air is supplied to the AIRRS, high-velocity air jets are produced through the holes. These jets are directed across the opening to be sealed in order to impede movement of dust particles through the opening.

The AIRRS is installed underneath the drill deck at the location where the drill steel penetrates the drill deck, positioned so that the drill bit and steel penetrate the inside diameter of the AIRRS. The diameter of the AIRRS should be as small as possible, but maintaining clearance for the drill bit and table bushing is an important consideration to minimize damage when raising the drill steel to remove the table bushing and drill bit. It is also very important that the drill's dust collector be maintained to achieve a collector to bailing airflow ratio of at least 2 to 1; otherwise more dust will be generated from the shroud area when using the AIRRS.

On surface mine drills equipped with the AIRRS, it successfully and significantly reduced respirable dust levels, as well as helped to keep the drill deck clean [Page 1991]. Solid-mounting the AIRRS flush with the deck virtually eliminates all large cuttings as well as the visible dust above the deck. The amount of air supplied to the AIRRS would be limited to what is available from the drill's compressor. An air pressure of 30 psi with 1/16-inch holes in the AIRRS was successful in reducing respirable dust levels [Page 1991]. However, this amount of air would be highly dependent upon the drill configuration.

In addition to using the AIRRS, installing a low-clearance deck bushing can minimize drill cuttings through the drill deck by reducing the gap size through which cuttings can escape. Replacement of the deck bushing as necessary can also enhance the performance of the air ring seal.

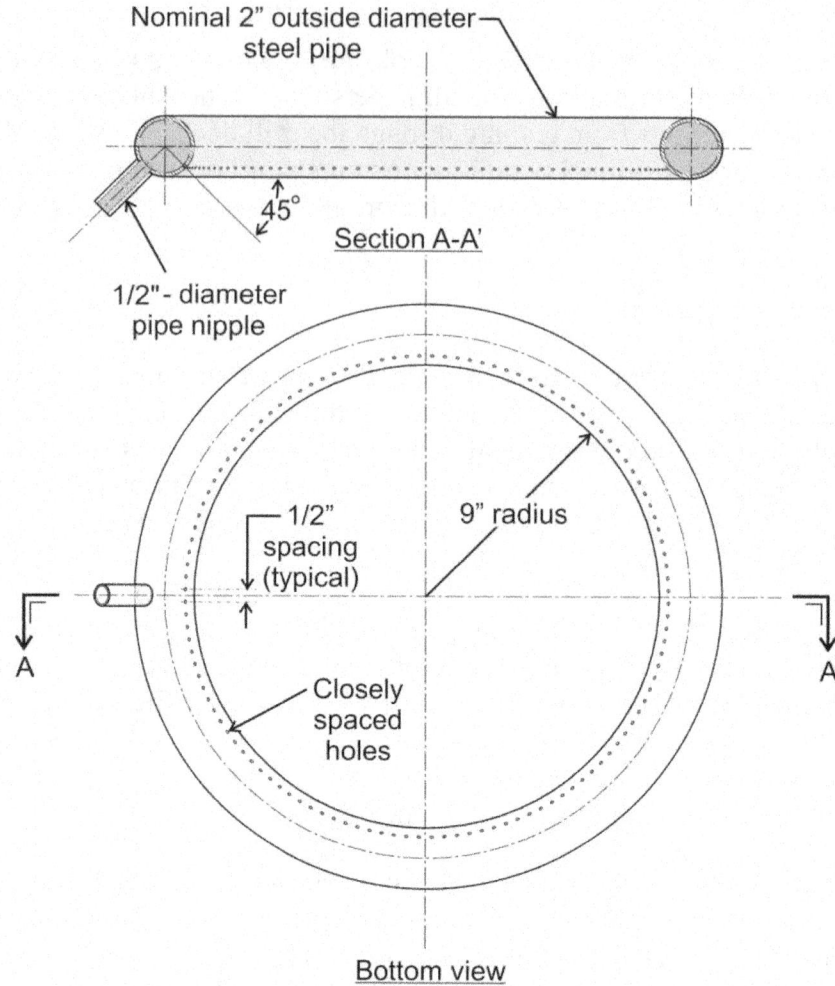

Figure 3.15. Air ring seal used to impede the movement of dust particles through an opening.

Small Diameter Drill Dust Collection Systems

A dust collection system used for small- to medium-diameter drills, small surface crawler, or "buggy" drills is shown in Figure 3.16. A schematic shows the operation of this type of system. The difference from the large diameter drill dust collection system is that this system collects all the drill cutting material, sends it to a large separation cyclone on the drill boom to drop out the large diameter material, and then sends the remainder to the dust collector at the back of the drill where the fine-sized material is discharged.

Maintenance of the dust collection system is important in sustaining efficient dust control with these systems. Two important maintenance issues are:

- filters should be replaced before they become clogged; and
- dust leakage from the bushing at the drill steel and the shroud device can be severe if not properly maintained (this has been shown to be even more important to dust control than the seal between the shroud device and the rock) [USBM 1956].

These drills may also contain a water misting system for dust control. This system includes a tank containing a dust suppressant mixture which is injected into the primary and fine collectors to reduce dust from their discharge [Sandvik 2005]. Monitoring the dust suppressant is important to maintain the integrity of the dust collection system.

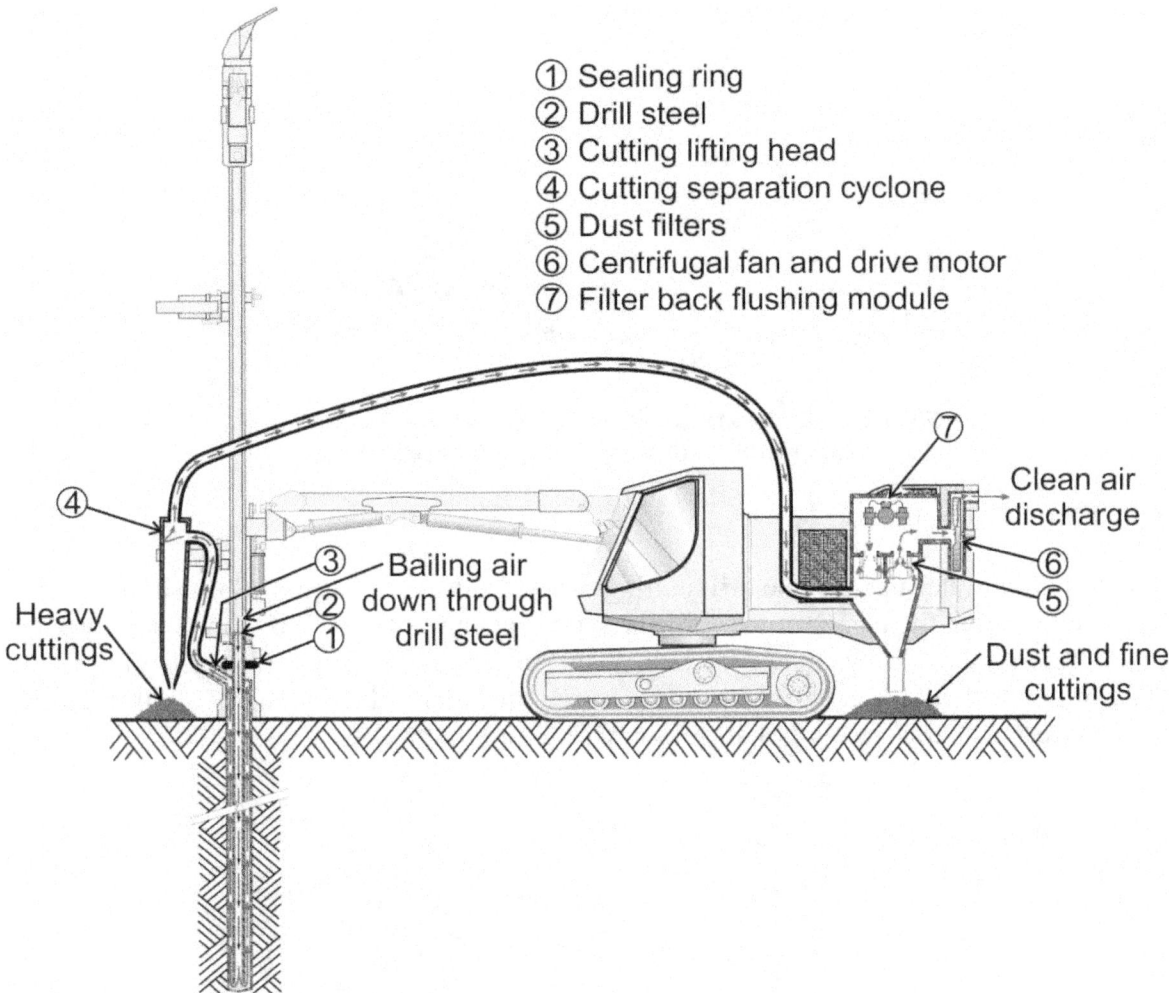

Figure 3.16. Typical dust collection system used by small crawler or "buggy" drills.

UNDERGROUND DRILLING DUST CONTROL

Underground drilling is generally accomplished using percussion drilling with small diameter holes (up to 3 inches). Depending upon the type of underground mining method used, these holes can be oriented in almost any direction. Generally the holes for a blast are consistently oriented horizontally or vertically and are drilled in a symmetrical pattern. The type of drilling equipment used can range from jackleg drills to stopers to jumbos which operate two to three drill booms, as shown in Figure 3.17. The most common method of dust control for underground drilling is using wet drilling techniques. Dry collectors have been used in the past, but are not commonly used due to the bulkiness of the collectors and their associated maintenance issues [Westfield et al. 1951].

Drilling and Blasting 97

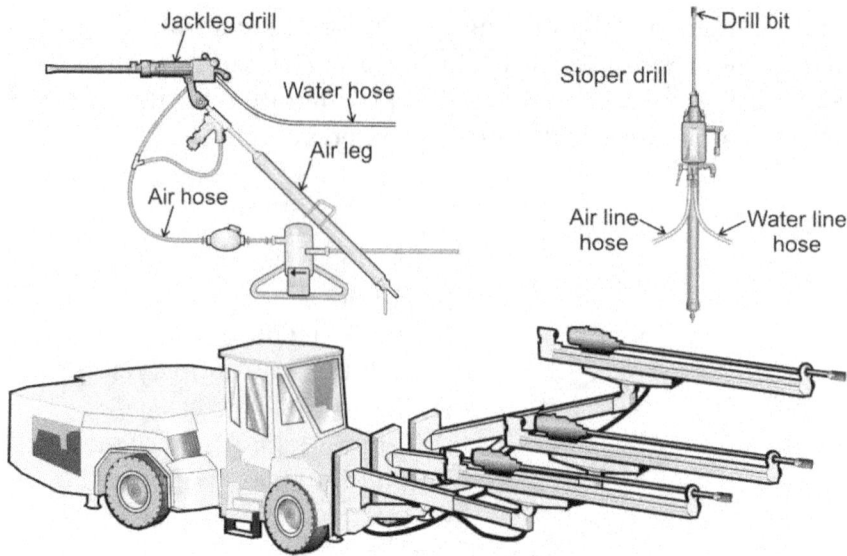

Figure 3.17. Illustrations showing a typical jackleg drill, stoper drill, and 3-boom jumbo, respectively.

Wet Drilling

Wet drilling uses water to flush the drill cuttings from the hole. Figure 3.18 illustrates the water being forced through the center of the drill steel, out the end of the drill bit, and back through the drill hole, forcing the cuttings out of the hole. The water pressure required to accomplish flushing of the drill hole is equal to the air pressure required for drilling, or, at a minimum, at least 10 psi less than this amount of air pressure [Hustrulid 1982].

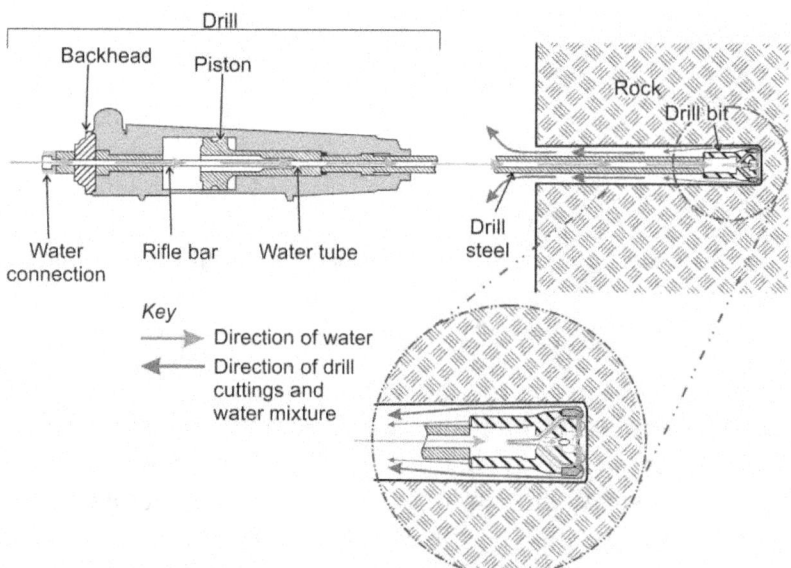

Figure 3.18. Water flow during a drilling operation to demonstrate water flushing of the drill cuttings. The water flows through the drill into the center of the drill steel and out the end of the drill bit to remove the cuttings from the drill hole.

98 Drilling and Blasting

Wet drilling has been found to be the best method of dust control, with dust reductions ranging from 86 to 97 percent depending upon the type of drilling involved [USBM 1921; USBM 1939]. The angle of the drill hole directly impacts wet drilling's effectiveness, with the amount of dust reduction being less for overhead vertical holes. Angle holes from the vertical were found to produce 50–60 percent less dust than overhead vertical holes [USBM 1938a]. The reason the overhead vertical holes produce more dust is due to the short amount of time that water contacts the drill hole face and the increase in sludge dropping through the air and running down the drill steel. The sludge running down the drill steel gets slung into the air from the rotation of the steel. Methods used to reduce dust from overhead vertical holes include [USBM 1938c]:

- increasing the amount of water used in wet drilling;
- creating a trap to capture the sludge from the drillhole and divert it away from the drill steel;
- designing drills that reduce the amount of air leaking from the front head along the drill steel; and
- preventing the exhaust air generated by the drill from dispersing sludge into the air.

The relationship of water flow to dust concentrations was tested in both stoper and drifter drills. For stoper drills, as the water flow increased to 1.3 gpm, it was found that the dust concentrations decreased rapidly. Above 1.3 gpm the dust concentrations decreased at a slower rate. This phenomenon was also seen in drifter drills, but at a water flow rate of 1.0 gpm. Therefore, it was recommended that the optimal water flow should be 1.3 gpm for stoper drills and 1.0 gpm for drifter drills [USBM 1938c]. Additionally, there are advantages gained through wet percussion drilling such as less drill steel breakage and greater penetration rate with wet drilling than with dry drilling [USBM 1921].

To characterize the typical dust emissions during wet drilling, it was shown through testing that the first 1 or 2 feet of drill hole was the dustiest when wet drilling [USBM 1938b]. After the first 1–2 feet was completed, the dust concentrations dropped rapidly and maintained a constant level. This study also demonstrated that collaring a hole was not responsible for the high dust concentrations, since the holes were precollared prior to testing [USBM 1938b].

Water Mists and Foams

Water mists, which use a mixture of compressed air and water, and foams injected through the drill steel, have also been shown to reduce dust concentrations by 91–96 percent, respectively [USBM 1982a,b]. However, the limited improvement in dust reduction by comparison to wet drilling does not justify the cost involved with using these methods.

Wetting Agents

The use of "wetting" agents or additives to the water used in drilling has also been evaluated, without much success. Results showed that wetting agent solutions applied at varying flow rates provided better dust control [USBM 1943]. However, the dust control provided by wet drilling alone was so good that the additional dust reduction from the use of wetting agent solutions was insignificant. The claim that wetting agents can improve penetration rate was shown to be unsubstantiated due to the inconsistent drilling rates (ranging from -24.9 to +57.5 percent) and the dependency upon drilling circumstances [USBM 1943].

External Water Sprays

External water sprays, depicted in Figure 3.19, have been used in the past, but when dry drilling they only produced dust reductions up to 25 percent compared with dry drilling alone—i.e., drilling without the use of any water [USBM 1921]. In an attempt to improve dust reductions, an external water spray device, a ring which contained spray holes on the inside diameter for dust control during dry drilling, was developed and tested. This device was slipped over the drill steel and sprayed water on the drill steel outside of the hole. It produced dust reductions that were variable, ranging from 75 to 88 percent, over dry drilling [USBM 1939]. Shrouding of the drilling area was tested through the addition of a rubber shroud mounted at the area at the end of the boom to surround the external spray and drill bit. It was shown to improve dust reduction to 53 percent [USBM 1982a]. Although these devices were an improvement over external water sprays, results showed that these devices along with external water sprays could not be used to replace wet drilling and maintain proper dust control during drilling operations. They could be used to enhance dust control in conjunction with wet drilling.

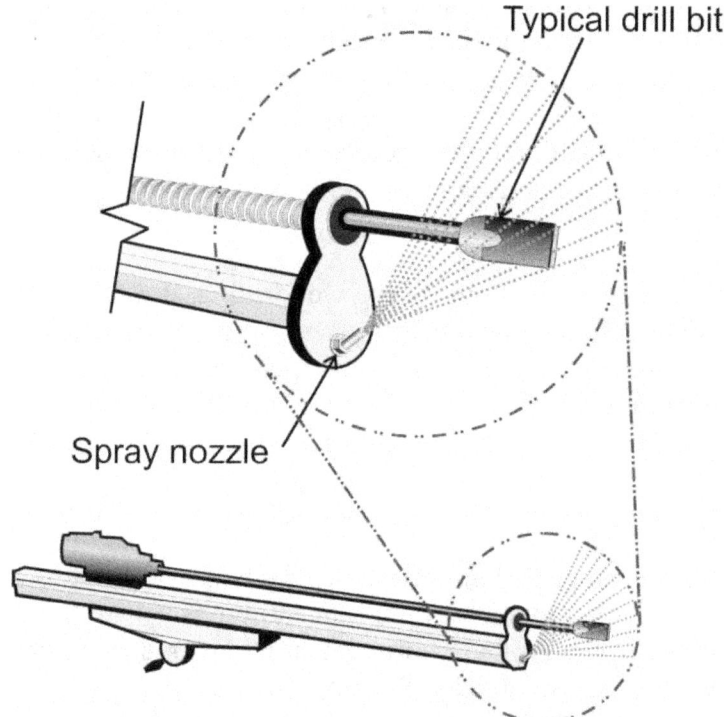

Figure 3.19. Drill boom showing external water sprays directed onto the drill steel.

RESPIRABLE DUST CONTROL FOR BLASTING

Blasting is a controlled operation which generally occurs intermittently at mining operations and is used to fragment the rock being mined. This is an important step in industrial mineral production as it is required to break up the mineral product, and the effectiveness of blasting may be a significant factor when optimizing the fragmentation and cost of the crushing and grinding circuit [Hustrulid 1999]. While blasting seemingly generates large amounts of dust, the operation occurs infrequently enough that it is not considered to be a significant contributor to particulate matter less than 10 micrometers (μm), or PM_{10} [EPA 1991; Richards and Brozell 2001]. Testing has been conducted to create an estimator for the amount of dust generated from blasting. The estimator is shown as the following equation:

$$e = 0.00050 A^{1.5}$$

where e = total suspended particulate emission factor (lb/blast); and
A = area blasted (m^2) as long as hole depth does not exceed 70 feet.

To estimate the PM_{10} emissions from blasting, the following equation is used:

$$PM_{10} = e \times 0.5$$

where PM_{10} = pounds per blast of particulate matter less than 10μm; and
e = total suspended particulate emission factor (lb/blast).

It must be noted here that the U.S. EPA determined that this estimator should only be used for guidance. More reliable emission factors for particulate emissions should be based upon site-specific field testing in order to determine the amount of particulate matter generated from blasting [EPA 1998].

Blasting Dust Control Measures

As a result of blasting being considered an insignificant dust source, there is very little documentation for dust control of blasting operations. There are five methods of dust suppression that can be used for the control of dust during blasting, many of which are effective for underground mining only [Cummins and Given 1973]:

- wetting down the entire blasting area prior to initiating the blast;
- the use of water cartridges alongside explosives;
- the use of an air-water fogger spray prior to, during, and after initiating the blast;
- the use of a filtration system to remove pollutants from the air after the blast; and
- dispersal and removal of the dust and gases using a well-designed ventilation system.

Wetting Down Blasting Area

A common method of dust control for blasting operations is to wet down the entire blasting area prior to initiating the blast. This procedure minimizes dust being entrained into the air from the blasting activity by allowing it to adhere to the wet surfaces [Cummins and Given 1973]. This method has been shown to be effective for dust control during blasting in underground mines. It could also be effective for surface mining depending upon the time interval between watering and blast initiation, with the possibility of exposure to the surface atmospheric conditions causing the moisture to quickly evaporate and rendering the watering ineffective. However, safety considerations may preclude wetting the blast area during the "tying in" phase of the blast where all the individual hole charges are connected together prior to blasting.

Water Cartridges

Water ampoules or cartridges which are inserted into the blasthole with the explosive have been used successfully for dust reduction in past underground coal mining blasting operations [ILO 1965]. The water cartridges consist of a properly sized plastic bag which is prefilled with water or can be filled in the hole. The cartridges can be placed in front of, alongside, or behind the explosive without causing any adverse effects to fragmentation. There is another type of cartridge which can be used in the place of stemming as shown in Figure 3.20. This cartridge uses a PVC bag which is inserted into the hole after the explosive and is then filled with water to maintain a tight seal with the blasthole. In coal mining operations, the use of these cartridges is claimed to have reduced dust by 40–60 percent.

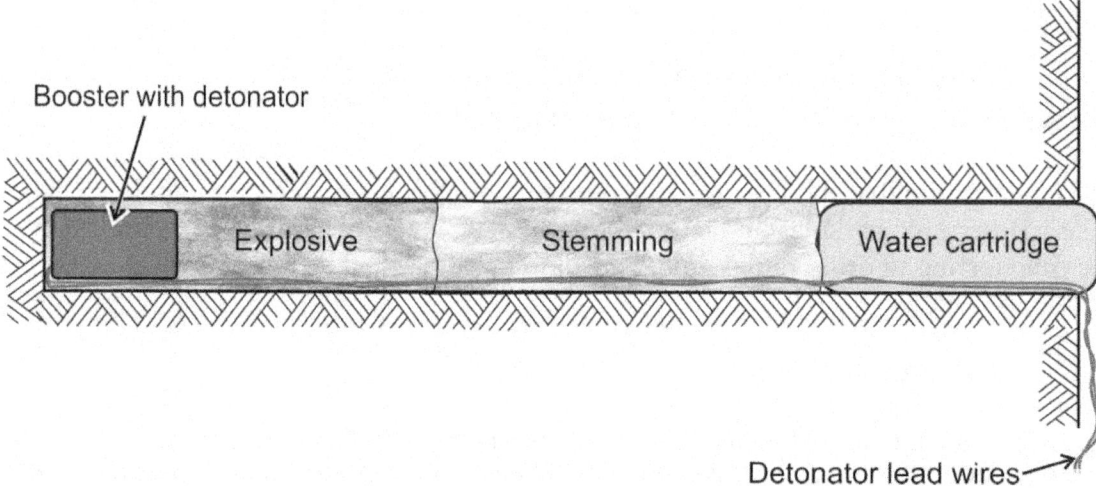

Figure 3.20. A typical blasthole containing an explosive charge utilizing a water cartridge to suppress dust generated during blasting.

Fogger Sprays

In the past, many underground operations used a fogger type spray in the heading where the blasting is conducted, shown in Figure 3.21. This setup uses both compressed air and water forced through a nozzle to create a fogged-in blast area. The nozzle is located at an approximate distance of 100 feet from the face, is turned on prior to blast initiation, and remains operational for 20 to 30 minutes after the blast [ILO 1965]. While a fogger type spray is effective for underground blasting, it is not known if such a system would be viable for surface blasting.

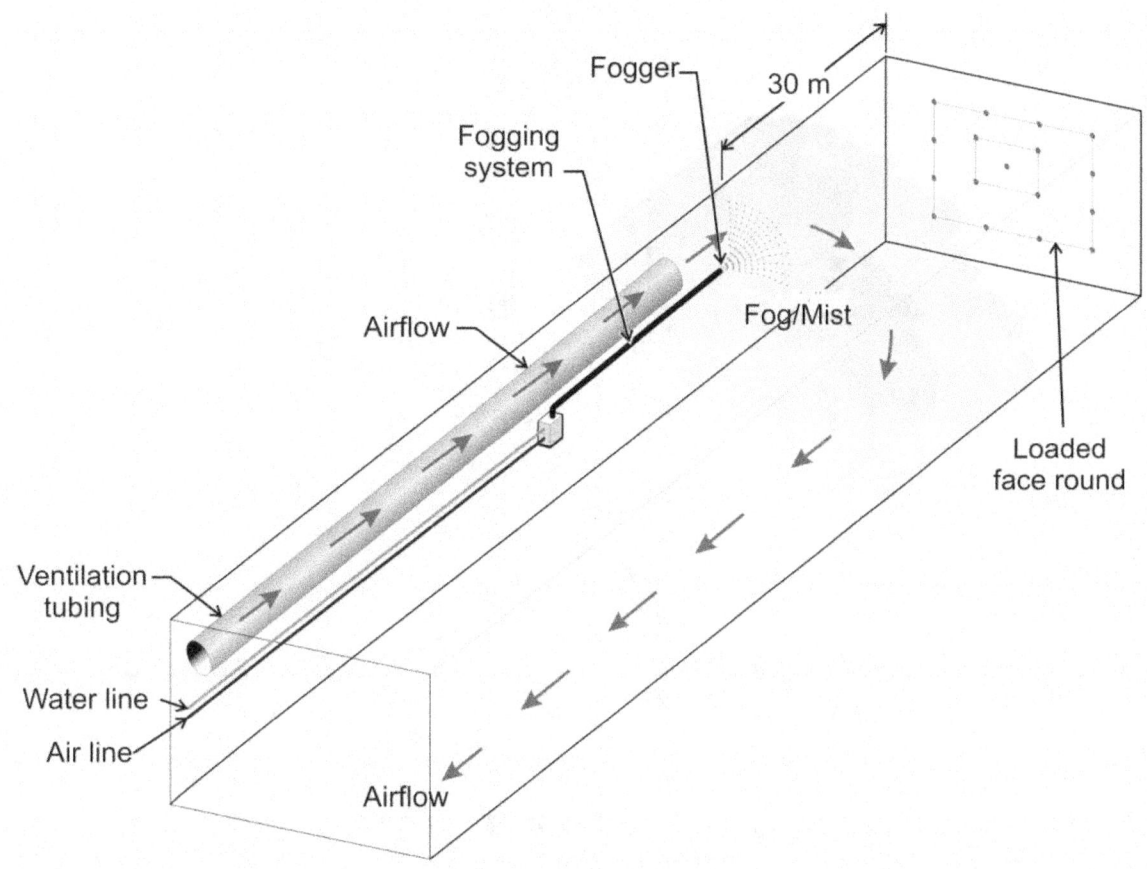

Figure 3.21. A fogger spray used to create a mist for dust suppression in the heading where blasting will occur.

Air Filtration Systems

Another dust control method that has been used in underground mining operations in the past is to filter the return air of the ventilating air in the blasting area as seen in Figure 3.22. One such filtration unit, used in South Africa, consists of a filter and a bed of vermiculite treated with sodium carbonate and potassium permanganate which removes the dust and nitrous fumes from blasting [Cummins and Given 1973]. Figure 3.23 shows another method, which is to place filters inside the exhausting ventilation duct with water sprays, which spray on the filters and are oriented in the same direction as the airflow. The filters are only used during blasting, and the duct containing the filters is approximately twice the diameter of the ventilation tubing [ILO 1965]. Also, dry filters have been used successfully in the past for the same purpose.

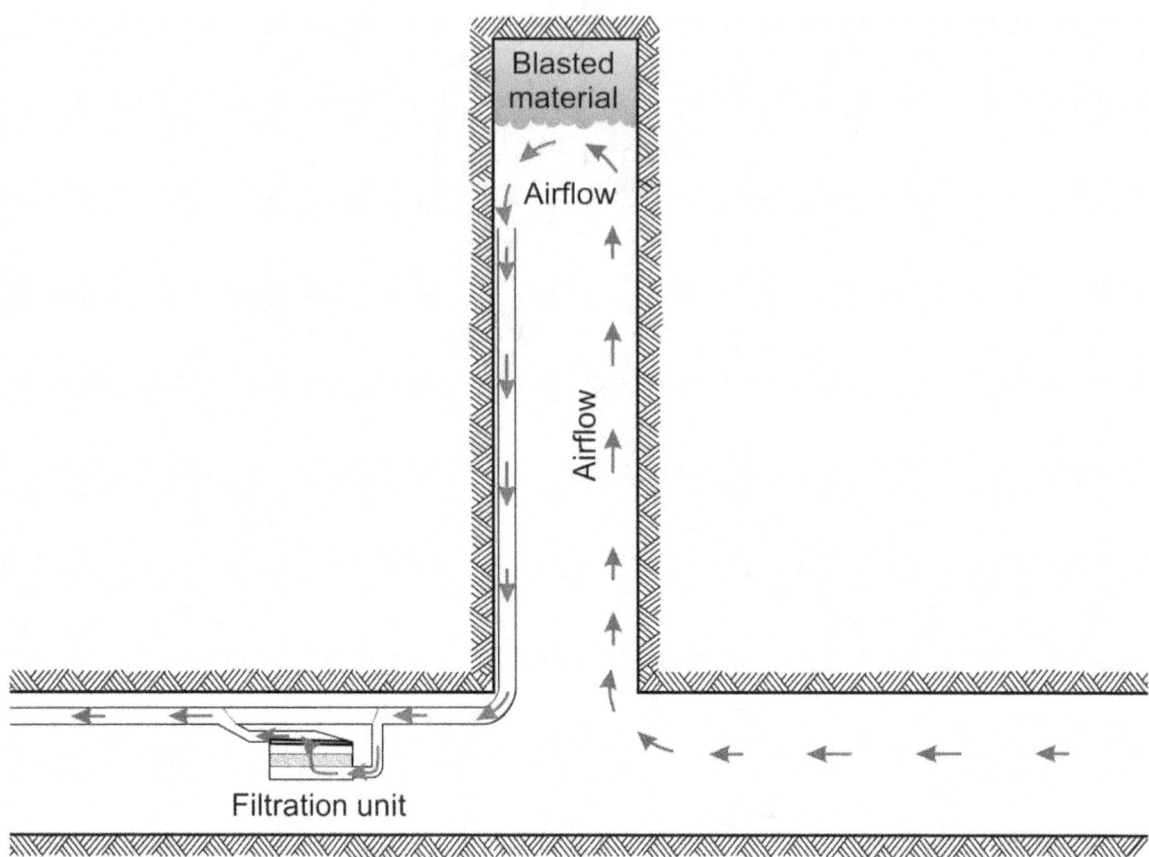

Figure 3.22. A filtration unit, located adjacent to the blast heading, used for filtering the contaminated ventilation air from the blast heading after blasting occurs.

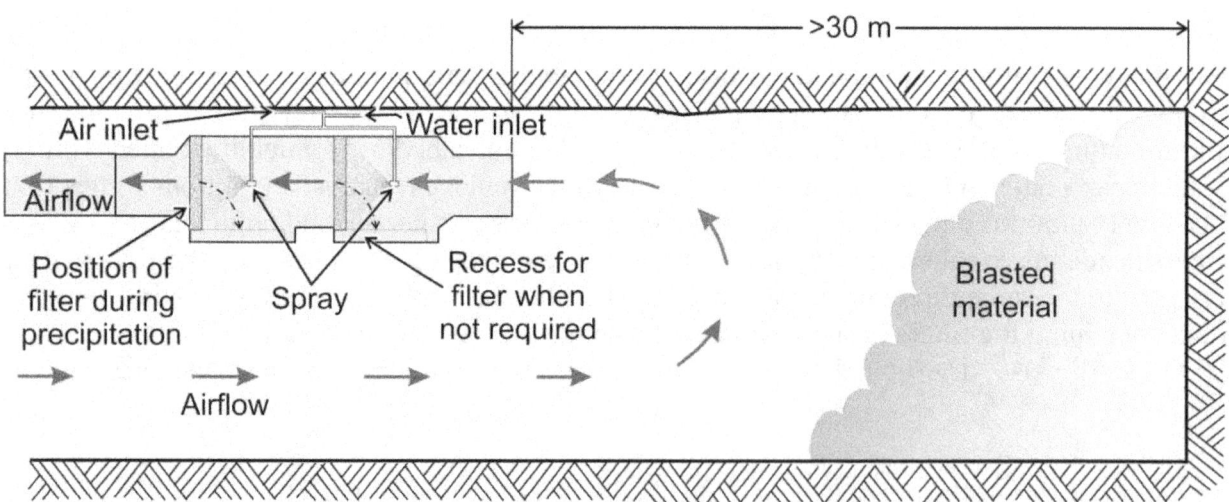

Figure 3.23. A filtration unit, located in the blast heading, used for filtering the contaminated ventilation air from the blast heading after blasting occurs.

104 Drilling and Blasting

Ventilation System Dispersal

The most common method of dust control is to allow the dust and gases from blasting to be dispersed and removed through the ventilation system in the case of underground operations, or through atmospheric dispersion in the case of surface operations. Underground operations generally schedule the blasting during off-shift times to allow sufficient time for the area to ventilate, to disperse, and to remove the dust and gases from the blasting [Cummins and Given 1973]. If off-shift blasting is not feasible, then areas affected by the blasting should be cleared and work should not commence until the dust and gases are removed.

The amount of time for dust and gas dispersal is mine-site specific, depending upon the efficiency of the mine's ventilation system. Therefore, it is important that the mine's ventilation system be maintained in good operating condition to optimize dust and gas removal and to minimize the time for the removal.

At surface operations, the area is cleared of personnel just prior to blasting. Scheduling the blast to take into consideration the meteorological conditions, i.e. low wind speed and low inversion potential, can be used to minimize the impacts of dust generation from blasting. Generally the dispersion of dust and gases occurs quickly after the blast, depending upon the wind speed and direction, and work is not allowed in the affected area until dispersion is completed. Additionally, it has been noted that the use of multi-delay detonators to initiate the individual explosive charges in millisecond time intervals may reduce dust generation from blasting, but this has not been verified [Miller et al. 1985].

REFERENCES

Cummins AB, Given IA, eds. [1973]. SME Mining engineering handbook, Vol. 1. New York: Society of Mining Engineers of the American Institute of Mining, Metallurgical, and Petroleum Engineers, Inc.

EPA [1991]. Review of Surface Coal Mining Emissions Factors. U.S. Environmental Protection Agency, Office of Air Quality Planning and Standards, Research Triangle Park, North Carolina, July.

EPA [1998]. Review of Emission Factors for AP—42 Section 11.9 Western Surface Coal Mining. U.S. Environmental Protection Agency, Office of Air Quality Planning and Standards, Research Triangle Park, North Carolina, September.

Hustrulid WA, ed. [1982]. Underground mining methods handbook. New York: Society of Mining Engineers of the American Institute of Mining, Metallurgical, and Petroleum Engineers, Inc.

Hustrulid WA [1999]. Blasting principles for open pit mining, Vol. 1—General design concepts. Rotterdam, Netherlands: A.A. Balkema.

ILO [1965]. Guide to the prevention and suppression of dust in mining, tunnelling, and quarrying. International Labour Office, Geneva, Switzerland; Atar, South Africa.

Miller S, Emerick JC, Vogely WA [1985]. The secondary effects of mineral development. Economics of the mineral industries, chapter 5.15B. Littleton, Colorado: Society of Mining, Metallurgy, and Exploration.

NIOSH [2006]. Technology news 512: Improve drill dust collector capture through better shroud and inlet configurations. Pittsburgh, PA: U.S. Department of Health and Human Services, Centers for Disease Control and Prevention, National Institute for Occupational Safety and Health Publication No. 2006–108.

Organiscak JA, Page SJ [2005]. Development of a dust collector inlet hood for enhanced surface mine drill dust capture. Inter J of Surface Min, Rec, and Env. *19*(1):12–28.

Page SJ [1991]. Respirable dust control on overburden drills at surface mines. Coal mining technology, economics, and policy. Session papers from the American Mining Congress Coal Convention, Pittsburgh, Pennsylvania, June 2–5, pp. 523–539.

Page SJ, Organiscak JA [2004]. Semi-empirical model for predicting surface coal mine drill respirable dust emissions. Inter J of Surface Min, Rec, and Env. *18*(1):42–59.

Page SJ, Listak JM, Reed WR [2008a]. Drill dust capture-Improve dust capture on your surface drill. Coal Age.

Page SJ, Reed WR, Listak JM [2008b]. An expanded model for predicting surface coal mine drill respirable dust emissions. Inter J of Surface Min, Rec, and Env. *22*(3):210–221.

Potts JD, Reed WR [2008]. Horizontal air blocking shelf reduces dust leakage from surface drill shroud. Transactions of the Society of Mining, Metallurgy, and Exploration. Littleton, Colorado: Society for Mining, Metallurgy, and Exploration, Inc. *324*:55–60.

Potts JD, Reed WR [2010]. Field evaluation of air-blocking shelf for dust control on blasthole drills. Submitted to Mining Engineering, 2010.

Reed WR, Organiscak JA, Page SJ [2004]. New approach controls dust at the collector dump point. Coal Age, June, pp. 20–22.

Reed WR, Potts JD [2009]. Improved drill shroud capture of respirable dust utilizing air nozzles underneath the drill deck. Transactions of the Society of Mining, Metallurgy, and Exploration. Littleton, Colorado: Society for Mining, Metallurgy, and Exploration. *326*:1–9.

Richards J, Brozell T [2001]. Compilation of National Stone, Sand and Gravel Association Sponsored Emission Factor and Air Quality Studies 1991–2001. Arlington, Virginia: National Stone, Sand and Gravel Association.

Sandvik [2005]. Tamrock ranger drills feature rock pilot, dust minimizor systems for enhanced and cleaner operation. Atlanta, Georgia: Sandvik Mining and Construction. [http://www2.sandvik.com/_C1256A1E004428BE.nsf/0/7C249CA00F258C2D85257062000D57EC].

USBM [1921]. Dust reduction by wet stopers. By Harrington D. U.S. Department of the Interior, Bureau of Mines Report of Investigations 2291.

USBM [1938a]. Effect of angle of drilling on dust dissemination. By Brown CE, Schrenk HH. U.S. Department of the Interior, Bureau of Mines Report of Investigations 3381.

USBM [1938b]. Relation of dust concentration to depth of hole during wet drilling. By Littlefield JB, Schrenk HH. U.S. Department of the Interior, Bureau of Mines Report of Investigations 3369.

USBM [1938c]. Relation of dust dissemination to water flow through rock drills. By Brown CE, Schrenk HH. U.S. Department of the Interior, Bureau of Mines Report of Investigations 3393.

USBM [1939]. Dust produced by drilling when water is sprayed on the outside of the drill steel. By Johnson JA, Agnew WG. U.S. Department of the Interior, Bureau of Mines Report of Investigations 3478.

USBM [1943]. Use of wetting agents in reducing dust produced by wet drilling in basalt. By Johnson JA. U.S. Department of the Interior, Bureau of Mines Report of Investigations 3678.

USBM [1951]. Roof bolting and dust control. By Westfield J, Anderson FG, Owings CW, Harmon JP, Johnson L. U.S. Department of the Interior, Bureau of Mines Information Circular 7615.

USBM [1956]. Dust control in mining, tunneling, and quarrying in the United States. By Owings CW. U.S. Department of the Interior, Bureau of Mines Information Circular 7760.

USBM [1982a]. An evaluation of three wet dust control techniques for face drills. Page SJ. U.S. Department of the Interior, Bureau of Mines Report of Investigations 8596.

USBM [1982b]. Evaluation of the use of foam for dust control on face drills and crushers. By Page SJ. U.S. Department of the Interior, Bureau of Mines Report of Investigations 8595.

USBM [1985]. Quartz dust sources during overburden drilling at surface coal mines. By Maksimovic SD, Page SJ. U.S. Department of the Interior, Bureau of Mines Informational Circular 9056.

USBM [1986]. Investigation of quartz dust sources and control mechanisms on surface mine operations, Vol. I—Results, analysis, and conclusions. By Zimmer RA, Lueck SR. U.S. Department of the Interior, Bureau of Mines Final Contract Report, NTIS: PB 86–215852, 72 pp.

USBM [1987]. Optimizing dust control on surface coal mine drills. By Page SJ. U.S. Department of the Interior, Bureau of Mines Technology News 286.

USBM [1988]. Impact of drill stem water separation on dust control for surface coal mines. By Page SJ. U.S. Department of the Interior, Bureau of Mines Technology News 308.

USBM [1995]. The reduction of airborne dust generated by roof bolt drill bits through the use of water. By Sundae LS, Summers DA, Wright D, Cantrell BK. U.S. Department of the Interior, Bureau of Mines Report of Investigations 9594.

Zimmer RA, Lueck SR, Page SJ [1987]. Optimization of overburden drill dust control systems on surface coal mines. Inter J of Surface Min *1*(2):155–157.

CHAPTER 4: CRUSHING, MILLING, AND SCREENING

CHAPTER 4: CRUSHING, MILLING, AND SCREENING

Mineral crushing, milling, and screening operations can be major sources of airborne dust due to the inherent nature of size reduction and segregation processes. Control of dust generated by these operations can be achieved with proper analysis of the sources, identification of appropriate control technologies, and consistent application and maintenance of selected controls.

Worker exposure may be managed through engineering controls to suppress or enclose the dust sources—as described below—or by isolating the worker from the dust source, as discussed in Chapter 9—Operator Booths, Control Rooms, and Enclosed Cabs. Administrative controls such as operating procedures, work practices, and worker training are also commonly applied to supplement engineering controls. Until feasible engineering and administrative controls are installed, or when they do not achieve the desired level of exposure reduction, and during maintenance, repair, and other unusual operating conditions, personal protective equipment may also be necessary.

As with any process component, installation of the selected dust controls, even if they are appropriate, will not guarantee continuing effective performance. The performance of installed dust control systems, which often represent large capital expenditures, should be periodically evaluated, maintained, and, when necessary, modified to maximize performance. For example, the effectiveness of dust control systems installed to protect worker health can only be demonstrated by collecting personal air samples for comparison to the occupational exposure limit for the substances in question.

Different mineral processing equipment generates different amounts of dust emissions. Relative dust emission rate ratios (setting the primary crusher emission rate as the baseline) for common mineral processing equipment are presented in Table 4.1. This ranking is based on EPA-estimated particulate emissions for crushed stone operations, and is presented only to illustrate the relative magnitude of the various dust sources [EPA 2003]. It can be seen that emissions increase as the size of the material processed decreases, as one would expect.

Table 4.1. Relative emission rate ratios of crushing and screening equipment

Equipment	Relative Emission Rate Ratio
Primary crusher	1
Secondary crusher	(No data, tertiary crusher rate would be an upper limit)
Tertiary crusher (dry)	51
Tertiary crusher (wet)	2
Screen (dry)	214
Screen (wet)	12

PREVENTION AND SUPPRESSION APPLICATIONS

Wet Control Methods

Wet dust control systems can be very effective and are less costly to install and operate, but they may not be feasible due to characteristics of the mineral, subsequent processing steps, or customer specifications. Also, when operations are in northern climates, freeze protection is necessary, and ice buildup can create additional safety hazards for workers.

As discussed more fully in Chapter 2—Wet Spray Systems, the use of water to control dust may be classified into prevention applications and suppression applications. Prevention is the application of water to prevent dust from becoming airborne. Suppression is the use of water to wet dust particles which have already become airborne, increasing their mass and causing them to settle more rapidly.

In general, prevention is more effective than suppression [NIOSH 2003; USBM 1978]. However, when the wetted material is subject to further size reduction as in crushing operations, effective prevention requires application of additional water to the dry—and larger—surface area of the material exposed by the size reduction process. This may create a complication because additional application of water to improve prevention may cause the material to become too wet, interfering with efficient handling, subsequent processing operations, or product specifications. As a result, some trade-off between wet dust control and process efficiency is often unavoidable. This trade-off can necessitate the use of other control approaches to achieve acceptable conditions.

Suppression of respirable airborne dust using water, usually through sprays directed into the dust cloud, is not always highly efficient. It is difficult with hydraulically atomizing spray nozzles to produce water droplets small enough to suppress respirable particles effectively. Appropriately sized water droplets can be produced with air-atomizing nozzles, but this method requires a source of compressed air to atomize the water, and the very small nozzle orifices are subject to clogging, requiring increased maintenance attention. Additionally, air-atomized spraying, due to the volume of the air/water mixture released, typically requires an enclosed area to avoid spreading dust into surrounding areas. All of these factors increase operating costs.

Dry Control Methods

Control of dust generated by transport (see Chapter 5—Conveying and Transport) and processing of minerals can be achieved through containment, exhaust, and cleaning of dusty air. This approach, termed dry control, can be more costly to install and operate than wet control methods, but can be very effective. Additionally, dry dust control may be necessary when the product is adversely affected by the addition of water, such as trona or clayey materials.

Dry dust control must create conditions that will prevent the escape of dusty air from the controlled space to areas occupied by workers. This control is achieved by using exhaust ventilation to create a negative air pressure inside the controlled space relative to outside of the controlled space (see Chapter 1—Fundamentals of Dust Collection Systems). The amount of ventilation necessary to achieve control is affected by:

- the degree of enclosure of the controlled space, with required exhaust volume increasing as the area of unsealed openings into the enclosure increases;
- the airflow created by the movement and processing of the mineral, including air that is entrained within moving material, airflow induced by moving material and equipment, and air displaced by the material flowing into or out of an enclosure;
- the effect of ambient wind speed and direction that can overcome the pressure differential between the interior and exterior of the controlled space.

CRUSHING

Size reduction processes will always contain at least one crushing circuit, and many times will have multiple crushers, often of different types. Selection of crushers is based primarily on material size-reduction and throughput requirements. Secondary selection considerations include the composition, hardness, and abrasiveness of the feed material(s). The most common primary crusher type is the jaw crusher (Figure 4.1) which operates by compressing the feed material between a fixed and moving plate or jaw. Cone (Figure 4.2 left) or gyratory (Figure 4.2 right) crushers may be used as primary or secondary crushers and also operate through compression.

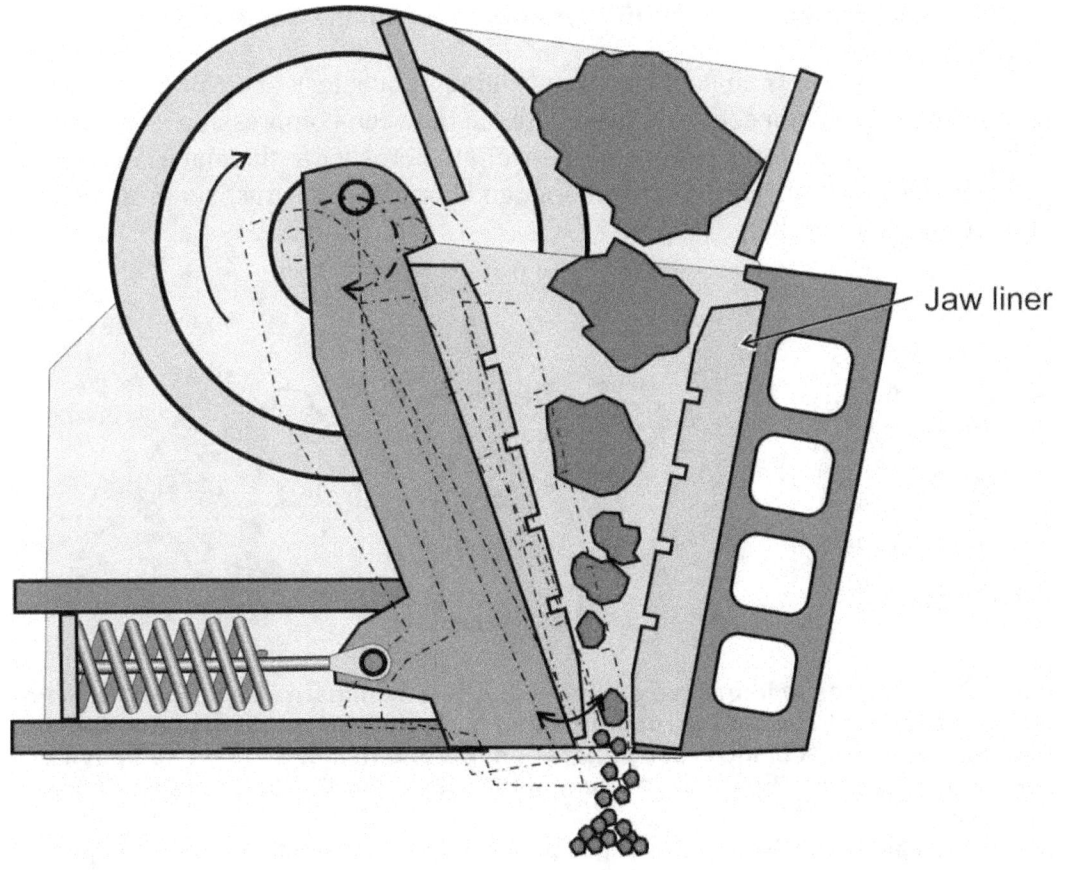

Figure 4.1. Illustration of a jaw crusher showing material being crushed between fixed and reciprocating jaws.

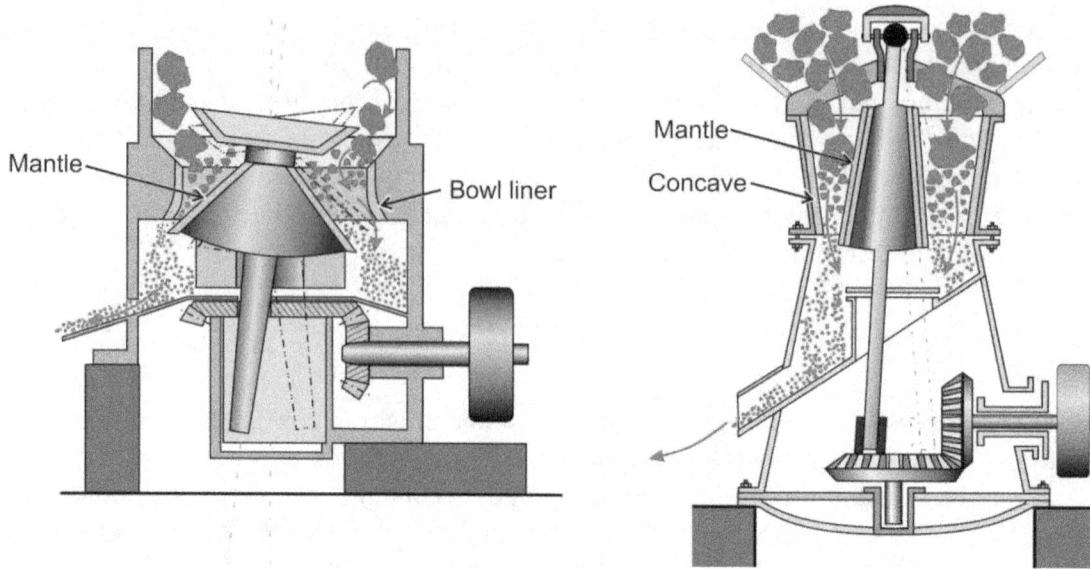

Figure 4.2. Examples of compressive crushers. Left—illustration of a cone crusher showing material being crushed between the cone mantle and bowl liner. Right—illustration of a gyratory crusher with material being crushed between the gyratory head and the frame concave.

Hammermills (Figure 4.3 left), impact breakers (Figure 4.3 center), and roll crushers (Figure 4.3 right) operate primarily by impaction. The difference between compression and impaction is the speed at which the breaking force is transmitted from the crusher to the material. Compression involves a slower energy transfer than impaction, and compressive crushers produce somewhat less dust than impactive crushers.

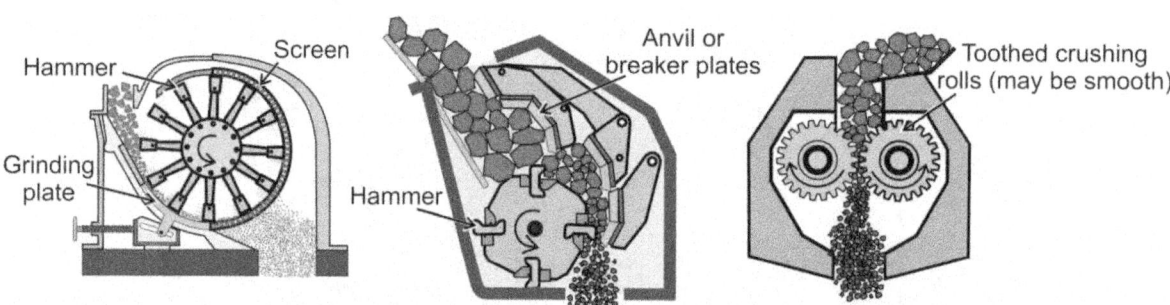

Figure 4.3. Examples of impactive crushers. Left—a hammermill-type crusher showing material crushed between the rotating hammers and fixed grinding plate. Center—illustration of material being crushed between rotating hammers and fixed anvil plates in an impact breaker. Right—illustration of the size reduction action of a roll crusher.

Dust control at crushers may be achieved through application of water, by enclosure of the dust source with or without exhaust ventilation, or a combination of wet and dry methods. Generally, the upper portion of the crusher (feed side) should be enclosed as completely as possible [USBM 1974]. Enclosures should be constructed of physically robust materials appropriate for the operating environment and climate conditions.

Figure 4.4 illustrates a wet dust control system at a crusher dump loading operation. The frontal open area of the enclosure may be reduced by installing hanging curtains and installing a high curb or berm at ground level. To conserve resources, operators should consider installing a presence-sensing control with a timed duration to activate the water spray system.

As described in Chapter 2—Wet Spray Systems, water sprays used for dust prevention typically use solid spray patterns or full cone spray patterns. Sprays used for dust suppression typically use hollow cone spray patterns. Sprays should overlap slightly to provide good coverage.

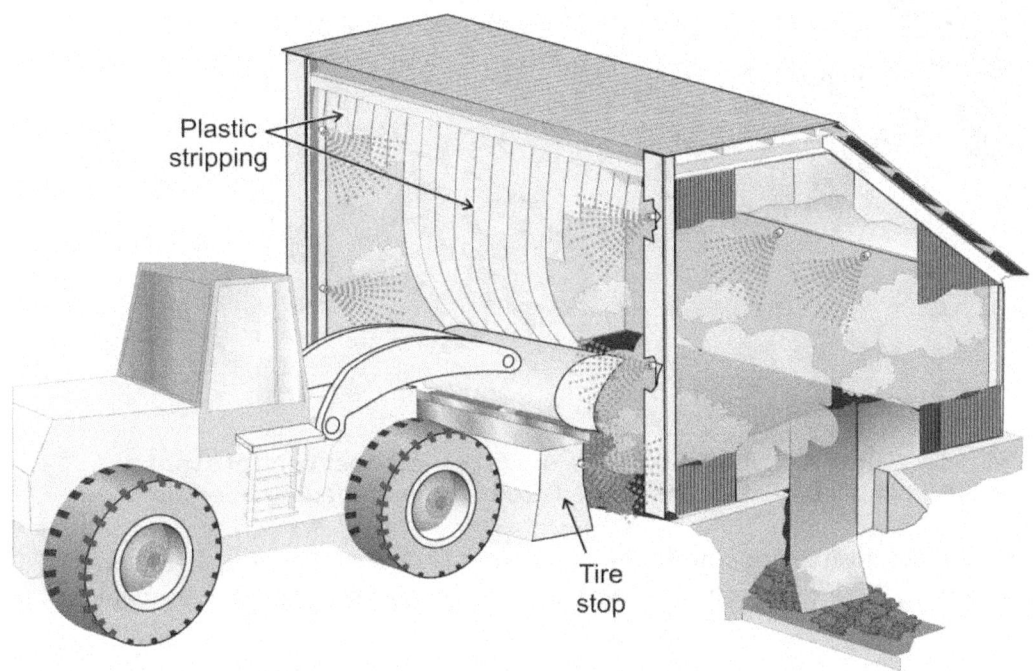

Figure 4.4. Illustration of a wet dust control approach with partial enclosure at a crusher dump loading operation. Note the blue "fan patterns" signifying water sprays.

A local exhaust ventilation dust control system at a crusher dump loading operation is illustrated in Figure 4.5.

Enclosure of the volume to be controlled should be maximized; in other words, the number and area of openings should be minimized as much as possible. Initial exhaust rates for the type of enclosure depicted in Figure 4.5 can be estimated by Equation 4.1, which accounts for air displaced by a dumping operation:

$$Q_E = 33.3 \times \left(\frac{600T}{G}\right) \quad (4.1)$$

where Q_E = exhaust air volume, cubic feet per minute;
T = weight of material dumped, tons per minute; and
G = bulk density of material, pounds per cubic foot.

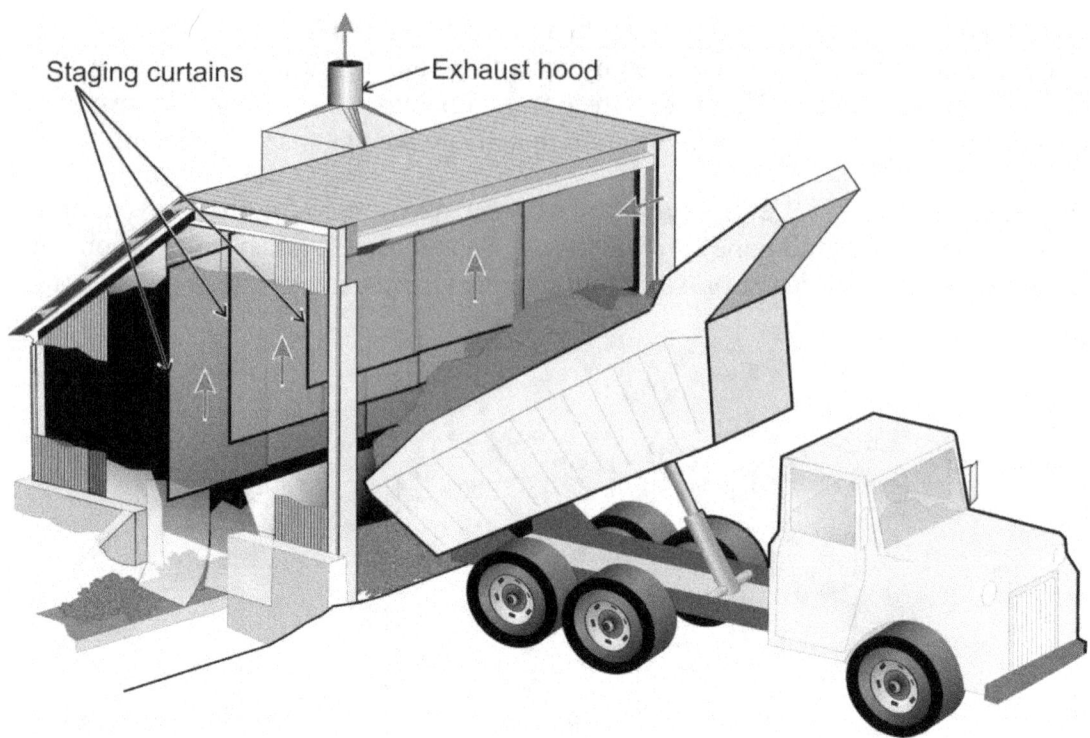

Figure 4.5. Illustration of a dry (exhaust) dust control system with a partial enclosure at a crusher dump loading operation.

If the dumping operation is not continuous, operators should consider utilizing a presence-sensing control with a timed duration to activate the exhaust system fan and water sprays as dumping occurs.

Whenever local exhaust ventilation is selected as a means of dust control, maintenance and clean-out openings should be designed into the enclosure and equipped with tight-fitting closures. Properly located and sized maintenance access openings will allow workers to perform their assigned tasks without having to modify the enclosure. This will reduce the potential for dust to escape through unintended openings.

A minimum air capture velocity of 200 feet per minute (fpm) is recommended to be maintained at all openings into enclosures [USBM 1974]. Where feed or discharge belt openings penetrate dust control enclosures, it is recommended to add the belt speed in fpm to the 200 fpm design velocity to account for the process material volumetric flow and belt-induced air movement [Yourt 1969].

As an example, for an enclosure with an initial design capture velocity (V_{INIT}) of 200 feet per minute (fpm) penetrated by a belt moving at 250 fpm (V_{BELT}), the recommended capture velocity (V_{CAP}) at openings into the enclosure becomes:

$$\text{Capture velocity } (V_{CAP}) = 200 \text{ fpm } (V_{INIT}) + 250 \text{ fpm } (V_{BELT})$$

$$\text{Capture velocity } (V_{CAP}) = 450 \text{ fpm}$$

Enclosure exhaust air volume (Q_E) in cubic feet per minute (cfm) can be estimated using the total area (in square feet) of all openings into the enclosure, with the capture velocity as:

$$Q_E \text{ (cfm)} = V_{CAP} \text{ (fpm)} \times \text{open area (sq ft)}$$

Again, these initial velocity and volume estimates should be verified as (1) providing acceptable dust control, and (2) not excessively consuming energy. Exhaust ventilation concepts are discussed more fully in Chapter 1—Fundamentals of Dust Collection Systems.

When possible, dust control enclosures should be large in volume to allow settling of coarse dust. Removing large airborne dust particles by this method will reduce the load on subsequent air handling and cleaning systems.

Falling material at transfer points will generate airflow both by induction and displacement [Anderson 1964; Goldbeck and Marti 1996]. When materials have an average size greater than ⅛ inch (0.01 foot), this airflow can be estimated by Equation 4.2:

$$Q_E = 10 \times A_U \sqrt[3]{\frac{RS^2}{D}} \tag{4.2}$$

where Q_E = exhaust volume, cubic feet per minute;
A_U = enclosure upstream open area, square feet;
R = rate of material flow, tons per hour;
S = height of material fall, feet; and
D = average material size, feet.

Determining exhaust volume through the use of Equation 4.2 is illustrated in Figures 4.6 through 4.9. In each case the height of material fall (variable S in Equation 4.2) is depicted. The value for the upstream open area variable—A_U—for a transfer point is the cross-sectional area of the chute transferring the material and any unsealed area between the chute and the enclosure structure. When chutes are not used, an estimate of the open area around penetrations (e.g., belt conveyor structure) into the enclosure is used for the A_U variable.

Figures 4.8 and 4.9 depict transfer from the discharge of a crusher to a conveyor. For jaw crushers, which have variable cross-sectional area, the upstream open area value (A_U) is the area of the bottom of the jaw crusher opening into the enclosure structure (Figure 4.8). For crushers with constant cross section, such as a cone crusher, the upstream open area is the crusher feed throat area (Figure 4.9).

More information on dust control at transfer points is found in Chapter 5—Conveying and Transport.

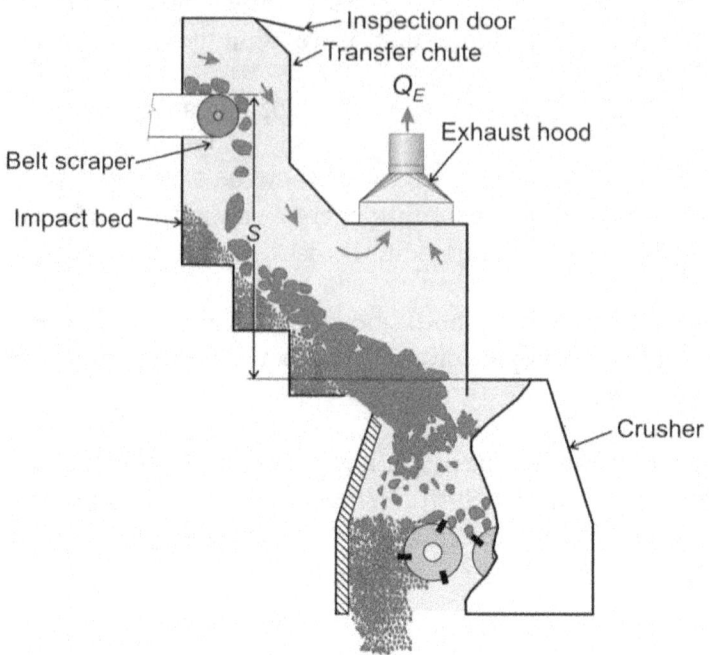

Figure 4.6. Illustration of a dry (exhaust) dust control system at the transfer point of a conveyor discharge to a crusher feed hopper.

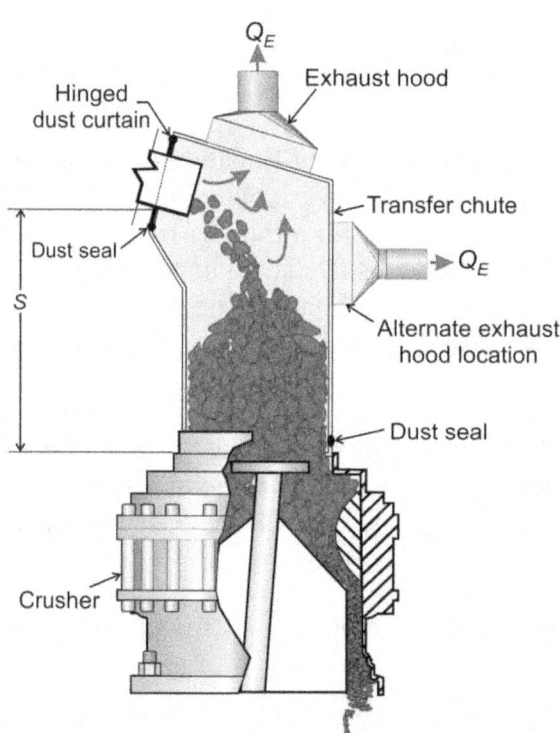

Figure 4.7. Illustration of a dry (exhaust) dust control system on a feed chute into a transfer chute feeding a crusher.

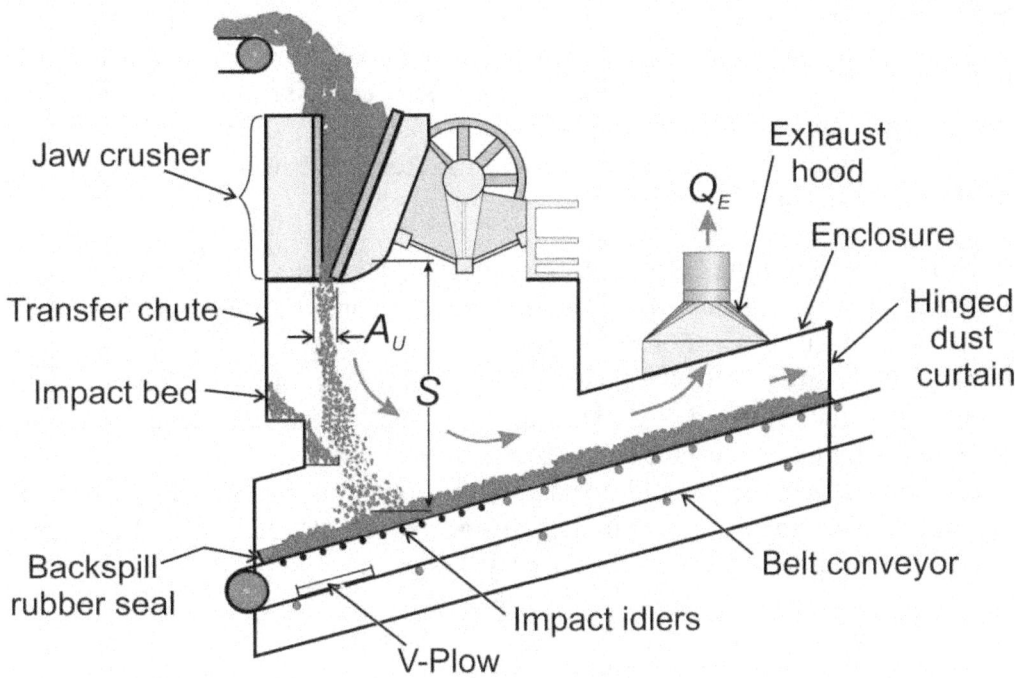

Figure 4.8. Illustration of a dry (exhaust) dust control system at the discharge of a jaw crusher onto a belt conveyor.

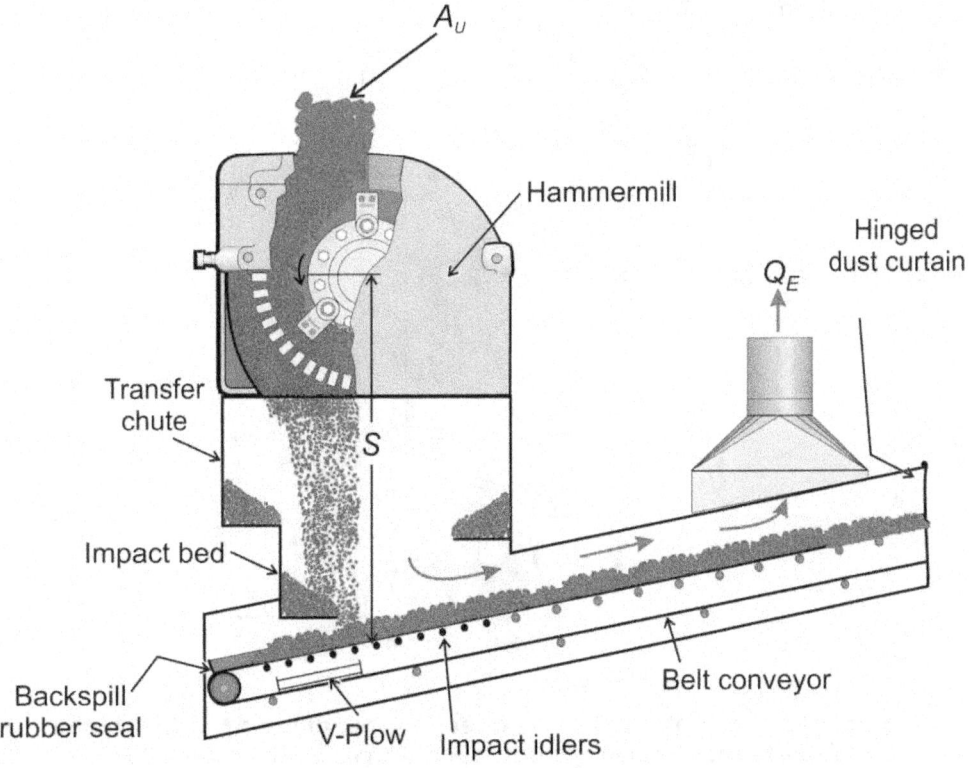

Figure 4.9. Illustration of a dry (exhaust) dust control system at the discharge of a hammermill crusher onto a belt conveyor.

Wet dust control methods for crushing operations, illustrated in Figures 4.10 and 4.11, involve wetting the process material before crushing, after crushing, or both. This is essentially treating the crushing operation as two transfer points, the feed side and discharge side. Crushing creates smaller sized material with an attendant increase in surface area, which will be dry. Thus, material wetted prior to crushing for dust control will likely have to be wetted again to address the additional dry surface area.

Where dust control through water application is selected, recall that solid spray nozzles, or full cone spray nozzles, produce larger drop sizes than hollow cone nozzles. Hollow cone nozzles should be used when dust suppression (reduction of airborne dust) is desired, with spray patterns arranged to cover the entire area of the dust cloud. When prevention of airborne dust is desired in a static application such as a bin or hopper, full cone nozzles should be used with spray patterns overlapping the entire surface of the material to be wetted. In a moving application such as a conveyor, fan spray nozzles should be used with spray patterns oriented perpendicular to conveyor travel and overlapping approximately 30 percent of the fan width. Further information on wet methods is found in Chapter 2—Wet Spray Systems.

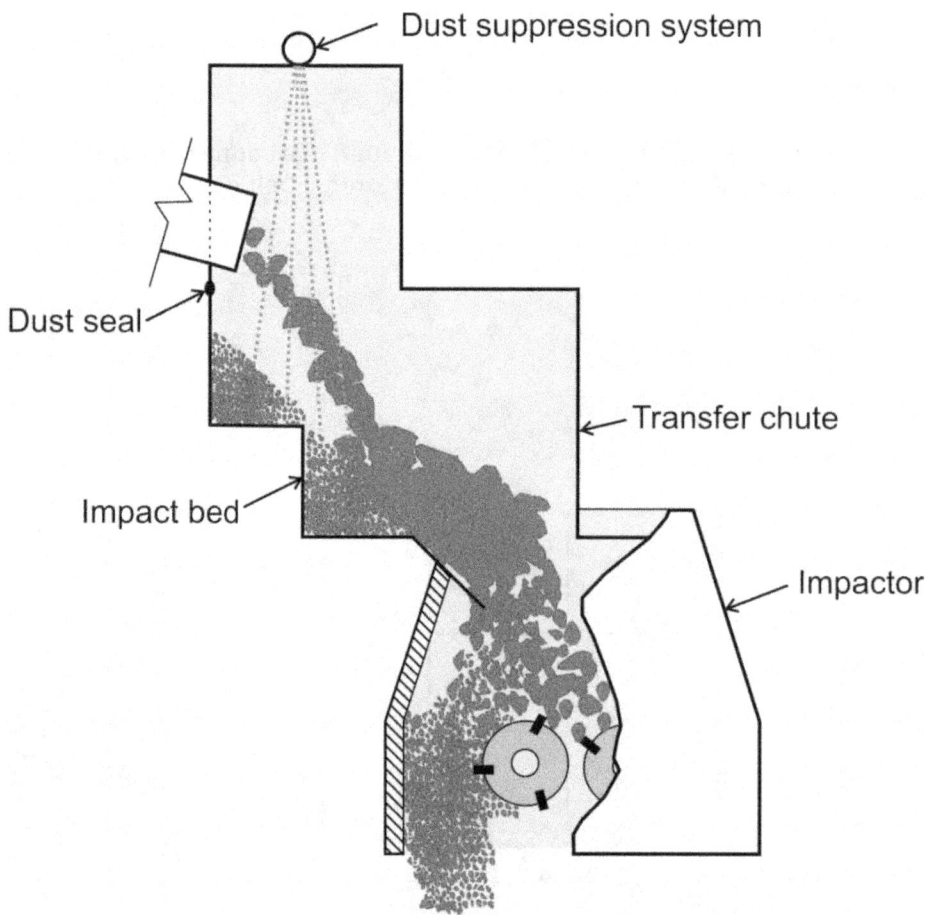

Figure 4.10. Illustration of a wet dust control approach with a transfer chute/impact bed enclosure at a crusher loading operation.

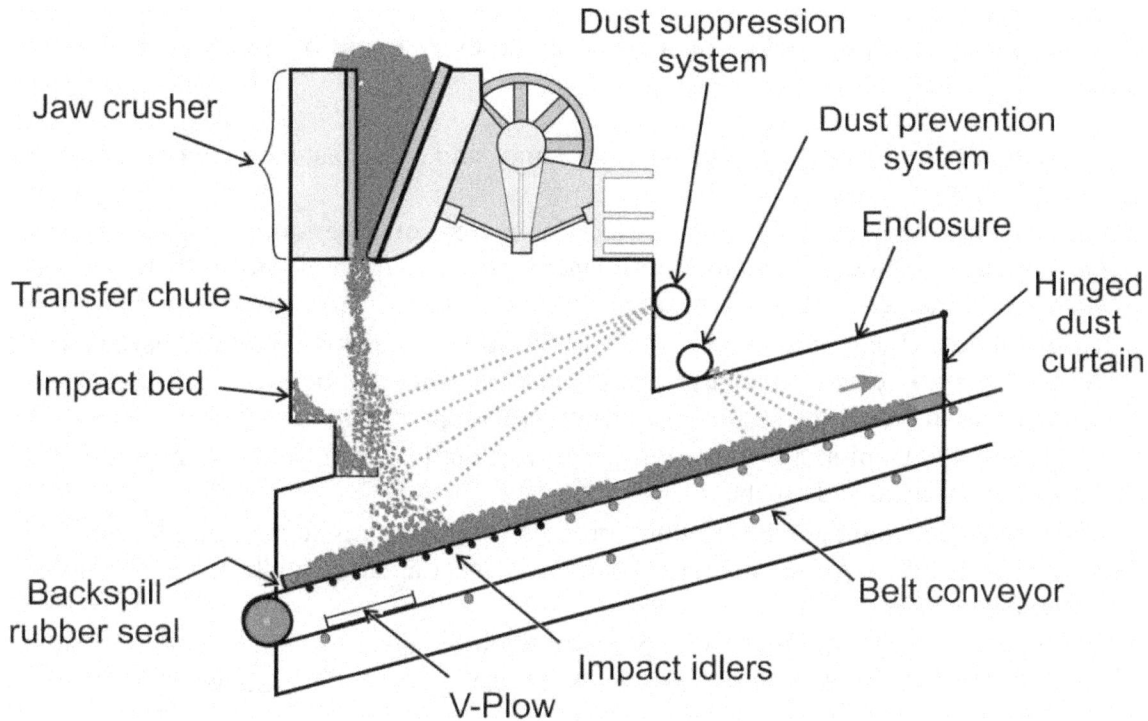

Figure 4.11. Illustration of a wet dust control approach on a crusher discharge/belt loading operation.

The mechanical action of crushers can generate air movement; i.e., jaw crushers can have a bellows-type effect, and cone or gyratory crushers can act as fans, although neither of these classes of crushers operate at high speeds. In contrast, hammermills operate with higher speed components and can act as centrifugal fans. A method to estimate maximum generated airflow from this type of crusher has been described [Burton 1999]. The maximum volume (cubic feet) of air generated per shaft revolution can be estimated by treating the hammer/shaft assembly as a fan using Equation 4.3:

$$A_{GEN\;REV} = \frac{\pi}{4D^2W} \quad (4.3)$$

where D = diameter of the hammer assembly from the tip to tip, feet; and
W = the width of the hammers, feet.

The volume of air generated per minute (cfm) is then the product of $A_{GEN\;REV}$ and shaft RPM.

A more general method to evaluate this issue for any type of crusher is to measure inflow velocity at permanent openings into the dust control enclosure with the crusher on, then off. If the inflow velocity is not satisfactory when the crusher is operating, it will be necessary to increase exhaust volume or install baffles inside of the enclosure.

Work Practices to Minimize Dust Exposure from Crushers

The following practices can serve to reduce operator exposure to airborne dust at crusher operations.

- Maintain closure/locking devices such as clamps and other fasteners. Fasteners are only effective when used.
- Where compatible with the process, wash down areas on a periodic basis. Periodic washing prevents the accumulation of material which can eventually become too large for wash down, necessitating a dry cleanup problem with the generation of dust.
- Keep makeup air and recirculating air dust filtration systems on operator booths, front-end loaders or other mobile equipment in place. Ensure that only correct filters are utilized, and maintain the systems to manufacturer specifications.
- Wear approved respiratory protection when working in dust collectors, mills, air classifiers, screens, and crushers.
- Where feasible, automate the crushing process using sensing devices and/or video cameras. This removes the operator from the crusher area and reduces the potential for dust exposure.
- Where crusher operation must be supervised continuously, provide an enclosed booth with a positive-pressure, filtered air supply for the operator. See Chapter 9—Operator Booths, Control Rooms, and Enclosed Cabs for more information.

MILLING

Milling, often called grinding, is a process by which granular minerals are reduced in size by compression, abrasion, and impaction. Mills may be classified into two broad types based on how they operate: *tumbling mills* (Figure 4.12) and *stirring mills* (Figure 4.13). Tumbling mills generally operate in a horizontal orientation, and the shell of the mill rotates to impart motion to the contents or charge. Stirring mills may be horizontal or vertical and motion is imparted to the charge by an internal stirring element.

Tumbling mills employ some type of medium to perform the size reduction, often rods or balls, usually manufactured from iron or a steel alloy, or high density ceramic material when metal contamination of the product is a concern. *Autogenous mills* use the feed stock (incoming ore) itself as the medium, and *semi-autogenous mills* use a combination of the feed stock and balls. Stirring mills reduce the size of the feed stock between fixed and rotating mill components. Materials discharged from mills can range in size from less than 40 to 300 micrometers (μm) for tumbling mills and from 40 to less than 15 μm for stirring mills [Wills 2008].

Although mineral processing milling is often done wet, the industrial minerals industry primarily utilizes dry milling. At some operations, the grinding mill is fed with coarse rejects or one of the products from their dry screening operation. These milling operations are fed dried granular minerals, usually less than 12 mesh (1.7 millimeters), and produce products in the 50-micrometer (μm) range, but some fine-grind products are less than 10 μm in size, which means that all of the fine-grind product is in the respirable size range.

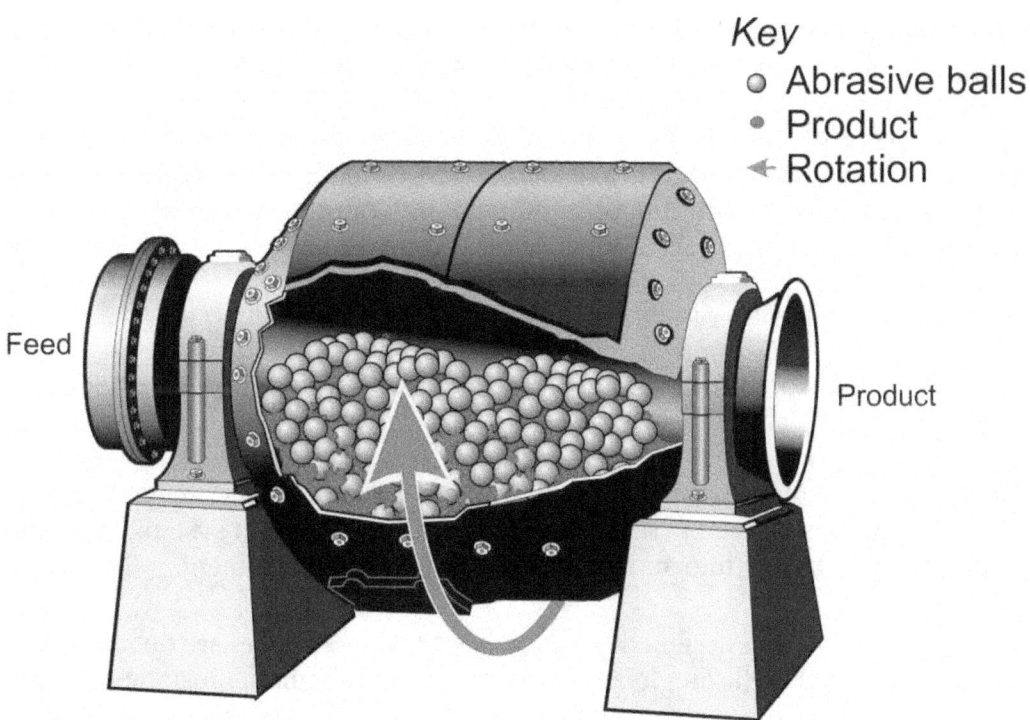

Figure 4.12. Cut-away illustration of a ball mill showing the charge consisting of the material being processed and balls.

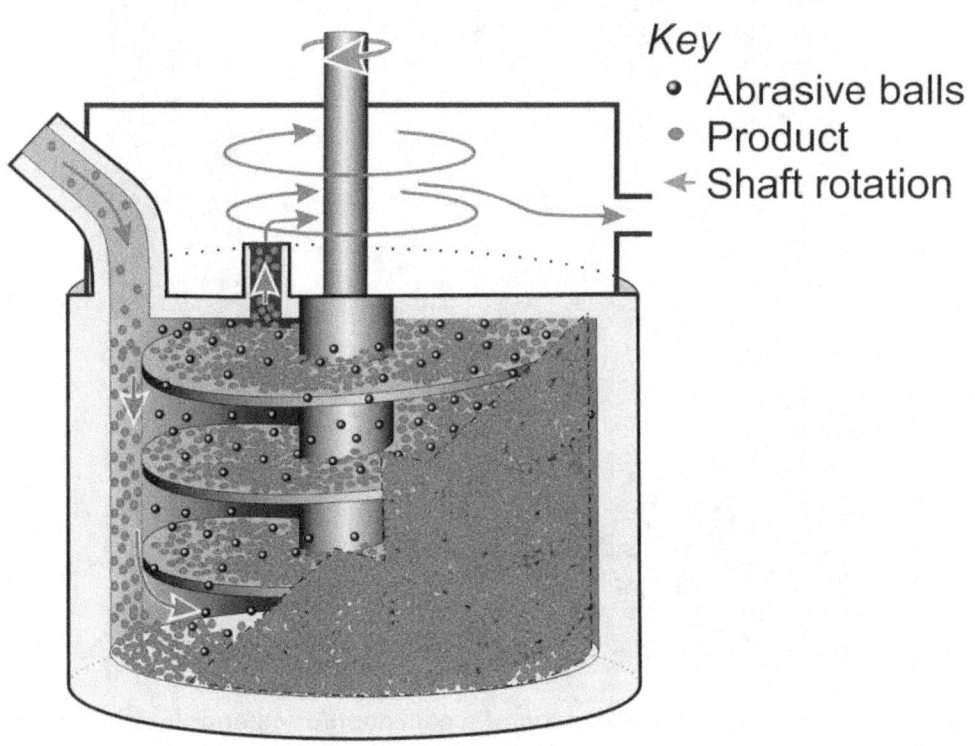

Figure 4.13. Illustration of a stirring mill.

Crushing, Milling, and Screening 123

Some specialty grinding operations purchase damp granular minerals and dry them before they are fed into the milling process. Regardless, all the process equipment involved in preparing and feeding the grinding mill have their respective dust collection systems.

The most important consideration for dust control within a grinding/classification/product storage circuit is dust containment, and the second is dust collection. All of the material conveying equipment related to a milling or grinding circuit must be enclosed. In addition, all transfer points must be fully enclosed.

Design and Work Practices to Minimize Dust Exposure from Milling

The following design and work practices can serve to reduce operator exposure to airborne dust around milling operations.

- Grinding circuit equipment must be tightly closed and maintained under a slight negative pressure 100 percent of the operational time by a dust collection system. This means that any air leakage will flow into and not out from the equipment, keeping the dust contained.
- Sampling points must be designed so that when the access doors are opened, the milled material stays inside. All sample and inspection doors should be installed at least 45 degrees off the vertical or greater, with horizontal doors being preferable for easy access and lower risk of spillage (see Figure 4.14).

Figure 4.14. Sampling doors located on top of a collection/transfer point to prevent leakage.

- Grinding circuits and buildings should be designed so that they can be washed down for cleaning.
- A vacuum system should be installed and used for cleanup. Equipment should not be dry swept or brushed off because of the possibility of dust liberation. Never use compressed air to clean equipment or work areas.
- Upper levels around process equipment should have solid floors. If a spill or leak were to occur over a solid floor, it immediately piles up instead of falling towards the next level where it will become airborne and contaminate the mill.
- Sections of the milling process should be isolated to improve containment efforts. Examples include uncovered storage, bagging areas, and bulk load out areas.
- Above freezing conditions, floors should be kept wet so that any falling dust will immediately hydrate and be trapped.
- Whole room ventilation with at least 10 air changes per hour (acph) is recommended.
- For classifier circuits, vendor guidelines should be followed for ducting and airflow to achieve optimum classification results. With other components of the grinding/loadout operation, the containment, pickup points, and ducting should be designed so that areas with high air movement do not occur at transfer points and other potential disturbance areas where the air stream will entrain product.
- When new or recently lined tumbling mills are started up for the first time, the air leaving the mill is fairly humid until the grinding action builds up enough heat to drive off the moisture out of the curing grout. This can lead to a buildup of moist material in ducting and may cause blinded bags and cartridges in the dust collectors. These problems will typically be resolved once the mill reaches operating temperature and the grout has cured.

SCREENING

Screening operations can produce high levels of dust because smaller sized material is handled. Airborne dust is generated from the vibrating screen decks that accomplish the size separation. Additionally, material must fall some distance as part of the separation process, and some dust will be suspended by this process.

Dust control for screening systems is similar to that for crushers, although wet systems are generally not used due to blanking of the screen openings by the wet material. Also, screens are not normally subject to large surges of material flow as are crushers. This is due to the fact that screens operate most efficiently within a designed flow range. Overloading screens, whether through excessive feed rate or large surges of material, can cause accelerated wear of components, reducing the efficiency of the operation and potentially increasing dust emissions.

Screens should be totally enclosed, and water suppression systems (when compatible with the process) or dust collection and exhaust systems should be incorporated. Necessary openings in the screen enclosure must be minimized, and inspection and maintenance openings must be provided with tight-fitting closures.

Flexible materials (e.g., rubber or synthetic sheeting) must be used to seal openings between the moving screen components and stationary equipment and structures. Because the seal between moving and stationary components is under dynamic stresses whenever the screen is operating, it should be frequently inspected for signs of failure, on a schedule that incorporates data from its operating history. Some screen original equipment manufacturers offer parts and materials such as covers and flexible sealing components designed specifically to fit their equipment well, enabling creation of an effective dust control enclosure. Equipment vendors should be consulted to explore available control options.

Initial design exhaust air volume may be estimated using the induced air method via Equation 4.2. Figures 4.15 and 4.16 illustrate exhaust ventilation dust control on screens. The dimension for material fall (Equation 4.2, variable S) is indicated in the figures. The upstream open area dimension (Equation 4.2, variable A_U) is the estimated area of the transfer chute plus any unsealed open area around the chute penetration into the dust control enclosure.

Alternatively, an exhaust volume of 50 cfm per square foot of screen area may be used as an initial value for flat deck screens [ACGIH 2010]. This value is not increased for multiple level screens because the stacked construction of multiple level screens presents no more "source area" than does a single screen. For cylindrical rotary screens, an exhaust volume of 100 cfm per square foot of screen cylindrical cross section with an in-draft velocity minimum of 400 fpm is recommended [ACGIH 2010].

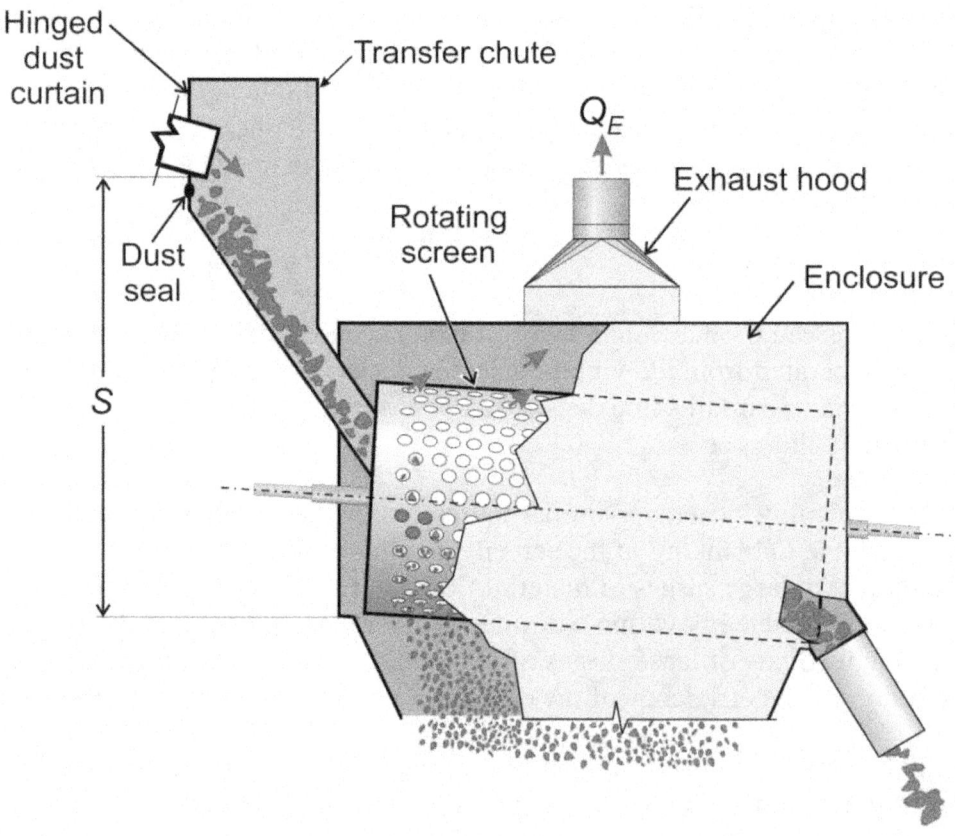

Figure 4.15. Illustration of a dry (exhaust) dust control system on the feed to a rotary screen with enclosed transfer chute.

126 Crushing, Milling, and Screening

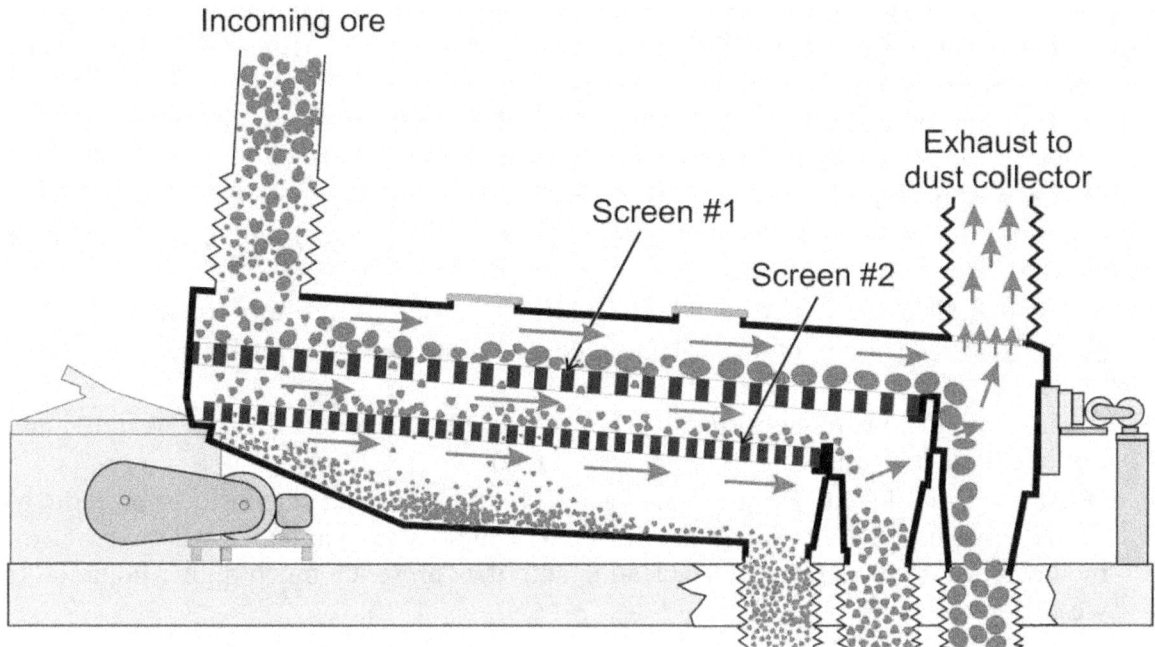

Figure 4.16. Illustration of a dry (exhaust) dust control system on a vibrating screen.

Exhaust air take-off location must be carefully considered so that desirable fines (product) are not collected. Locating the take-off closer to the discharge (away from the feed end) of the screen will help avoid the loss of desirable fines. Also, collection of undersized material can be reduced by using take-off openings large in area to reduce the air entry velocity.

Work Practices to Minimize Dust Exposure from Screens

The following practices should be observed where compatible with installed equipment and existing operating procedures.

- Clean equipment and area before and during work, when needed, preferably by washing or vacuuming.
- Stop material flow before cleaning screens, if operationally feasible. The absence of flowing material minimizes the potential for dust emissions.
- When cleaning, only open the top or the bottom part of screen decks at a time. Do not open both at the same time. The larger the size of the opening, the less effective the ventilation system can be.
- Only open one screen at a time during cleaning. Do not open multiple screens simultaneously. Again, numerous openings decrease the capture effectiveness of a ventilation system.
- When cleaning screens, only use long-handled brushes that will provide more distance between the worker and the dust source.
- Consider variable speed vibrators that significantly reduce the frequency of manual cleaning of screens. The less time cleaning screens, the less the potential for dust exposure.

- Open screen decks slowly to allow internal exhaust system to function at the deck periphery. Opening decks rapidly can cause a swirling effect with dust being released into the work area, potentially exposing personnel.
- Do not slam screen decks closed. Any material, no matter how slight, has the potential of becoming airborne due to the shock of deck closure.
- Decks must be kept clear of dust. A clean deck lessens the potential of dust emissions resulting from opening and closing decks, as well as from environmental factors such as wind.
- Maintain seals on screen decks where an airtight closure is desired. Decks that are properly sealed should remain sealed when all fasteners and sealing materials are in place.
- Closure and locking devices such as clamps and other fasteners must be well maintained. Fasteners are ineffective unless they are utilized.
- Where compatible with the process, wash down areas around screens on a periodic basis. Periodic washing removes material before it accumulates to a degree where dry cleanup methods become necessary. Dry cleanup activities present a much higher potential for worker exposure.

MAINTENANCE

Maintenance of dust control systems is critical to ensure continuing worker protection [USBM 1974]. Crushers and screens are subject to constant vibration when in use and this condition can cause accelerated wear on all components. Even small openings from worn seals, missing fasteners, and damaged flexible connectors can cause unwanted air infiltration, degrading capture velocity at necessary penetrations [Goldbeck and Marti 1996]. This could potentially result in worker overexposure, particularly when exposure limits are low.

Operators should document baseline parameters of an effectively functioning dust control system, including for example:

- inflow velocity at enclosure openings,
- static pressure in ducts,
- velocity pressure in ducts and at hood entry points, and
- water pressure and flow in wet systems.

Results of subsequent periodic checks should be compared to the baseline values. Significant deviations from the baseline values should be explored, and corrective action implemented when necessary.

A preventive maintenance schedule should be established based on manufacturer recommendations, observed component wear, system performance measures, and exposure sampling results.

REFERENCES

ACGIH [2010]. Industrial ventilation: a manual of recommended practice. 27th ed. Cincinnati, Ohio: American Conference of Governmental Industrial Hygienists.

Anderson DM [1964]. Dust control design by the air induction method. Industrial Medicine and Surgery. *33*:68–72.

Burton DJ, ed. [1999]. Hemeon's plant & process ventilation. 3rd ed. Boca Raton, Florida: Lewis Publishers.

EPA [2003]. AP 42, Fifth Edition, Compilation of air pollutant emission factors, volume 1: stationary point and area sources. United States Environmental Protection Agency. Chapter 11: Mineral Products Industry; Background Information for Revised AP-42 Section 11.19.2, Crushed Stone Processing and Pulverized Mineral Processing [online] available at URL: [http://www.epa.gov/ttn/chief/ap42/ch11/]. Date accessed 09/04/2008.

Goldbeck LJ, Marti AD [1996]. Dust control at conveyor transfer points: containment, suppression and collection. Bulk Solids Handling *16*(3):367–372.

NIOSH [2003]. Handbook for dust control in mining. By Kissell FN. Pittsburgh, PA: U.S. Department of Health and Human Services, Centers for Disease Control and Prevention, National Institute for Occupational Safety and Health, NIOSH Information Circular 9465, DHHS, (NIOSH) Publication No. 2003–147.

USBM [1974]. Survey of past and present methods used to control respirable dust in noncoal mines and ore processing mills—final report. U.S. Department of the Interior, Bureau of Mines Contract No. H0220030. NTIS No. PB 240 662.

USBM [1978]. Improved dust control at chutes, dumps, transfer points and crushers in noncoal mining operations. U.S. Department of the Interior, Bureau of Mines Contract No. H0230027. NTIS No. PB 297–422.

Wills BA [2008]. Wills' Mineral Processing Technology, 7th ed. Oxford, UK: Elsevier Ltd.

Yourt GR [1969]. Design for dust control at mine crushing and screening operations. Canadian Mining Journal *90*(10):65–70.

CHAPTER 5: CONVEYING AND TRANSPORT

CHAPTER 5: CONVEYING AND TRANSPORT

There are numerous types of equipment used at mineral processing operations to transfer material from one location to another. The material being transferred can range from raw unprocessed ore to fully processed, finished product. Each material transfer methodology has its own strengths and weaknesses. Proper selection of the correct type of equipment is a function of the specific application, taking into account the material to be transferred, transfer distance, and nature of the transfer (i.e. horizontal, vertical, incline or decline). This chapter will discuss some of the most commonly used transport systems and the methods available for controlling dust and spillage.

BELT CONVEYORS

Belt conveyors are among the most commonly used piece of equipment at mineral processing operations. A conveyor, and the associated transfer points, can generate significant quantities of respirable dust and be one of the greatest sources of fugitive dust emissions within an operation. Operations must control these emissions by containing, suppressing, or collecting dust mechanically, either before or after it becomes airborne, giving special attention to transfer points. A conveyor belt consists of many different parts as seen in Figure 5.1.

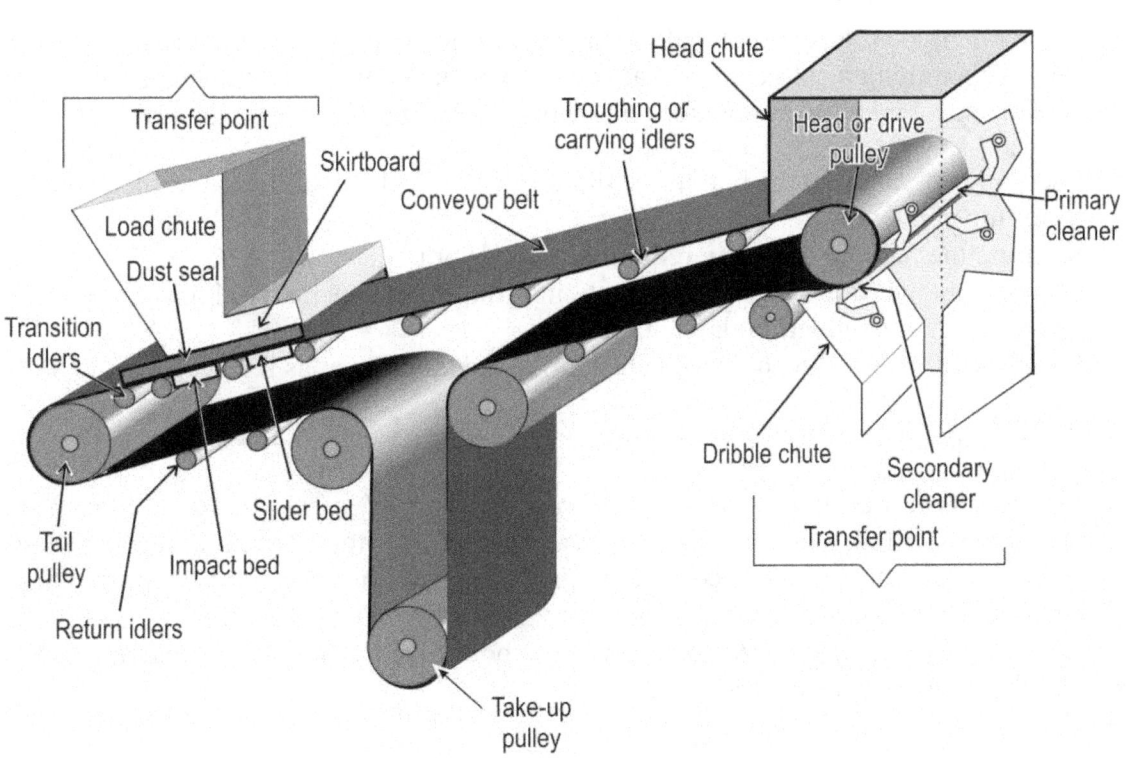

Figure 5.1. Basic components of a conveyor belt.

Conveying and Transport 133

There are three primary root causes for fugitive dust emissions associated with conveyor belts: spillage, carryback, and airborne dust (Figure 5.2). Control of all three primary dust sources is necessary to eliminate fugitive dust emissions.

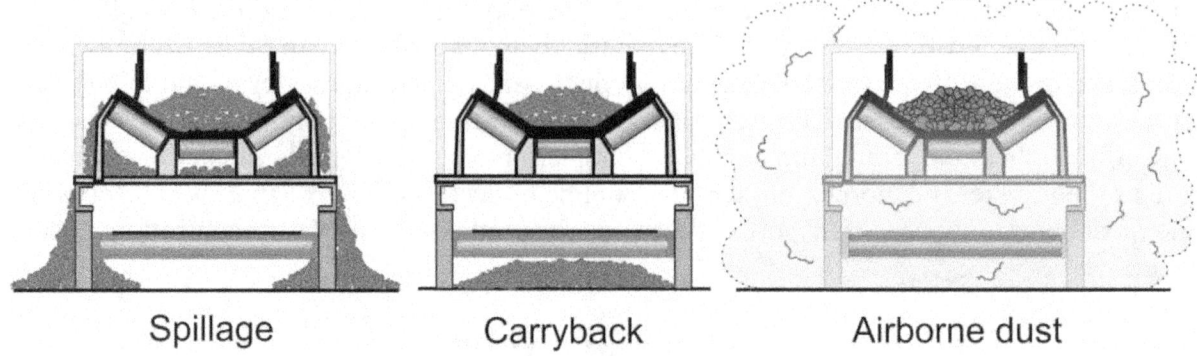

Spillage   Carryback   Airborne dust

Figure 5.2. Types of fugitive dust emissions from conveyor belts.

Controlling Material Spillage

Material spillage from a conveyor belt is caused by a lack of material control, either at a transfer point or along the transfer route. Spillage along the transfer route is generally associated with carryback, a condition discussed later in this chapter.

The most prevalent location for material spillage is at transfer points. Material may be deposited virtually anywhere along a conveyor, but always leaves the conveyor at the head pulley. In order to effectively control material at transfer points, the following four steps are used.

1. Feeding material onto the belt in the direction of travel and at the same speed as the receiving conveyor.
2. Keeping the fed material centered on the conveyor belt.
3. Minimizing the transfer distance which, in turn, minimizes material impact on the belt and air entrainment within the material.
4. Using skirtboard ("skirting") to manage the loaded material as it travels on the belt.

Each of these steps will be discussed in detail below.

Step 1. Feeding material onto the belt in the direction of travel and at the same speed as the receiving conveyor. Figure 5.3 shows a conveyor head pulley (top conveyor) feeding material onto a conveyor tail pulley (bottom conveyor). This figure depicts the simplest of transfer points, where the feeding and receiving conveyors are in-line with each other. The material is being fed in the same direction and the speed of the bottom conveyor should match, as closely as is possible, the top conveyor's speed.

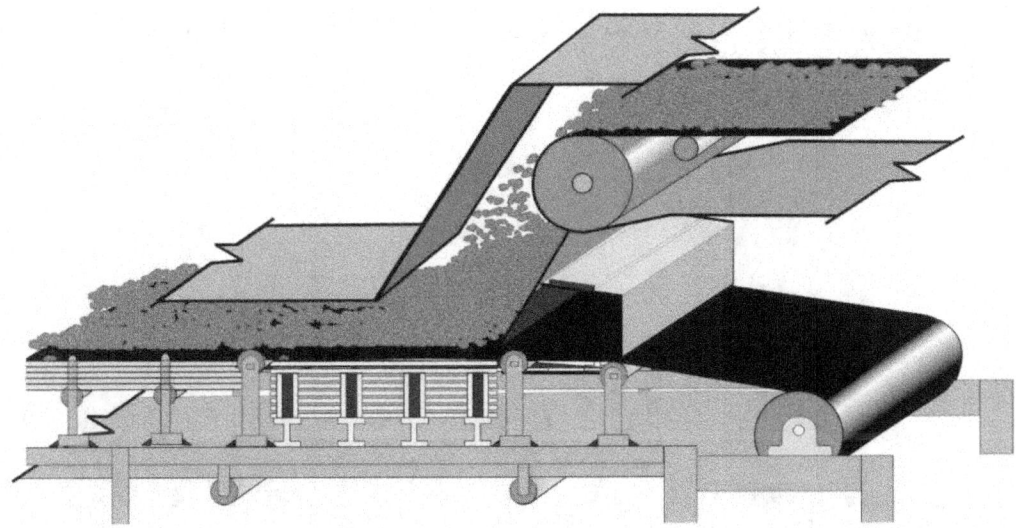

Figure 5.3. Basic depiction of material transfer from a top conveyor to a bottom conveyor.

Achieving effective transfer of material in the same direction and at the same speed when conveyors are, for example, perpendicular to one another, is more challenging. Impact beds are often used to allow ore to fall onto ore rather than directly onto the belt. This is done to control the direction and speed of the material being transferred and to minimize wear on the inner surfaces of the chute and skirting (Figure 5.4).

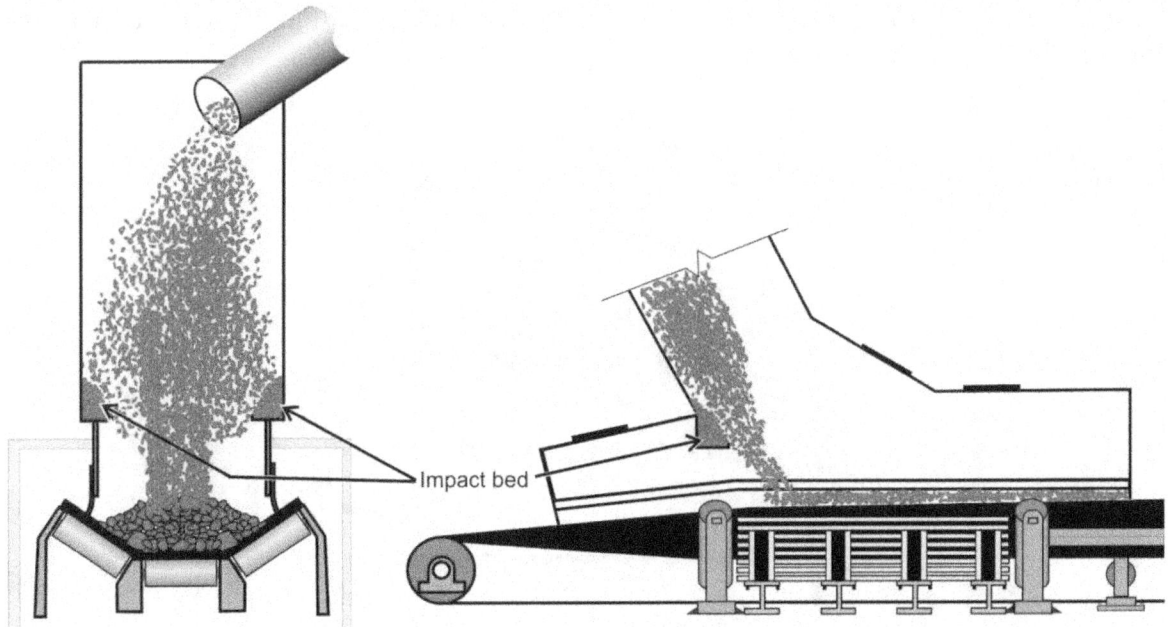

Figure 5.4. Depiction of impact bed used to control material direction and speed.

At times, curved plates are used near the surface of the receiving belt to help steer the material being fed. Plates and impact beds can help to encourage the material being fed to flow in the direction of the receiving conveyor as well as control the placement of the material on the receiving conveyor (Figure 5.5).

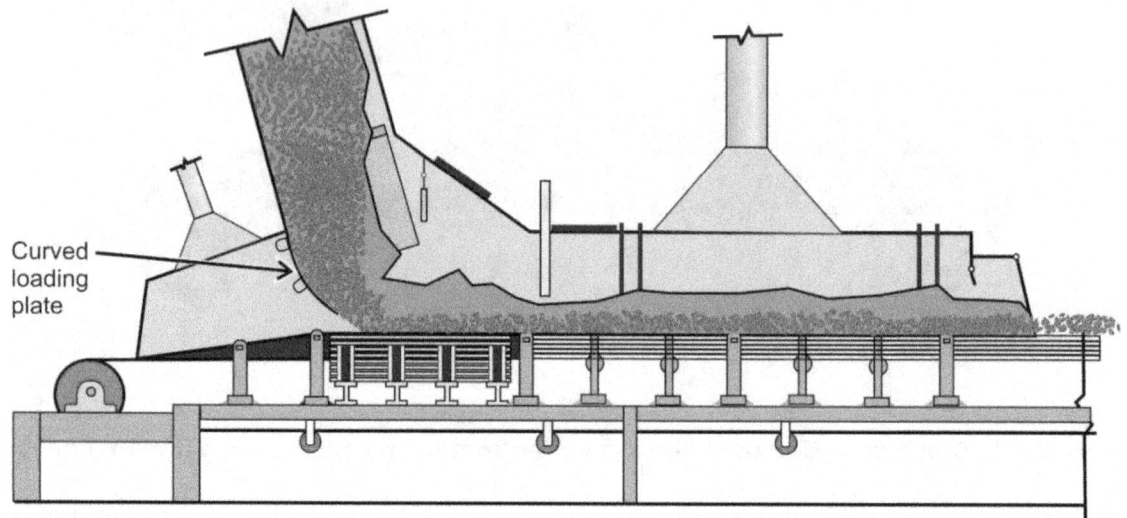

Figure 5.5. Depiction of curved loading plate to help steer loaded material in a particular direction.

Step 2. Keeping the fed material centered on the conveyor belt. Keeping the fed material centered on the conveyor belt is extremely important in controlling material spillage at transfer points. Material fed more to one side of a belt than another (Figure 5.6) can cause the belt to shift sideways and run off-center. A belt that runs off-center can spill material over the edge of the belt outside of the transfer point.

Figure 5.6. Spillage as a result of material being improperly fed onto a belt.

Step 3. Minimizing the transfer distance which, in turn, minimizes material impact on the belt and air entrainment within the material (see Figure 5.7). The transfer distance is the distance that the material falls before landing on the belt. Longer transfer distances permit fine particles to separate from the material being transferred, while shorter transfer distances reduce the time available for fine material to separate from the material being transferred, thus reducing dust emissions. Additionally, longer transfer distances also contribute to belt sagging. When belts sag, they allow material to become trapped between the idlers, which in turn allow the material to release fugitive dust, as well as directly damaging the belt. To address this issue, some operations use impact cradles (often called impact beds). An impact cradle system may include a slick top cover to allow smooth belt movement as well as underlying rubber layers to absorb impact (Figure 5.8). In areas of low impact or no-impact, even lighter duty side-support cradles can be used (Figure 5.9). For effective impact cradle design, refer to Conveyor Equipment Manufacturers Association (CEMA) Standard 575–2000 (Table 5.1 and Figure 5.7) [CEMA Standard 575–2000].

Table 5.1. Impact cradle ratings to be used based on impact loading [CEMA Standard 575–2000]

Code	Rating	Impact Force Lbf	$W \times h$ (ref.) lb-ft
L	Light duty	<8,500	<200
M	Medium duty	8,500 to 12,000	200 to 1,000
H	Heavy duty	12,000 to 17,000	1,000 to 2,000

Key
Q = Weight of material conveyed — tph
W = Weight of largest lump — lb
h = Height of drop — ft
k = Spring constant for specific equipment (consult equipment manufacturer) — lb/ft
F = Impact force — lbf

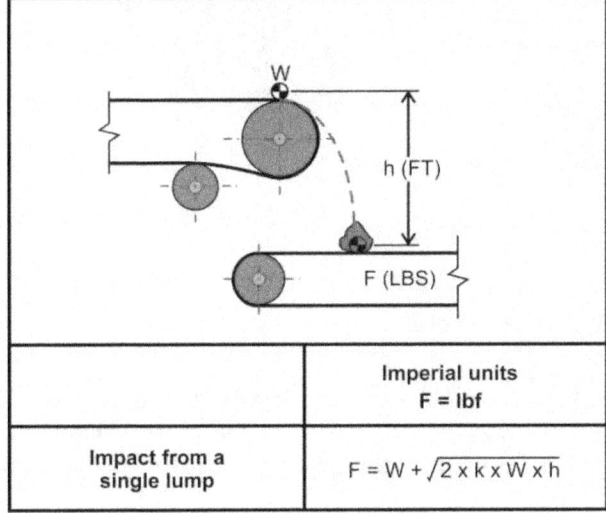
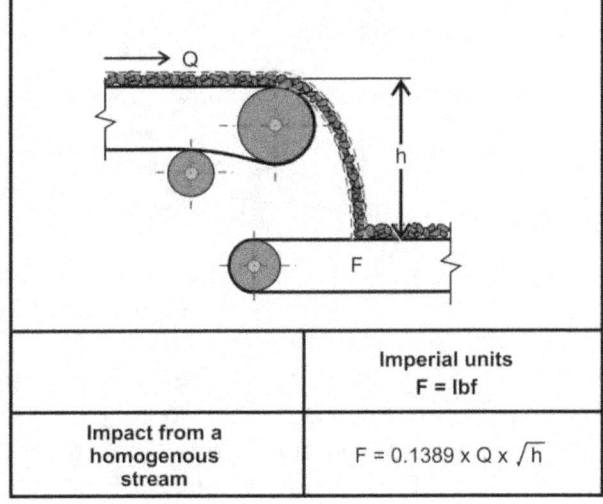

Figure 5.7. Calculations for determining material impact loading at a conveyor transfer point [CEMA Standard 575–2000].

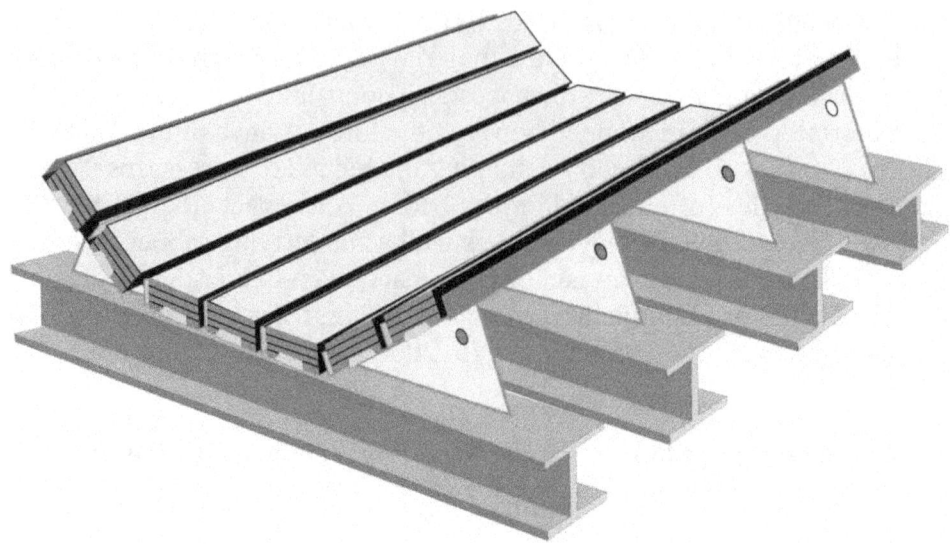

Figure 5.8. Depiction of an impact cradle with underlying rubber layers.

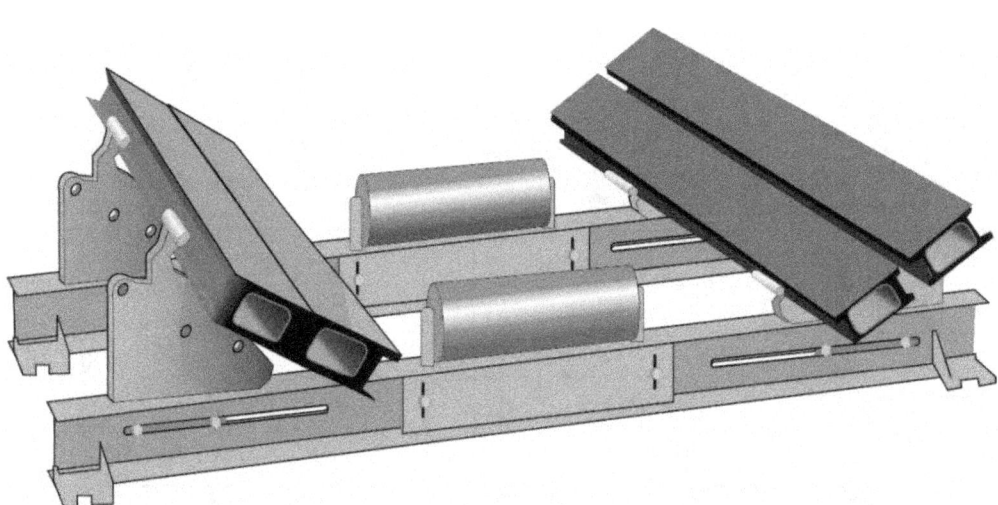

Figure 5.9. Side support cradle.

Step 4. Using skirtboard (skirting) to manage the loaded material as it travels on the belt.
Skirting (Figure 5.10) is the horizontal extension of the loading chute used to contain fines (very fine pieces of ore) and dust within the transfer point and to shape the material until it has settled into the desired profile on the center of the belt.

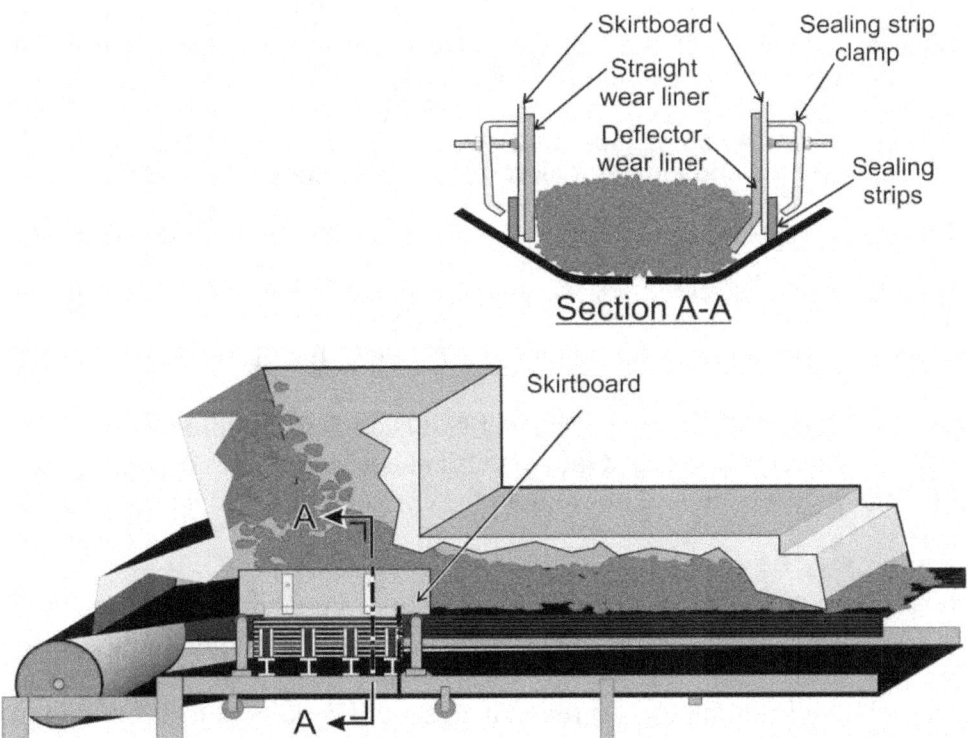

Figure 5.10. Skirtboard used within the transfer point to help manage loaded material.

Skirtboards are employed to retain material on the belt after leaving the loading chute and until the material reaches belt speed. Skirtboards are generally made of steel with the lower edges positioned some distance above the belt. The gap between the boards and the belt is in turn sealed by a flexible elastomer sealing strip attached (clamped or bolted) to the outside wall of the boards (see Figure 5.10).

Skirtboard length should, in theory, extend in the belt's direction beyond the point where the material has settled into the profile that it will maintain for the remainder of its journey to the head pulley. CEMA notes that the length of skirtboarding is "a function of the difference between the velocity of the material at the moment it reaches the belt and the belt speed" [CEMA Standard 575–2000]. In essence, CEMA recommends two feet of skirtboard for every 100 feet per minute (fpm) of belt speed, with a minimum length of not less than three feet. While this rule of thumb is generally sufficient for easily conveyed materials, it is insufficient for materials that typically require dust collection or materials that roll on the conveyor, making their management more complex. Other rules of thumb also exist from various organizations. For example, the U.S. Army Corps of Engineers technical manual *Design of Steam Boiler Plants* recommends that skirting extend 5 feet from material impact and extend an additional 1 foot for every 100 fpm of belt speed [Swinderman et al. 2002].

As one demonstration, using the rules of thumb from CEMA and the Army Corps of Engineers, Table 5.2 compares the two recommendations using a belt speed of 600 fpm. Based on this scenario, CEMA recommends 12 feet of skirting while the Army Corps of Engineers recommends 11 feet of skirting. While the 1-foot difference may not seem significant, in operational conditions this could mean the difference between no material spillage and significant material spillage. Therefore, each circumstance must be evaluated individually with the behavior of the material under conditions of conveyance as a primary consideration.

Table 5.2. A comparison of CEMA and Army Corps of Engineers recommendations for skirting for a belt speed of 600 fpm

CEMA	Army Corps of Engineers
2 ft of skirting for every 100 fpm of belt speed	5 ft plus 1 ft of skirting for every 100 fpm of belt speed
2 ft x (600 fpm/100 fpm) = 12 ft of skirting	5 ft + (600 fpm/100 fpm) = 11 ft of skirting

An inclined skirting design, in which the skirting belt is angled at approximately 30 degrees from vertical, is more advantageous over a standard vertical design because of wear issues. This skirting design improves the loading of ore onto the conveyor and reduces the amount of dust generated.

Dust curtains (Figure 5.11) are an effective and inexpensive containment device that should be installed at the entrance and exit of the chute enclosure. Some design guidelines include:

- Exit air velocity from the enclosure or chute should be kept below 500 fpm to minimize the entrainment of large particles [Yourt 1990].
- Dual rubber curtains should be hung roughly 18 inches (450 mm) apart from each other to form a "dead" area where dust can settle [Goldbeck and Marti 2010].
- Entrance curtains should extend to the conveyor belt to maximize the containment area.
- Exit curtains should extend to approximately one inch (25 mm) below the top of the product pile [Goldbeck and Marti 2010].

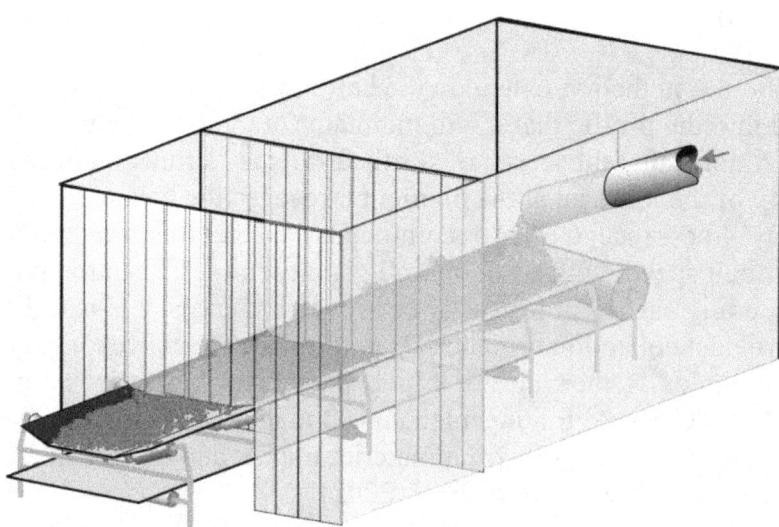

Figure 5.11. Typical dust curtain used at the entrance and exit of the chute enclosure.

Carryback

Material that sticks or clings to a conveyor belt after passing over the head pulley is called carryback. Carryback tends to fall from the belt as it passes over return idlers. This creates piles of material that require clean-up, which can increase worker dust exposure. Also, respirable portions of carryback can become airborne and increase fugitive dust exposure levels. The goal is to remove carryback before it is released into the air and becomes a source of contamination to the workers or creates piles of material that require clean-up.

The primary means of controlling carryback is to clean the belt as it passes over or past the head pulley (i.e. shortly after material is discharged from the belt). The two most common means of cleaning a conveyor belt of carryback are to mechanically "scrape" the belt via scrapers or brushes or to wash the belt.

Belt Scrapers

There are a myriad of different types and configurations of scrapers and addressing them all is beyond the scope of this handbook. Instead, the basic principles and types of scrapers will be addressed. Although belt cleaner systems come in many different styles, types, and trade names sold by numerous commercial manufacturers, their function remains the same: to reduce the amount of carryback on the belt once the ore is discharged.

Material can be either scraped or peeled from a belt. When material is scraped from a belt, the scraper is angled with the direction of belt travel (Figure 5.12).

Figure 5.12. Typical method for scraping material from a belt.

Multiple scraper belt cleaner systems provide a way to address carryback, with the use of a precleaner on the face of the head pulley to remove most of the material, followed by secondary cleaners to perform final cleaning. The secondary scrapers can be installed either where the belt leaves the head pulley or further along the conveyor return. To be effective as well as to reduce belt wear, multiple cleaners should be incorporated with low blade-to-belt pressure [Goldbeck and Marti 2010]. When dust levels are high, it is not uncommon to use two or three belt scrapers at different locations in an effort to further reduce the amount of carryback material on the belt [Roberts et al. 1987].

It is most desirable to return the scrapings to the primary material flow through chutes (Figure 5.13). Chutes should therefore be designed large enough to capture the scrapings and their walls steep enough to prevent scrapings from building and stacking up within the chute.

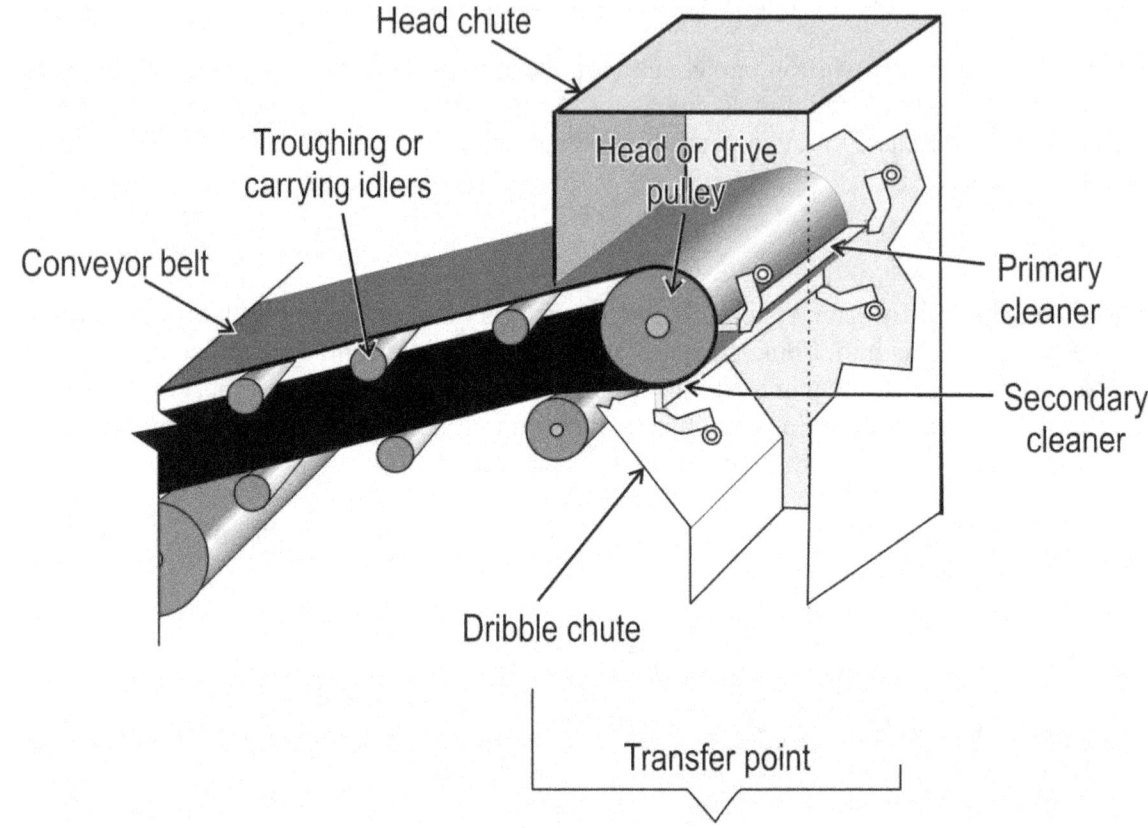

Figure 5.13. Belt conveyor discharge chute used to return scrapings to the primary material flow.

Belt Washing

Some studies have shown that oversized material is more easily removed through scraping, but that the smaller respirable size particles tend to remain adhered to the conveyor. Washing is an alternative when belt scraping does not remove particles adhering to the belt. A belt wash system sprays the conveyor belt with water while simultaneously scraping it to remove the product. A handbook from Martin Engineering, *Foundations: The Practical Resource for Cleaner, Safer, More Productive Dust & Material Control* [Swinderman et al. 2009], devotes a chapter to a thorough discussion of belt washing systems.

CONVEYOR DESIGN AND MAINTENANCE ISSUES

The great challenge involved with conveyors is the number of belts involved and the total distance traveled throughout a mineral processing plant. Some belts are located outside where dust liberation is not as critical as when they are located within a building. Another challenge particular to conveyors is their ability to generate or liberate dust whether they are loaded heavily with ore or nearly empty. Controlling dust from conveyors requires a constant vigilance by the maintenance staff to repair and replace worn and broken parts, including conveyor belting. Basic maintenance and inspection are required to ensure that all parts of the system are performing to their capacity. Material can escape through chutes worn from rust or abrasion, and even small holes created by missing bolts or larger holes created from open access doors can be a pathway for fugitive dust. In some cases, it may become necessary to replace an entire loading chute to ensure proper containment.

Belt Splices

All conveyor belts require at least one splice to connect the ends together and thus close the loop. Due to belt failures, conveyors can often have more than one splice. Mechanical and vulcanized splices, similar to the styles shown in Figure 5.14, are commonly used in the mining industry.

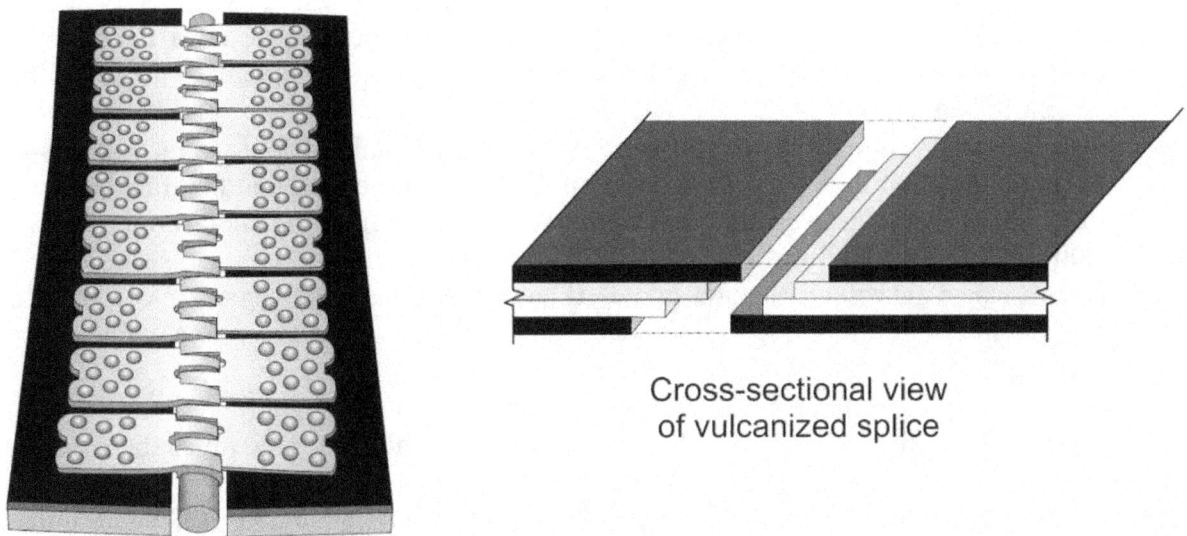

Figure 5.14. Illustration of a mechanical belt splice (left) and a vulcanized belt splice (right).

Mechanical belt splices can often be a source of significant belt spillage because product and dust can fall through small openings in the splice. Vulcanized belt splices can be used to eliminate this source of spillage. Vulcanized splices provide a method of joining the ends of conveyor belts without interrupting the continuity of the belts, and usually without altering the geometry or dimensions of the belts. It is recommended that vulcanized belt splices be used, where feasible, to reduce dust and spillage.

DESIGN CONSIDERATIONS FOR TRANSFER POINTS

Enclosures at both the head and tail ends of a conveyor are a very common practice because they are effective at controlling dust at these locations and provide guarding safety. While the elimination of dust generation at conveyor transfer points is not feasible, effective dust control is. Designing the proper size enclosure is a critical factor because as the ore is dumped onto the conveyor, it entrains a measurable amount of air (venturi effect), and this can pressurize the enclosure if it is undersized. Enclosures for both conveyor and transfer points can be either full or partial, depending on the various components of the system.

Although this seems like a very simple process, significant dust generation and liberation can result from transfer points if they are not properly designed and installed. The following are some important design considerations for an effective transfer point or chute used with an exhaust ventilation system (also see Figure 5.15):

- It has been shown that the exhaust port of the transfer point enclosure should be a least 6 feet from the dump point to minimize the entrainment and pickup of oversized particles [MAC 1980]. The air velocity at the base of the exhaust port cone should also be kept below 500 fpm to avoid the pickup of larger particles [Yourt 1990].
- Transfer point enclosures should be designed to have a 200 fpm intake velocity at any unavoidable opening to eliminate dust leakage from the openings [USBM 1974]. The Mining Association of Canada (MAC) recommends adding 25 percent to the 200 fpm guideline [MAC 1980].
- It is important to minimize openings and thus preserve intake velocity by using plastic stripping and other types of sealing systems. It is recommended that the transfer point enclosure be made large enough so that the air velocity within the enclosure is below 200 fpm (1 m/sec) [Goldbeck and Marti 2010]. The larger enclosure can serve as a plenum where air velocity can dissipate. If an enclosure is undersized, air will be forced from the interior high pressure of the enclosure to the lower pressure area outside, carrying dust particles outside the system.
- Transfer chutes should be sized to allow ore to flow without clogging or jamming. A general rule of thumb is that the chute width should be at least three times the maximum lump size to avoid clogging [USBM 1987].
- The skirtboard for the transfer point should be high and long enough to serve as a plenum so that the dust has a chance to settle. On belts with minor air movement, a good rule-of-thumb is two feet per 100 feet/min of belt speed [Goldbeck and Marti 2010]. When belts have a greater calculated air movement, the transfer point should approach three feet per 100 feet/min of belt speed [Goldbeck and Marti 2010].
- The dump point of the ore should be designed to impact on a sloping bottom or an impact bed. Impact beds are designed to allow ore product to build up so that ore contacts ore at transfer locations or chutes to reduce wear and abrasion (Figure 5.16).
- Fall height of ore should be minimized whenever possible. Some methods to perform this are through the use of rock ladders, telescopic chutes, spiral chutes, and bin-lowering chutes.
- Any abrupt changes in product direction or flow should be avoided.

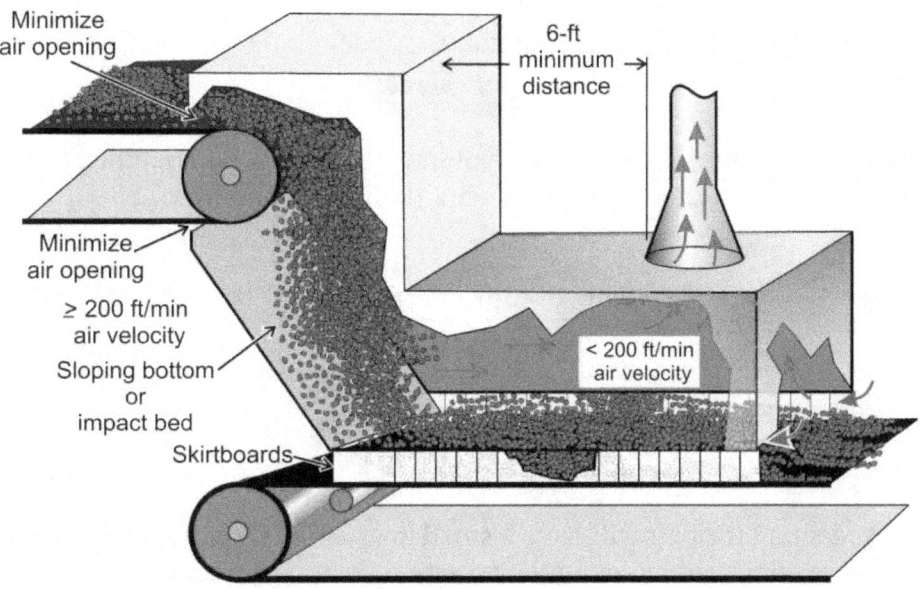

Figure 5.15. Conveyor transfer enclosure used with an exhaust ventilation system.

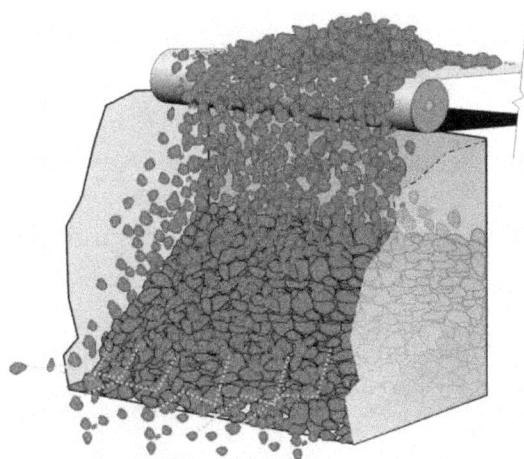

Figure 5.16. Typical impact bed designed for ore-on-ore contact at transfer locations or chutes.

Assessing the correct amount of the required air volume is essential to effective dust collection. Total airflow at a given transfer point can be calculated by Equation 5.1 [Goldbeck and Marti 2010]:

$$Q_{tot} = Q_{dis} + Q_{ind} + Q_{gen} \qquad (5.1)$$

where Q_{tot} = total air movement, cubic feet per minute;

Q_{dis} = calculated displaced air, cubic feet per minute;
Q_{ind} = calculated induced air, cubic feet per minute; and
Q_{gen} = determined generated air, cubic feet per minute.

The amount of displaced air increases as the volume of moving material increases. The amount of displaced air can be calculated by Equation 5.2 [Goldbeck and Marti 2010]:

$$Q_{tot} = \frac{\text{conveyed product (lb/min)}}{\text{bulk density (lb/ft}^3)}$$

(5.2)

$$\text{lb/min} = \frac{\text{t/h} \times 2000}{60}$$

Material conveyed along the belt will have a small amount of air entrapped in the product bed. As the product leaves the head pulley in the normal trajectory it increases in volume as each particle of material collects an amount of air. Once the product lands, this induced air is released, causing substantial positive pressure flowing away from the center of the load zone. The movement of this induced air can be calculated as follows in Equation 5.3 [Goldbeck and Marti 2010]:

$$Q_{ind} = 10 \times A_U \sqrt[3]{\frac{RS^2}{D}}$$

(5.3)

where Q_{ind} = induced air, cubic feet per minute;

A_U = enclosure upstream open area, square feet (where air is induced into the system by the action of falling material);

R = rate of material flow, tons per hour;
S = height of material free fall, feet; and
D = average material diameter, feet.

Other sources of moving air (Q_{gen}) may be devices feeding the load zone such as crushers, and these devices can create a fan-like effect. Equipment manufacturers can provide operators with additional information to help quantify the amount of air generated by these mechanical devices.

WATER SPRAYS FOR PREVENTION OF AIRBORNE DUST

As discussed more fully in Chapter 2—Wet Spray Systems, the use of water to control dust may be classified into prevention applications and suppression applications. When properly designed and installed, water sprays are a cost-effective method of controlling dust from conveyors. The most common and effective practice for conveyor sprays is to wet the entire width of product on the belt. The best practices for the application of water sprays include:

- High-volume, high-pressure sprays should be avoided. The energy of the spray transfers to the material particles, resulting in more dust production [Goldbeck and Marti 2010].
- The amount of moisture applied should be varied and tested at each operation to determine the optimum quantity, but a one percent moisture to product ratio is a good starting point. Excess moisture can promote slippage and adversely affect conveyor belt performance in cold weather conditions.
- Some studies indicate that wetting the return side of the conveyor belt also helps to minimize dust liberation. This practice reduces dust generation from the idlers as well as at the belt drives and pulleys.
- Water sprays located on the top (wetting the product) and the bottom (reducing dust from the idlers) at the same application point can be a very effective strategy [Courtney 1983; Ford 1973]. These locations are also beneficial from an installation and cost standpoint.
- Fan spray nozzles are the most commonly used because they minimize the volume of water added for the amount of coverage. It is more advantageous to locate the water sprays at the beginning of the dust source (i.e., the dump or transfer location) because as the water and ore continually mix together, the wetted surface of the ore grows in volume, increasing the suppression potential and reduction in dust liberation.
- Using more spray nozzles at lower flow rates and positioning them at locations closer to the ore is more advantageous than using fewer sprays at higher flow rates [NIOSH 2003].

Suppression of airborne dust involves the application of water, either untreated or chemically treated, in the form of spray, fog, or foam to prevent the dust fines from escaping into the air. Fog and foam are less used than water application alone, in particular because of the potentially higher costs and chemical contamination issues.

SCREW CONVEYORS

Screw conveyors are one of the oldest and simplest methods for moving bulk materials (Figure 5.17). They consist of a conveyor screw rotating in a stationary trough. Material placed in the trough is moved along its length by rotation of the screw. Screw conveyors can be mounted horizontally, vertically, and in inclined configurations.

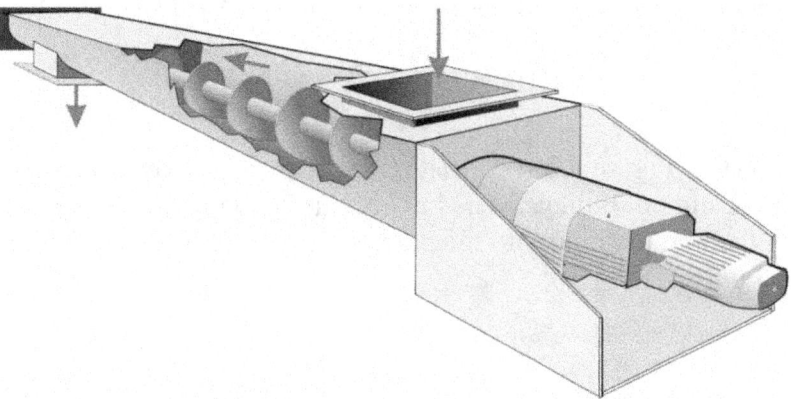

Figure 5.17. Depiction of a typical screw conveyor.

Normally, screw conveyors are totally enclosed except at the ends, where dust emissions can be controlled by proper transfer chute design. The screw conveyor cover is usually fastened by nuts and bolts. However, to maintain a proper dust seal, a self-adhesive neoprene rubber gasket should be installed.

Screw conveyors are essentially dust- and spillage-free as long as the casings, packing glands, and transfer chutes are properly designed and maintained. The only location where spillage typically may occur is where the screw shaft exits the screw conveyor housing. Pedestal-mounted pillow block support bearings (Figure 5.18) should be used and mounted outside the casing. Packing gland seals can be installed where the shaft exits the casing and before the bearing. This will prevent spillage and protect the bearings.

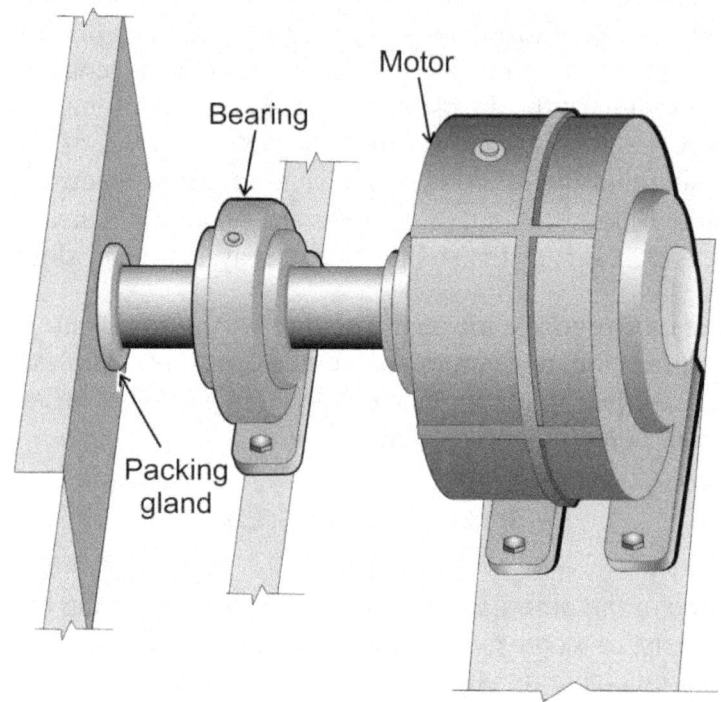

Figure 5.18. Packing gland and pillow block support bearing.

BUCKET ELEVATORS

A typical bucket elevator (Figure 5.19) consists of a series of buckets mounted on a chain or belt that operates overhead and tail pulleys. The buckets are loaded by scooping up material from the boot (bottom) or by feeding material into the buckets. Material is discharged as the bucket passes over the head pulley.

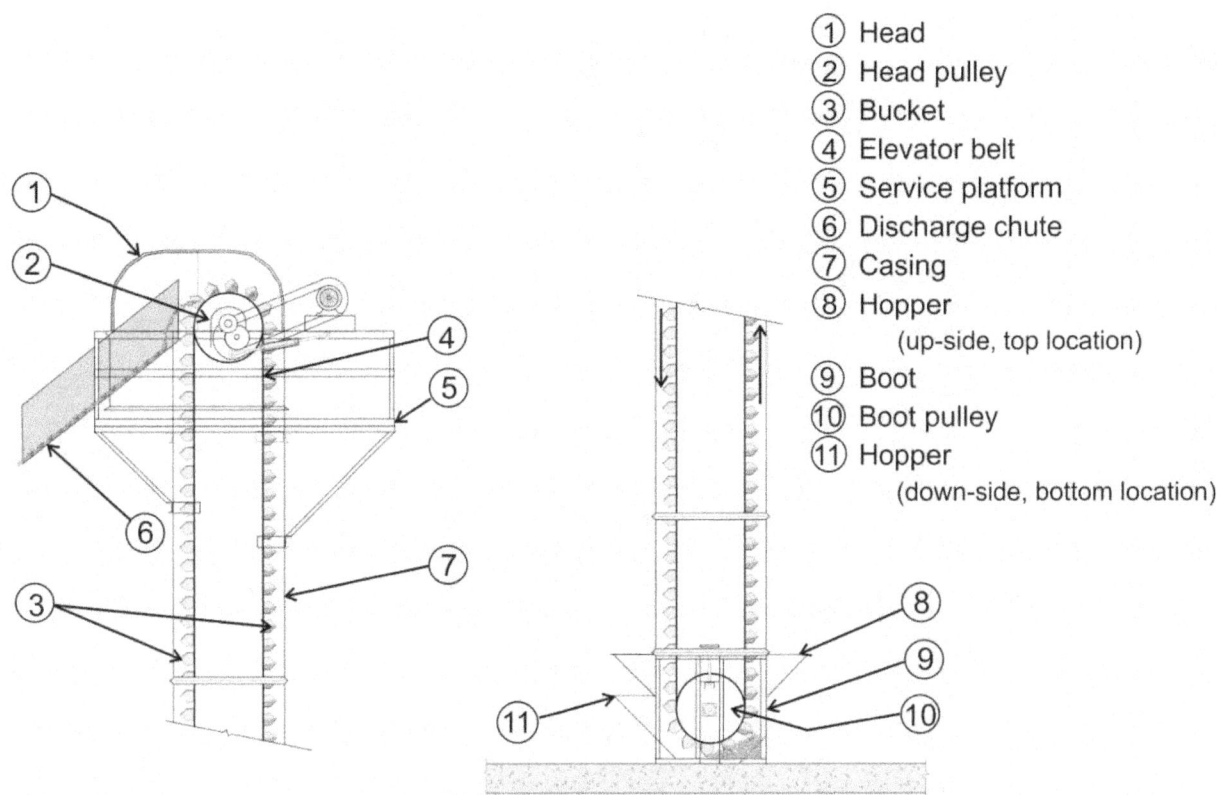

Figure 5.19. Depiction of a bucket elevator.

A steel casing usually encloses the entire assembly and effectively contains dust unless there are holes or openings in the casings. Dust emissions typically occur at the boot of the elevator where material is being fed into the elevator or at the head of the elevator where material is being discharged.

Emissions at the boot of the elevator can be controlled by proper design of a transfer chute (similar to belt conveyors) between the feeding equipment and the elevator. Dust can be reduced significantly by keeping the height of material fall to a minimum and by gently loading material into the boot of the elevator.

Controlling dust emission at the discharge end of the bucket elevator can be accomplished by proper venting to a dust collector (see Chapter 1—Fundamentals of Dust Collection Systems), as well as through the use of proper enclosures and chutes between the elevator discharge and the receiving equipment. It is recommended that at least 100 cfm of dust collection air be provided for every square foot of casing cross-sectional area [ACGIH 2010]. Dust collection pickup should be provided from the top of the casing, just above the head pulley (Figure 5.20). Additional ventilation may be needed for belts traveling over 200 fpm. If the elevator is over 30 feet tall a second pickup should be installed on the side of the casing just above the tail pulley. Dust collection ventilation should also be provided where the material is being discharged into the elevator.

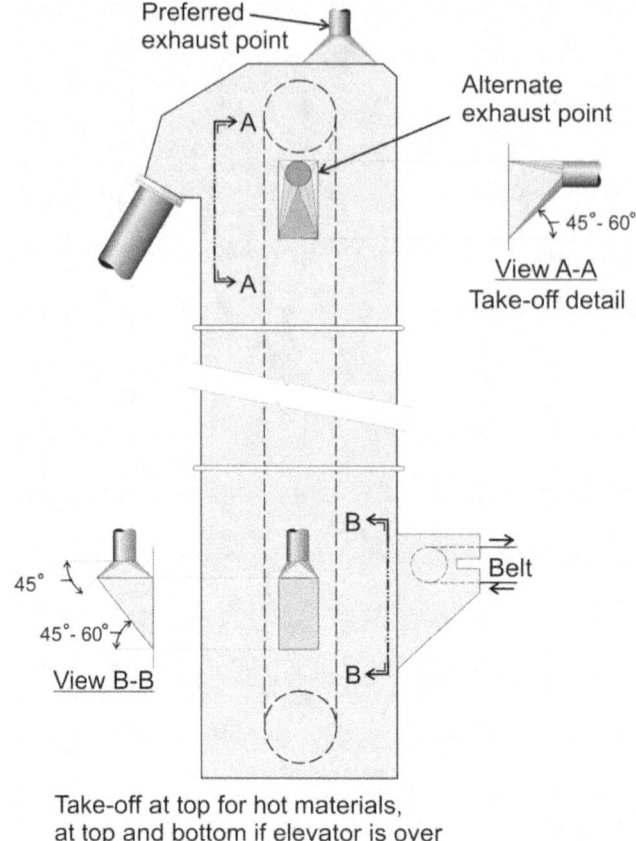

Figure 5.20. Depiction of a typical bucket elevator dust collection process.

PNEUMATIC CONVEYANCE

Pneumatic conveyors are tubes or ducts through which material is moved by pressure or vacuum (suction) systems. Positive pressure systems can be either *dilute phase* or *dense phase*. Dilute phase uses a low (dilute) product to air ratio for transport, while dense phase uses a high (dense) product to air ratio. Dilute phase flow is when the air velocity in the conveyor line is high enough to keep the product being conveyed airborne. Dense phase does not require the product to be airborne. Material being conveyed lies for periods of time in the bottom of a horizontal line and sometimes flows through the line in slugs. Dilute phase systems typically operate at pressures obtainable from a fan and dense phase systems use a high-pressure compressed air source.

When material is fed into a pressure system, the material is conveyed to a storage bin with dust collection, cyclone, or filter-type collector. The conveying air then escapes through the cyclone vent or a filter (Figure 5.21).

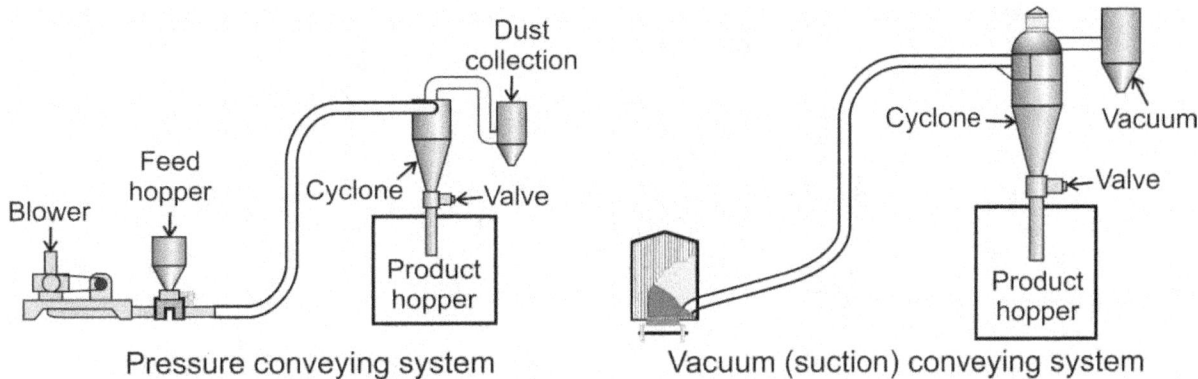

Figure 5.21. Depiction of two types of pneumatic conveying systems.

Since positive pressure pneumatic systems are totally enclosed, dust emissions do not usually occur unless the system has worn-out areas. Because maximum wear in the conveying ductwork occurs at elbows, long-radius elbows made of heavy gauge material should be used. Numerous styles of wear-resistant elbows are available (Figure 5.22). The elbows can also be lined with refractory or ceramic material to further reduce wear and abrasion. In low-pressure pneumatic systems, dust may also leak through joints. Self-adhesive neoprene gaskets should be used at all joints to provide a dust-tight seal.

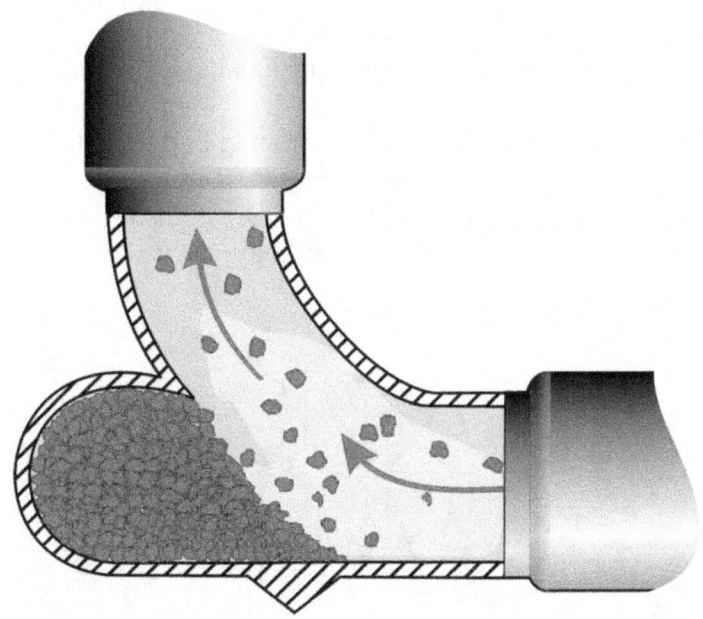

Figure 5.22. HammerTek wear-resistant elbow installed in conveying ductwork.

Dilute phase pneumatic conveying systems operate under pressure and require some mechanical means, such as a rotary valve or double dump valve, to provide an airlock and prevent transport air from short circuiting into the feed material storage bin (Figure 5.23). These types of devices employ moving parts and seals that are often subject to high wear and significant maintenance.

The dismantling of these airlock devices can also be a significant source of dust for mechanics and requires downtime for process equipment.

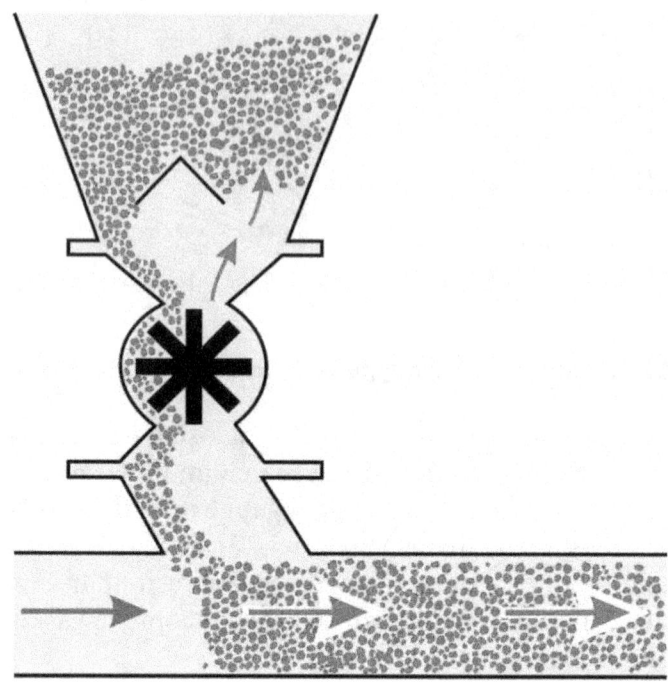

Figure 5.23. Rotary airlock feeding pneumatic conveying line.

Venturi eductors are often a good alternative to the mechanical airlock devices. Eductors convert the blower output into a suction that is used to draw the feed material into the conveying line (Figures 5.24 and 5.25). These devices have no moving parts and typically require minimal repair and downtime for maintenance activities. The venturi eductor can be used on dust collectors, grinders/mills, belt and vibratory feeders, etc. Systems utilizing this type of device are generally designed by the manufacturer due to variables associated with the process.

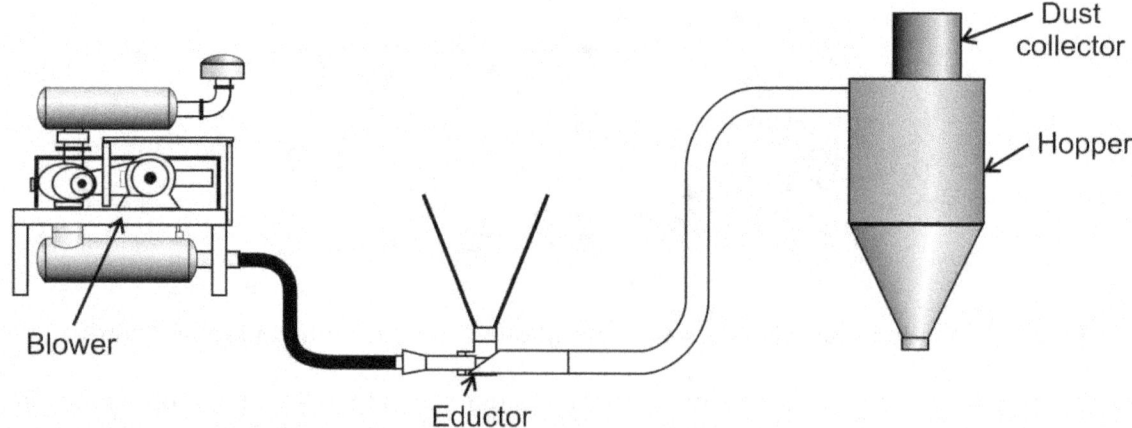

Figure 5.24. Conveyance system utilizing a venturi eductor device.

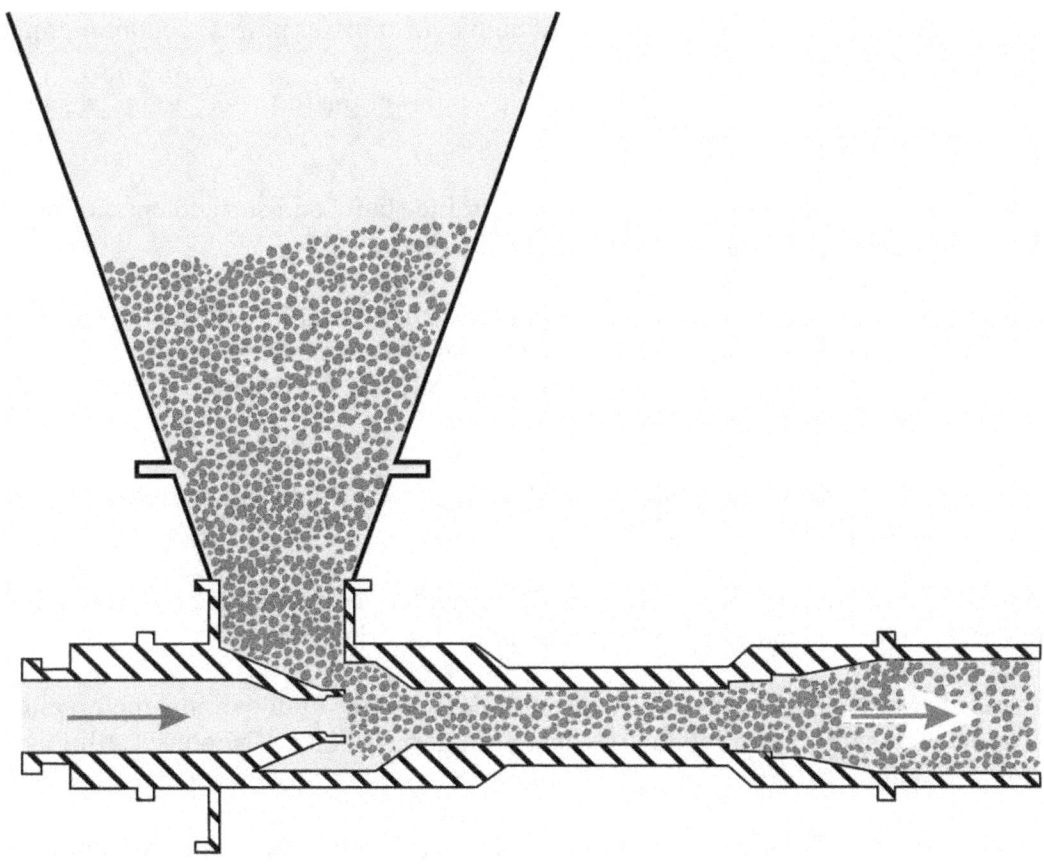

Figure 5.25. Venturi eductor utilizing suction to draw the feed material into the conveying line.

Vacuum systems offer clean, efficient pickup of material from rail cars, trucks, or bins and hoppers for unloading into other types of equipment. Since the systems are under a negative pressure dust leakage is not normally a problem. Cyclone receivers or filters are used at the end of such systems to separate the material. The level of separation desired will determine the type of separation equipment required.

REFERENCES

ACGIH [2010]. Industrial ventilation: a manual of recommended practice for design. 27th ed. Cincinnati, OH: American Conference of Governmental Industrial Hygienists.

CEMA Standard 575–2000. Bulk material belt conveyor impact bed/cradle selection and dimensions. [http://www.cemanet.org/publications/previews/CEMA%20Standard%20575%20-2000pv.pdf].

Courtney WG [1983]. Single spray reduces dust 90%. Coal Mining & Processing. June, pp. 75–77.

Ford VHW [1973]. Bottom belt sprays as a method of dust control on conveyors. Min Technology (United Kingdom) 55(635):387–391.

Goldbeck LJ, Marti AD [2010]. Dust control at conveyor transfer points: containment, suppression and collection. [http://www.ckit.co.za/secure/conveyor/troughed/transfer_points/transfer_ponits_dust_control.html]. Date accessed August 23, 2010.

MAC [1980]. Design guidelines for dust control at mine shafts and surface operations. 3rd ed. Ottawa, Ontario, Canada: Mining Association of Canada.

NIOSH [2003]. Handbook for dust control in mining. By Kissell FN. Pittsburgh, PA: U.S. Department of Health and Human Services, Centers for Disease Control and Prevention, National Institute for Occupational Safety and Health, NIOSH Information Circular 9465, DHHS, (NIOSH) Publication No. 2003–147.

Roberts AW, Ooms M, Bennett D [1987]. Bulk solid conveyor belt interaction in relation to belt cleaning. Bulk Solids Handling 7(3):355–362.

Swinderman RT, Goldbeck LJ, Marti AD [2002]. Foundations 3: the practical resource for total dust & material control. Neponset, Illinois: Martin Engineering.

Swinderman RT, Marti AD, Goldbeck LJ, Strebel MG [2009]. Foundations: the practical resource for cleaner, safer, more productive dust & material control. Neponset, Illinois: Martin Engineering Company.

USBM [1974]. Survey of past and present methods used to control respirable dust in noncoal mines and ore processing mills—final report. U.S. Department of the Interior, Bureau of Mines Contract No. H0220030. NTIS No. PB 240 662.

USBM [1987]. Dust control handbook for minerals processing. U.S Department of the Interior, Bureau of Mines. U.S. Bureau of Mines Contract No. J0235005. Martin Marietta Corporation and Marcom Associates, Inc.

Yourt GR. [1990]. Design principles for dust control at mine crushing and screening operations. Canadian Mining J 10:65–70.

CHAPTER 6: BAGGING

CHAPTER 6: BAGGING

This chapter discusses techniques used by mineral producers for controlling worker dust exposures while bagging and stacking product in various types of bags for shipment to customers. The loading of product into some type of container is normally called "bagging." The stacking of these bags of product onto pallets for shipment to customers is typically called "palletizing." There is a wide spectrum of different types of bags that are used to ship product to customers, ranging from 50-pound to over one-ton bulk bags. Only the smaller type bags (100 pounds or less) are typically palletized because the larger bulk bags are in most cases, individually shipped. Both the bagging and palletizing process can be performed manually or through some type of semi-automated or totally automated process. Dust generated by the manual bagging and palletizing of 50- to 100-pound bags directly affects a number of worker's dust exposure.

Workers performing bagging and stacking tasks, or working in and around these areas, typically have some of the highest dust exposures of all workers at mineral processing operations. A few of the difficulties with controlling dust in work processes associated with bagging and palletizing are the wide range of different equipment being used, as well as the variety of bag types. The bagging process can range from single-station manual bagging units to fully automated multi-station machines.

In addition to the potential for high respirable dust exposures from the bagging and palletizing process, there is also a significant risk for musculoskeletal disorder (MSD) from all the repetitive motion and the significant weight lifted by workers while performing tasks associated with these processes. The dust exposure and MSD concerns have motivated many operations to pursue the implementation of equipment that either semi-automates or totally automates these work tasks.

BAGS AS DUST SOURCES

To address problems associated with the bagging process, a number of different dust sources need to be addressed and controlled, specifically product blowback, product "rooster tail," and contaminated bags. When manually bagging and stacking 50- to 100-pound bags, these dust sources directly affect the worker's exposure. Two different types of bags that are used to transport product within this weight range are open-top bags and closed bags with an internal valve.

Open-top bags are slid up over the loading chute and filled with product in a mass loading technique. Open-top bags are typically used for larger particle sizes called whole-grain, which are normally greater than 120 mesh (0.125 mm) in size. The dust sources when loading open-top bags are leakage from the loading chute once the bag is removed and liberation from the open bag before it is sealed. Sometimes these open-top bags are moved into a cage device to properly position the bag before sealing; this task of moving the bag into the cage can also be a dust source. Since these bags are typically used with large particle sizes, dust generated or liberated is not normally as significant as with the valve-typed bags (closed-top).

For valve-type bags, three major dust sources need to be addressed for effective dust control. The first dust source is from product blowback, which occurs during bag filling and results from product spewing out of the bag valve. Product blowback occurs as excess pressure builds inside the bag during bag filling and is then relieved by air and product flowing out of the bag valve around the fill nozzle. The second major dust source is product spewing from the fill nozzle and bag valve as the bag is ejected from the filling machine. This is typically called a "rooster tail" because of the spray pattern of the escaping product. Both product blowback and product "rooster tail" occur because of the air pressure injected into the bag with product during filling. Both release dust into the air and contaminate the outside surface of the bag. The contaminated bags then become the third significant source of dust exposure for the bag stackers, or for any other individuals handling the bags, including the end user of the product. Figure 6.1 shows product blowback during bag filling, the product "rooster tail" as the bag is ejected from the fill nozzle, and dust contamination on the outside of the bag after loading is completed. If bags are undersized, they create a greater amount of product blowback and "rooster tail" during the bagging process, and this needs to be considered when evaluating these dust sources.

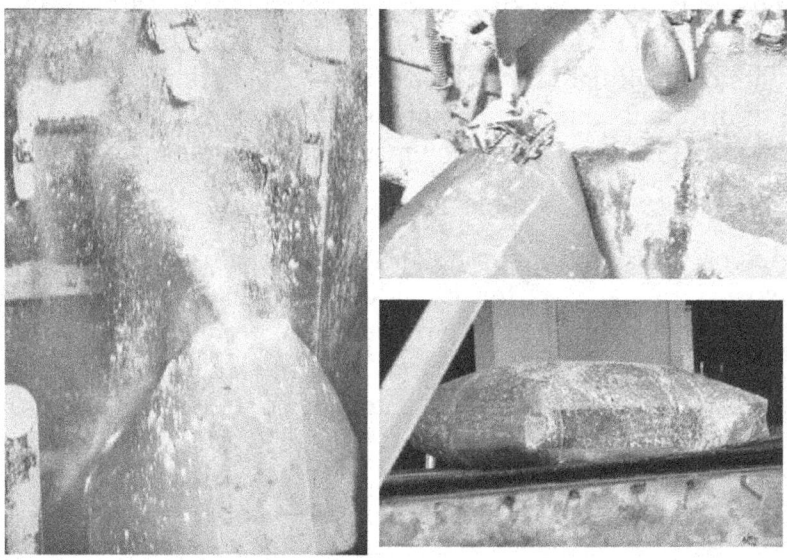

Figure 6.1. Product "rooster tail" shown on left, product blowback shown on top right, and dust-soiled bag shown on bottom right.

With 50- to 100-pound bags, another area that impacts the amount of dust generated is the bag valve. The valve is an internal sleeve in the bag through which the fill nozzle is inserted prior to the bag being filled with product (Figure 6.2). The type of bag valve used impacts the amount of product blowback and product "rooster tail." Once the bag falls from the fill station, the weight and volume of the product inside the bag is supposed to force the valve closed, sealing the area and keeping product from leaking from it during the conveying, stacking, and the transportation process. However, many times the bag valve is lined with product and does not seal properly, causing product and dust to leak from the valve during movement/transportation. This sealing ability also varies for different types of bag valves.

One last dust source at bagging operations is from broken, torn, or ruptured bags. When a bag breaks, a significant amount of product and dust is released into the mill atmosphere. Defects caused during manufacturing, including improper gluing and improper storage of empty bags, are just a few of the factors that can cause bag breakage. The location of the break has a significant impact on the amount of dust liberated. Bags that break during filling tend to be more violently affected than those that break during ejection or conveying. In all cases, bag breakage liberates a substantial amount of dust and significantly impacts workers in these areas.

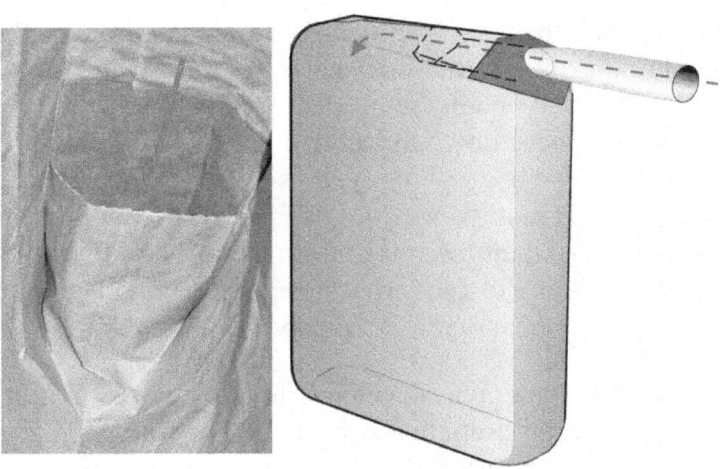

Figure 6.2. Typical bag valve—interior view (left). Artist concept of fill nozzle and bag valve (right). Fill nozzle enters through the bag valve and delivers product into bag.

BAG CONSIDERATIONS

A number of different factors ultimately impact the amount of dust generated or liberated from bags of product material: bag construction, bag perforations, bag fill type, and the effects from bag failures.

Bag Construction

The multi-wall pasted valve bag is by far the most common bag used for the packaging of mineral products and is the only style discussed below. Many of the considerations discussed for pasted valve bags are also applicable to open-top bags.

A very important element in controlling the amount of dust liberated during the bagging process is the design and construction of the paper bag itself. The composition of the paper, its porosity, the number of plies or layers used, the "basis weight/thickness" of the paper sheets which provides strength, the type of pasted valve, and the use of bag perforations, are all factors that affect and can minimize the amount of dust generated or liberated from the bag.

One area of significant advances by bag manufacturers over the years has been the improvements and advances in "paper" technology [Kearns 2004]. Currently, conventional natural kraft (NK) paper, also known as flat kraft (FK) paper, is the industry standard. Normal constructions for 50- to 100-pound bags are to use either three or four plies of NK paper. Due to the low porosity of NK paper, meaning that air does not flow through the paper material, bag perforations are normally required for most industrial mineral processing applications to keep the bag from rupturing/exploding.

Another development in paper technology is the use of extensible kraft (EK) paper, sometimes referred to as "high performance paper." The strength of this EK paper has been increased to such an extent that the number of plies can be reduced by approximately 30 percent in multi-layer bags when compared to conventional kraft papers. A savings in paper can be achieved in one of two ways. The first, and the most common approach, is to reduce the number of plies/layers in the bag. For instance, rather than using NK or FK paper that would normally require three or four plies/layers to obtain the needed strength, by using EK paper, the number of plies/layers can normally be reduced down to just two. For every ply/layer that is reduced, there is a benefit from an increase in the bag's porosity, thus allowing the bag to breathe better. Despite its popularity, extensible kraft paper does not allow for the proper ventilation required for many packaging applications; therefore, bag perforations are still required to provide the necessary air escape during bagging to avoid ruptures. The second way to reduce paper is by reducing the weight/thickness of each ply/layer of paper.

Airflow extensible kraft (AEK) paper was developed in the 1990s and offers the same cross-directional and machine directional strength as EK paper, but with an air-permeable (breathable) characteristic. A great benefit from this air-permeability characteristic is that it eliminates the requirement for void spaces in the bag construction, allowing for a smaller bag size. The porous paper is superior to perforated paper as far as strength and air evacuation (ability to relieve air pressure). No bag perforations are required when using this AEK paper type, and by eliminating perforations, bag strength can be increased by as much as 20 percent. The faster the air is evacuated from bags, the faster they can be filled and palletized, thus increasing production and profitability.

The AEK paper also allows the flow of air to depressurize the bag at a rate of three to four times faster than standard kraft (NK or FK) paper and at the same time retains bag strength. The paper acts as a filter during the bag filling process which allows the product to flow freely into the bag while containing dust that could be liberated into the work environment and expose workers. Because of this, the outer surface of the bag stays cleaner. When considering the various options available for bag construction, the use of AEK paper appears to be the best choice to minimize the amount of dust generated and liberated from the bag during the filling process.

Bag Perforations

As product enters the paper bag during the filling process, fluidizing air entrapped in the product must be evacuated in order to fill the bag with product. In order to remove the air as the product is introduced into the bag, perforations or vent holes are integrated in the paper used in the bag. For this purpose, manufacturers use multi-wall bag perforations, which are small vent holes punched through the walls of the paper bag to allow air to escape rapidly during the filling process.

There are two common types of perforations used in bags: overall perforations and undervalve perforations. Both types of perforations reduce the strength of a paper bag by as much as 20 percent.

Overall Perforations

The use of overall perforations is a critical factor to consider when evaluating the effectiveness of bag performance. Each ply of paper is individually perforated during the manufacturing process, creating staggered perforations throughout the bag. The perforations (pinholes) do not go directly through all layers (2, 3, or 4 plies/layers) and are not aligned with each other. The goal is to allow the fluidizing air to work its way out of the bag while still containing the product/material. Perforations allow the bag to depressurize, minimizing the possibility of it rupturing during bag filling. Since the perforations are staggered for each of the different plies/layers, it is uncommon for the product being bagged to escape through the perforations, however, this can sometimes occur for very fine product sizes—i.e., 325 mesh (0.045 mm) and finer. As previously mentioned, the bag strength can be lessened by up to 20 percent when using perforations. Standard perforation sizes are 1/8, 3/32, and 1/16 inches, and patterns range from 3/4 x 3/4 inch centers up to 2 x 3 inch centers (Figure 6.3).

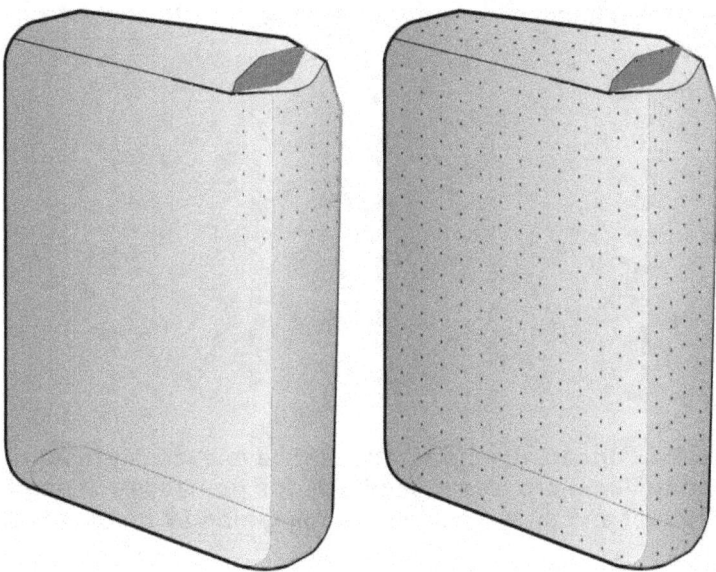

Figure 6.3. Undervalve bag perforations (left) and overall perforations (right).

Undervalve Perforations

Unlike overall perforations, undervalve perforations are created after all plies are positioned into the finished bag stage. They are punched through all 2, 3, or 4 plies/layers at the same time and are aligned. Since the fluidizing air is most likely to escape out of the bag valve around the fill nozzle, (product blowback), having perforations as close as possible to the valve area is most advantageous/effective. In addition, this location also minimizes the amount of bag contamination as the excess air and product escape. The perforation pattern can range from 1/4 x 1/4 inch centers up to 1 x 1 inch centers. Perforation sizes are the same as for the overall perforations, being: 1/8, 3/32, or 1/16 inches. Using an exhaust hood with a local exhaust ventilation (LEV) system is most beneficial in this application because of the likelihood of product escaping from the bag during the filling process. Figure 6.4 shows an example of undervalve perforations and how they depressurize the bag to keep it from rupturing or exploding during bag filling.

Because of the product and dust emitted from the perforations during bag filling, most filling operations require some type of ventilation system to capture this dust. Two common techniques to achieve this dust capture are to place exhaust hoods around the bag loading nozzles and/or to use an exhaust hopper located below the bag loading station with an LEV system incorporated. These techniques can be used individually or together, depending on the level of dust control needed. Both of these techniques are discussed in greater detail later in this chapter.

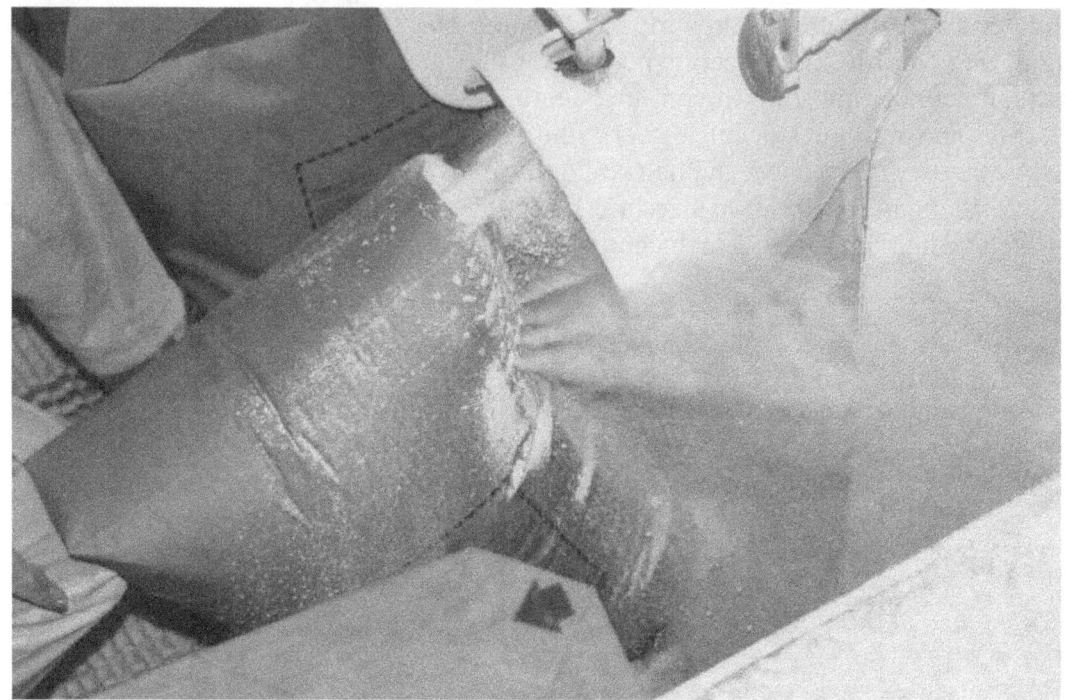

Figure 6.4. Undervalve perforations relieve excess pressure from bag during filling process to minimize the possibility of bag failure. Product and dust are seen escaping along with excess air.

Bag Fill Type

Two types of bags will be discussed in this section: open-top and valve bags. Figure 6.5 shows a typical open-top bag loading process. The bag operator slides an open-top bag up over a loading spout and then activates a start button to begin the loading process. The bag is then filled relatively quickly in a bulk loading fashion. During filling, the top of the bag is well-sealed and typically tied into an LEV system to exhaust any dust generated during the loading process. When filling is completed, the bag automatically releases from the loading spout and drops a short distance down onto a conveyor. The top of the bag is then sealed using either heat sealing/gluing or is sewn (stitched) by a machine with nylon thread. Both of these sealing techniques are effective and create minimal amounts of dust.

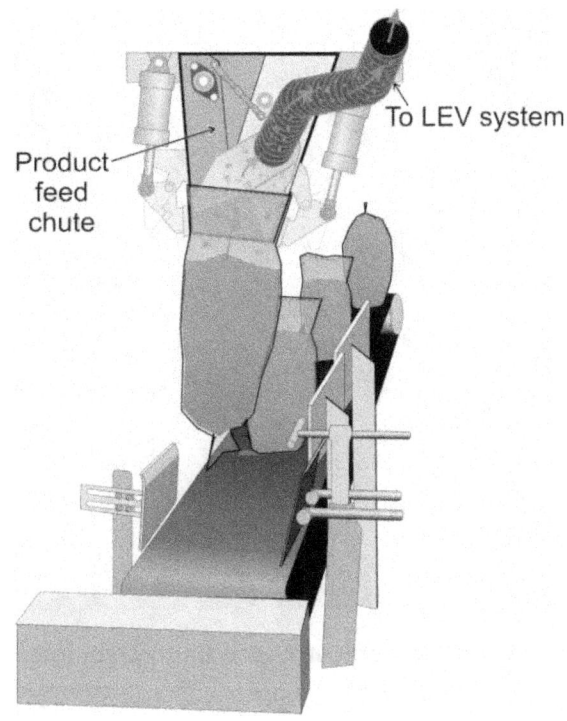

Figure 6.5. Open-top bag being loaded with product (normal whole-grain material, which is less dusty).

The second and most common bag type used in mineral processing operations is the pasted valve bag. The choice of valve type from among different commercially available bag valves can be a significant factor in the ability to effectively seal 50- to 100-pound bags of product and to minimize the amount of dust liberated during the bag filling, conveying, and stacking process. From a study performed a number of years ago, five different valves were tested to compare their effectiveness at sealing the bag and minimizing dust liberation. The valves were: standard paper, polyethylene, extended polyethylene, double trap, and foam [USBM 1986b; Cecala and Muldoon 1986]. By far the most effective valve was the extended polyethylene (Figure 6.6). This is simply a plastic valve approximately two inches longer than the standard paper or polyethylene valve.

In the study, two factors determined the effectiveness of the bag valve: valve length and valve material. The longer valves were more effective in reducing the amount of product blowback and bag-generated dust. However, it is speculated that when the valve length became too much longer than the fill nozzle, it began to negatively impact the bag filling performance. For the second factor, which is valve material, foam was the most effective material tested. The foam valve was the shortest valve tested for the evaluation at four inches in length, while the extended polyethylene valve was six inches long. The rankings of valve types from the most to least effective were as follows: extended polyethylene, foam, standard paper, polyethylene, and double trap. Figure 6.7 shows a comparison of the extended polyethylene and the foam valve compared to the standard paper valve. Respirable dust levels ranged from approximately 45 to 65 percent lower with the extended polyethylene as compared to the standard paper valve [USBM 1986b; Cecala and Muldoon 1986].

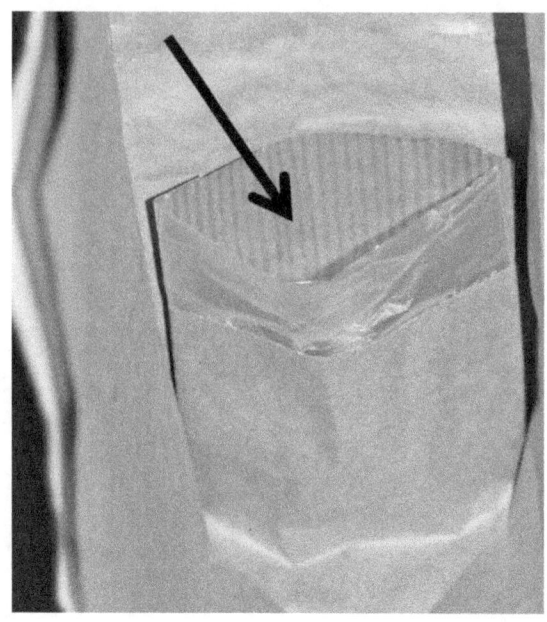

Figure 6.6. Extended polyethylene bag valve, interior view.

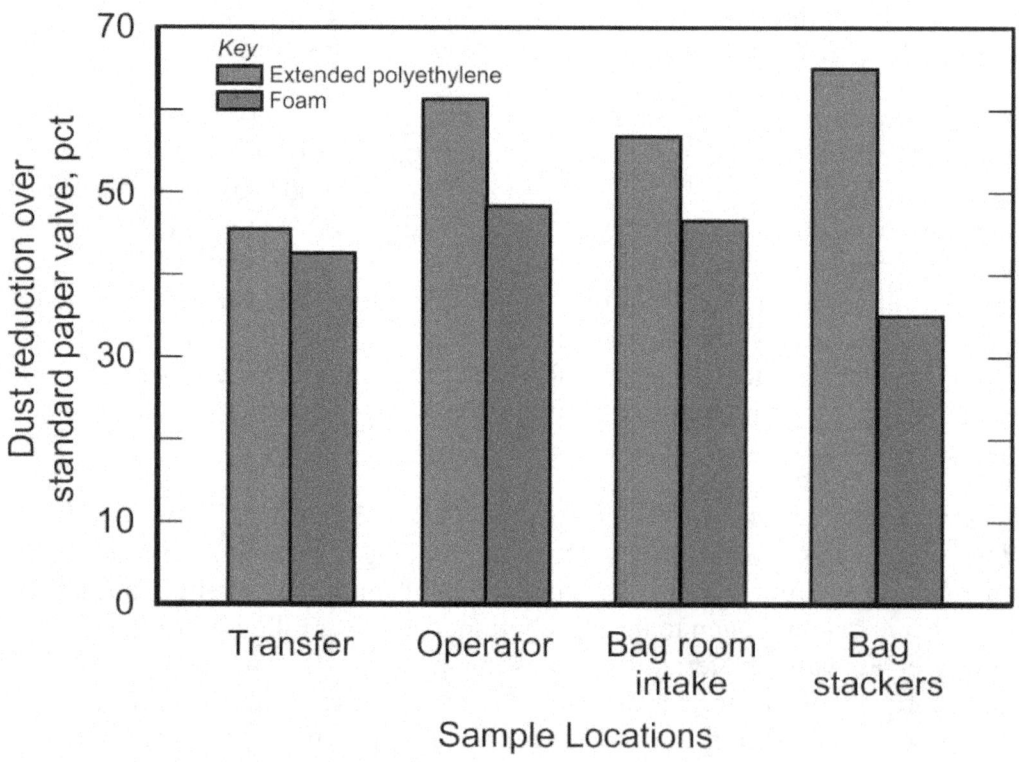

Figure 6.7. Dust reductions with extended polyethylene and foam valves as compared with the use of the standard paper valve.

Bagging operators should be aware that changes in product leakage and dust liberation are directly based on the type and effectiveness of the bag valve used in their bags. The extended polyethylene valve was the most effective valve with only a minor cost increase over the standard paper valve.

A recent review of bag valves revealed that the standard paper, polyethylene, and extended polyethylene valve types are still the main valves being used by industry. Based upon these choices, operators should definitely consider using the extended polyethylene valve type to minimize the amount of dust liberated during bag loading, conveying, and bag stacking process.

Another commercially available bag valve type is the sonic energy sealed bag valve, in which the valve is glued after filling. This valve construction has adhesive on the inner surface of the valve, activated by sonic energy applied either at the packer fill nozzle or from ancillary equipment downstream. Theoretically, heating/melting of the adhesive should provide improved sealing of the inner surface of the valve and thus minimize leakage.

A problem that sometimes occurs with this valve type is that the glue does not properly seal because of product in the valve area, which then leads to product spillage. Another drawback with this valve type is that the cost to perform this sonic energy sealing is typically much higher than for the other types of valves.

Effects of Bag Failures

In addition to the considerations mentioned above, bag failure from broken, torn, or ruptured bags can occur during the bag filling, transportation, or handling process and can have catastrophic effects on the amount of dust generated or liberated into the work environment. When a bag breaks during bag filling, it can be a violent occurrence which quickly releases a significant amount of dust and product into the mill atmosphere. Another concern from this type of bag failure is that most dust control equipment is not normally sized adequately to handle the large volume of dust released when this occurs. Gluing defects during manufacturing and the improper storage of empty bags (storing for long periods of time in high humidity and/or high temperatures, or damages to bags during storage) are just a few of the circumstances that can cause bag failure problems during filling.

In addition to bags rupturing during loading, another problem can be bag failures during the conveying or pallet loading process, which can also cause a significant amount of dust to be liberated into the work environment. Bag failures most often occur from rips and tears from sharp edges during conveying, or damage from mishandling during pallet loading.

Every operation needs to address how it will deal with bag failures because of the significant impact that they can have on overall mill respirable dust concentrations, and ultimately to workers in the plant. Some operations use a bag scale (weighing) system during conveying which identifies underweight (broken) bags and diverts them onto another conveyor line. This secondary conveyor then takes the bags and discards them into a disposal device, normally a dumpster unit so the bags do not need to be rehandled. It is recommended that this dumpster unit be located outside the plant to minimize the possibility that the dust liberated as the bags are discarded contaminates the plant air. Once the dumpster is full, it is taken away and the bags are disposed of in a safe manner. A new dumpster is then positioned at the dump point and the cycle is repeated.

CONTROL TECHNOLOGIES FOR FILLING 50- TO 100-POUND BAGS

The bag filling process can range from a single-station manual bagging unit to a fully automated multi-station machine. This section addresses various control techniques to reduce respirable dust during bag filling of 50- to 100-pound bags.

Exhaust Hood Placed around Bag Loading Nozzles

One of the earliest techniques developed to capture dust generated from bag filling machines was a very simple exhaust hood, which surrounds the bag to capture the dust generated and liberated during the bag filling process [USBM 1983]. If the fill machine is a multiple bag filling unit, an individual exhaust hood is placed around each fill nozzle. The exhaust hood is normally shop-fabricated from sheet metal and can be adapted to fit most valve type bag fill machines (Figure 6.8).

To minimize installation and maintenance time, the hood is normally constructed in quarter-piece sections to allow it to be more easily installed and bolted together. The bottom of the hood should be sloped to allow product to slide from the hood and fall down into a hopper, normally located below the bag filling station. The exhaust hood also needs to be designed to allow the bag clamp, fill nozzle, or pinch tube to enter through the back of the hood.

To accommodate the installation of an exhaust hood on retrofit systems, one item that sometimes needs to be relocated is the start/stop button which can easily be moved to a location outside the hood. The intent is for the hood to capture the respirable dust generated during the bag filling process (blowback and "rooster tail"), while allowing the oversized and heavier product to fall into the hopper and be recycled/reused.

Each exhaust hood should be connected to an LEV system to capture the dust generated and liberated during bag filling and bag ejection. The hood's exhaust volume requirement is a function of the open hood area, with tests showing that air velocities of 200 feet per minute (fpm) into the hood are adequate for dust capture/containment. A typical hood size is in the range of 4 ft^2 of open face area, which requires an exhaust volume of 800 cubic feet per minute (cfm). Exhaust ventilation hoods have been shown to be from 90 to 100 percent effective at reducing dust leakage into the plant [USBM 1981]. For information on tying these hoods into LEV systems, refer to Chapter 1—Fundamentals of Dust Collection Systems.

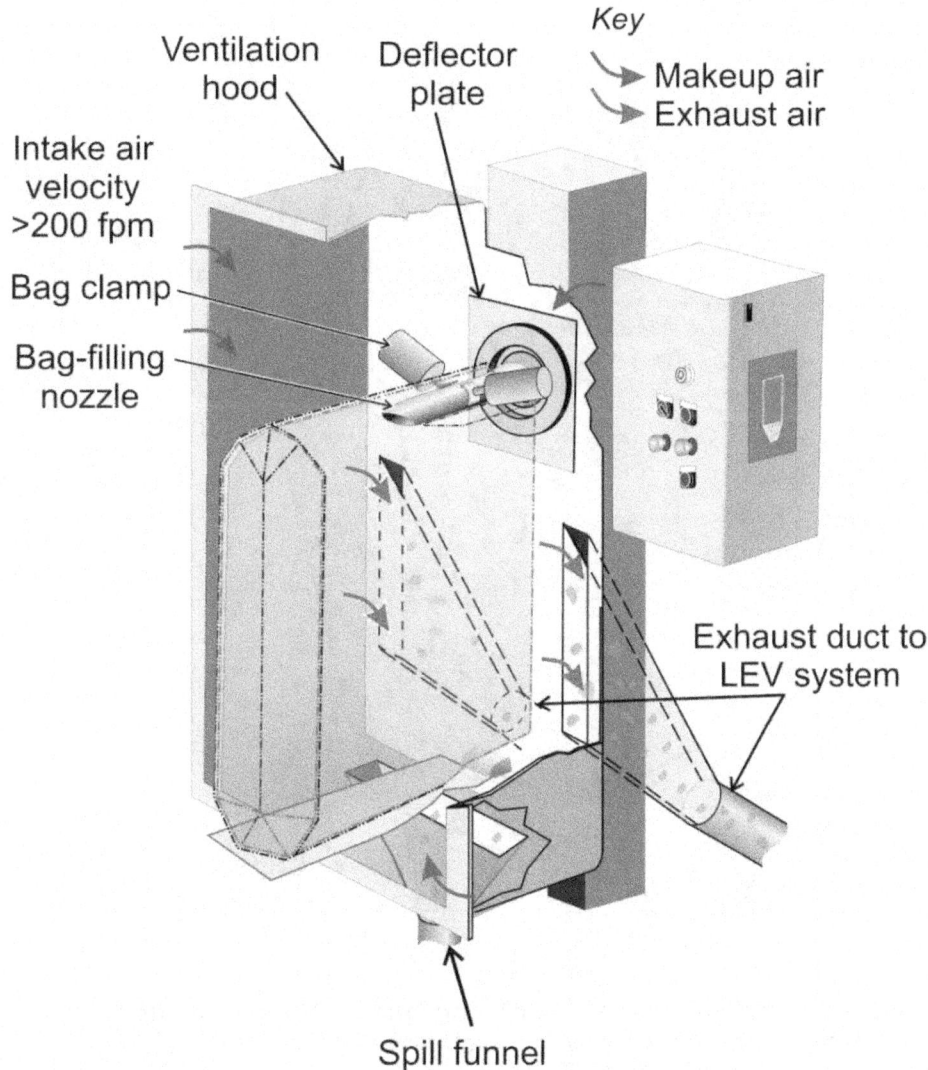

Figure 6.8. Exhaust hood to capture and exhaust respirable dust away from each fill nozzle.

Exhaust Hoppers (Below Entire Fill Station)

Similar to the above technique, which places an exhaust hood around each individual bag loading unit, a hopper can also be placed below the entire fill station to capture and recycle all the dust and product lost from the entire bag loading area or station. This is especially beneficial with three- and four-fill-nozzle bag loading systems. In these cases, incorporating the LEV into a single-hopper design is easier and more financially advantageous than trying to connect ductwork to individual hoods around each fill nozzle.

Figure 6.9 shows a typical design for a single-hopper LEV type system. One critical design component is to locate the exhaust system takeoff points high in the hopper unit and in close proximity to the actual bag loading area. The heavier dust and product that is not captured by the exhaust system simply falls into the base of the hopper and is recycled back into the product line using a screw conveyor.

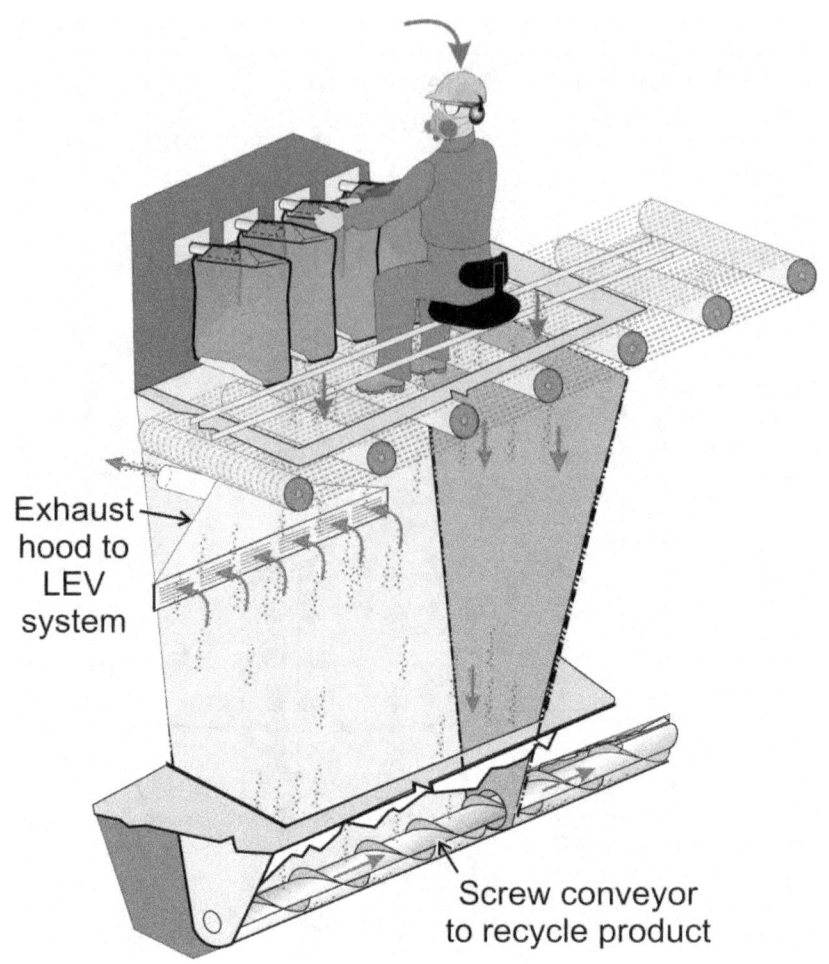

Figure 6.9. Exhaust hopper below bag loading station with integrated LEV system to capture dust generated during bag loading process.

One critical component for both the exhaust hood being placed around each bag loading nozzle and the exhaust hopper being below the entire fill station is to provide clean (dust-free) makeup air to the bag filling area. Any time an LEV system is used to capture dust generated at a bagging station, air will be drawn from the surrounding area as makeup air. Since this makeup air flows over the bag operator before being captured by the LEV system, it is important that it be dust-free so as to not increase the exposure of the bag operator [USBM 1986a; Constance 2004]. Outside air can be used as makeup air when there are no outside dust sources near the intake location and when outside air temperatures are acceptable.

Dual-Nozzle Bagging System

The primary dust sources for manual bagging have been identified as product blowback, product "rooster tail," and contaminated bags. The dual-nozzle bagging system is designed to reduce these major dust sources and lower the exposure of the bag operator. Figure 6.10 depicts the major components of the dual-nozzle bagging system. This system uses a two-nozzle arrangement with an improved bag clamp to reduce the amount of product blowback and product "rooster tail."

In the original design, a single nozzle is used to fill the bag with product. With the two-nozzle arrangement, the inner nozzle is used to fill the bag and the outer nozzle relieves excess pressure from the bag after it has been filled. Once filling is completed, depressurization of the bag is accomplished with the aid of an eductor, which uses the venturi principle (compressed air entraining additional exhaust air) to exhaust excess air from the bag at approximately 50 cfm. This exhaust is controlled by a pinch valve, which is opened when the exhaust is on. The bag is slightly overfilled and held in place by the bag clamps, then the pinch valve is opened and the exhaust is initiated to depressurize the bag. After a few seconds, the bag clamp opens and the bag falls from the fill station. The exhaust system also continues to operate as the bag falls away, cleaning the bag valve area. The exhausted material can then be recycled back into the system.

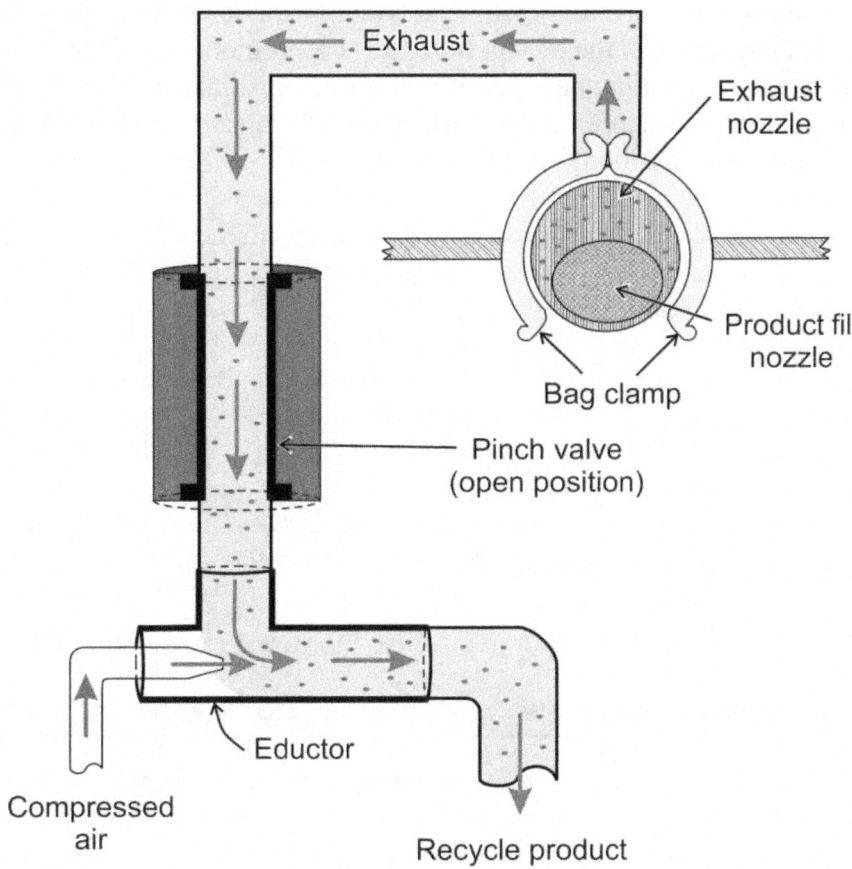

Figure 6.10. Dual-nozzle bagging system design.

Another key component of this dual-nozzle bagging system is an improved bag clamp. With the original bag clamp, there were only two small points of contact with the fill nozzle. The improved bag clamp makes direct contact with approximately 60 percent of the nozzle (concentrated at the top and sides), thus reducing the amount of product blowback during bag filling. A controlled amount of blowback is necessary so the bag does not rupture during filling, but this should occur at the bottom of the nozzle to minimize the amount of dust contamination to the outside of the bag.

The dual-nozzle bagging system can be very effective at lowering respirable dust levels to workers performing the bagging and palletizing tasks [USBM 1984a,b]. Figure 6.11 indicates an 83 percent reduction in the bag operator's respirable dust exposure before and after the installation of a dual-nozzle bagging system at one plant. A 90 percent reduction was also measured in the hopper below the fill station. This represents a substantial reduction in product blowback and product "rooster tail" during the bagging process. The dual-nozzle bagging system also resulted in less product and dust on the outside of the bags and accounted for a 90 percent reduction in the bag stacker's dust exposure while bags were being loaded into an enclosed vehicle.

The dual-nozzle bagging system is mainly recommended for operations with three- and four-fill-nozzle bag machines because there is a slight decrease in production due to the time needed to depressurize the bags after filling is completed. The system can be used on one- and two-nozzle filling machines, but the decrease in the production rate is even more significant because the bag operator must wait for each individual bag being depressurized, rather than waiting on a cycle of bags. Other similar systems that depressurize the bag after filling is completed have also been designed and sold [Case History 2006a].

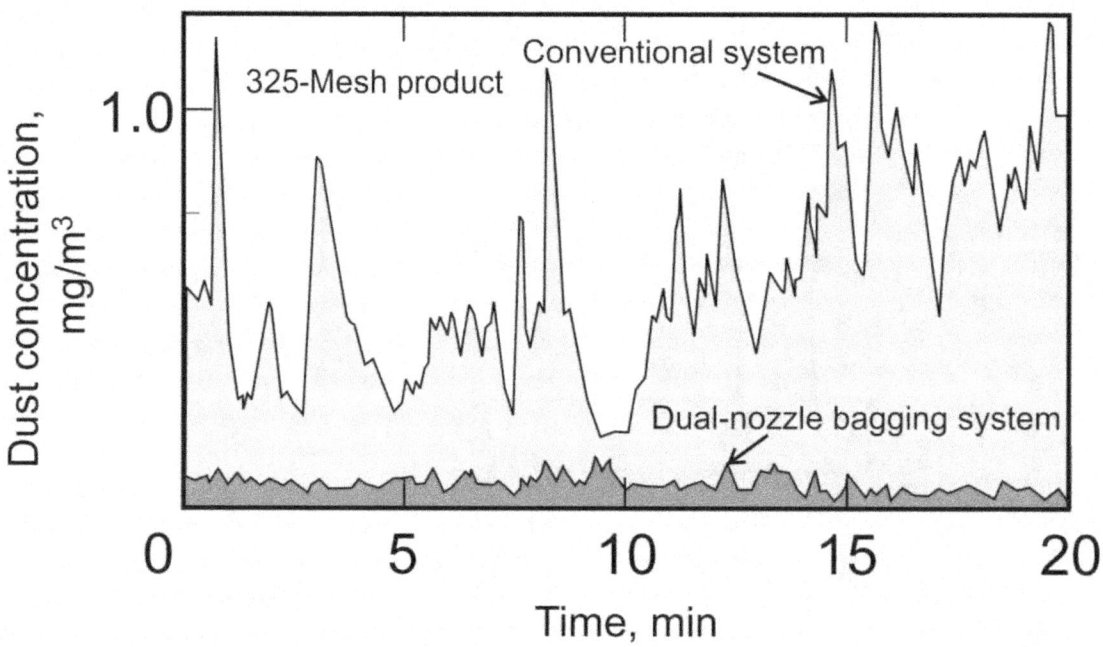

Figure 6.11. Bag operator's dust exposure with and without the use of a dual-nozzle bagging system.

Overhead Air Supply Island System

The overhead air supply island system (OASIS) can be used to provide an envelope of clean, filtered air to a worker at a stationary bagging or palletizing unit. One of the main advantages of the OASIS is that it is suspended over a worker and operates independently of any processing equipment. Figure 6.12 shows an OASIS located over a bag operator.

The OASIS is a relatively simple design and system. Mill air is drawn into the unit and passes through a primary filtering chamber, normally using HEPA quality filters. After the air exits the primary chamber, it passes through an optional heating or cooling area, which can be incorporated into the unit if temperature control is desired. The air then flows through a distribution manifold filter when exiting the unit. This provides an even distribution of clean filtered airflow down over the worker, while providing backup filtering in case there is a problem with the primary filtering chamber. If the bagging or palletizing process generates a significant amount of dust, it will contaminate the envelope of clean air from the OASIS after it flows down over the worker. In these cases, it is normally advisable to then capture this dust-laden air with an LEV system.

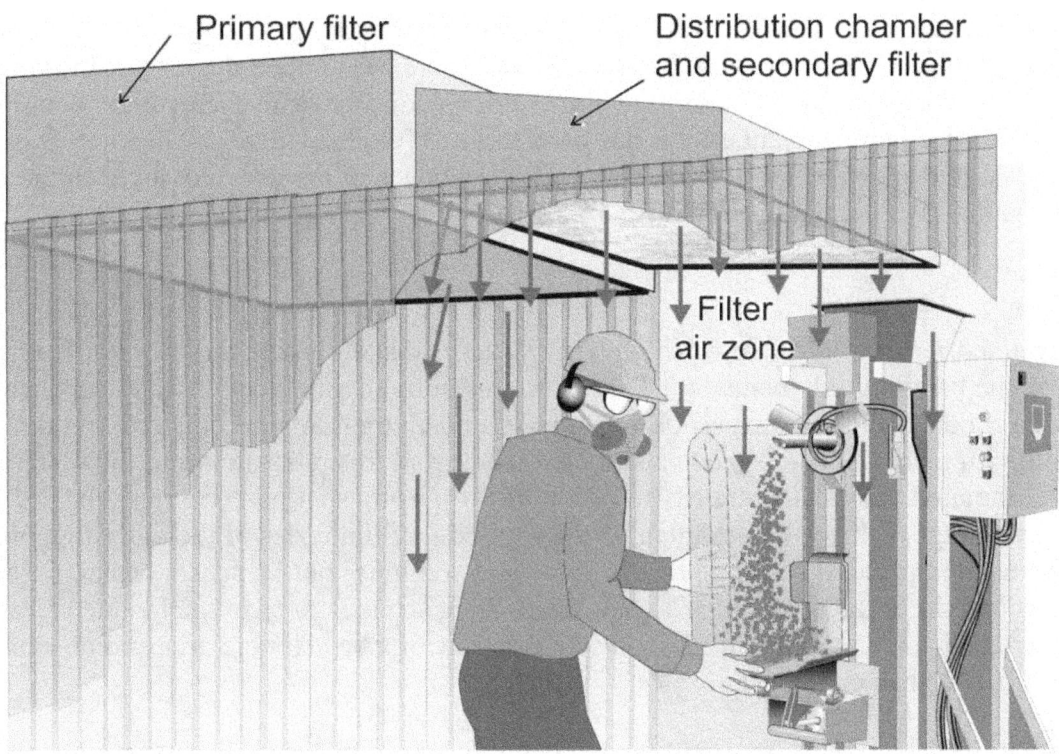

Figure 6.12. OASIS positioned over bag operator to deliver clean filtered air down over work area.

With the OASIS, the target is to have an average velocity of roughly 375 fpm flowing down over the worker, which normally keeps any dust-laden mill air from entering the clean air core [Volkwein et al. 1986]. One important feature with an OASIS system is to incorporate a self-cleaning primary filter design. There are a number of different self-cleaning techniques used to remove the dust cake from the filters. For example, the reverse air jet pulse collector is a very common and effective method to perform this task (refer to Chapter 1—Fundamentals of Dust Collection Systems).

The OASIS is very effective at lowering workers' respirable dust exposures. A bag operator's respirable dust exposure was reduced by 82 and 98 percent at two different operations using the OASIS, as compared to when the unit was not being operated [Robertson 1986]. At both of these operations, the dust concentration within the clean filtered air of the OASIS remained under 0.04 mg/m^3.

When the OASIS is used with standing workers, it can be advantageous to place clear plastic stripping down the sides of the device. This allows the workers to recognize the boundary of the clean air zone and the inability to be protected once exiting the physical barrier of the plastic stripping.

An additional benefit provided by the OASIS is a general improvement to the overall mill air quality. This occurs because the OASIS is drawing air from the plant environment, filtering it, and then blowing this clean air down over a worker. After flowing over the worker, it then becomes part of the general mill air again. At one site using the OASIS, there was a 12 percent reduction in respirable dust levels throughout the entire mill building [Volkwein et al. 1986]. The volume of clean air delivered by the OASIS is variable based upon the size of the unit, but it is normally in the range of 6,000 to 10,000 cfm. Other studies have also shown the benefit of recirculating air back into a plant after it has been cleaned [Godbey 2005]. The OASIS is generic in design and can be fabricated and installed in-house or through any local engineering company that handles ventilation and dust control systems.

Automated Bag Placer and Filling Systems

Many manufacturers provide automated bagging systems that eliminate the operator from performing the bagging task manually. These systems normally include both a mechanical bag placer and a product filling system. For most automated systems, a worker is only required to place empty 50- to 100-pound bags on a storage rack. From this point forward, the system becomes automated. Normally some type of robotic arm using suction cups takes the unfilled bags from a storage rack and places them on the individual fill nozzles of the bag filling unit. When the bag is in place, the filling process starts and continues until a sensor indicates that the bag has reached the correct weight. The bag saddle then actuates to eject the loaded bag from the fill nozzle/fill station. Figure 6.13 shows an example of a bag storage area and an automated device to place the bag on the fill nozzles.

For open-top bags, typically a semi-automated bagging system places, fills, and weighs the bags of product, then a worker manually feeds the bags through a closer device. For fully automated systems, all tasks are performed automatically and the operator/worker only needs to monitor the system's performance.

There are many different factors to evaluate when an operation is considering some level of automation (semi-automatic to fully automatic) [White 2006]. For facilities only bagging one shift per day, semi-automated systems are more commonly used. When operations bag many different product types, the cleanout or change over between batches is also easier with semi-automatic systems. Automated bagging systems require more care and maintenance than nonautomated systems. It is also important to stock some spare parts for these systems, especially if the vendor is not located nearby. There are fewer parts necessary for semi-automated systems; thus a smaller inventory of these critical parts is necessary.

Figure 6.13. Bag storage shelves shown on left; a robotic arm with suction cups to place empty bags on fill nozzle shown on right.

With a totally automated system, greater production can be achieved. In addition, savings can be achieved with automated systems because they eliminate the worker at the bagging process. Eliminating the need to have workers performing tasks manually reduces their respirable dust exposure, MSD, and lost-time injuries.

One aspect that does not change from manual to automated bagging systems is to have an LEV system incorporated into the bagging station to capture and remove the dust liberated from product blowback and product "rooster tail." Although workers are not located directly within the bag filling area, if the dust is not captured during bag loading and ejection from the fill station, it will flow out into the work environment and contaminate workers in and around the automated bag filling system. One very effective technique often used is to enclose the bag filling area with clear plastic stripping. This provides a physical barrier that indicates the potential dust-laden air zone to workers and allows the LEV a better opportunity to contain and capture the dust liberated from bagging.

The initial costs of automated systems are much greater than for manual systems. If a cost analysis is performed, costs must be evaluated on a long-term basis, considering the respirable dust exposure and MSD exposure to workers performing the tasks manually.

CONTROL TECHNOLOGIES FOR CONVEYING AND PALLETIZING 50- TO 100-POUND BAGS

Bag and Belt Cleaning Device

The bag and belt cleaning device (B&BCD) is designed to reduce the amount of dust escaping from bags as they travel from the bag loading station to the stacking/palletizing area. This device reduces the dust exposure of all workers involved in and around the conveying process, as well as anyone handling the bags once they are filled at the loading station.

These systems are applicable to any mineral processing operation that loads product into 50- to 100-pound paper bags. This system should be located in-line on the conveyor at the closest available position to the filling station.

In one particular application, the B&BCD was 10 feet long and used a combination of brushes and air jets to clean all sides of the bags (Figure 6.14). Initially, the bags enter the B&BCD through a curtain made of clear, heavy-duty flexible plastic stripping to seal it from the plant air. The bags then travel past a stationary brush on a swing arm that cleans the front and top of each bag. Next, the bags go through a second set of plastic stripping and into the main cleaning chamber. The bags then travel under a rotating circular brush that further cleans the top of each bag. The sides are cleaned by a stationary brush located on each side of the bag, and an air jet located at the end of these brushes provides for additional cleaning. The side of the unit facing the valve uses a higher volume (velocity) air jet to provide the maximum amount of cleaning of the valve area. After passing through the air jets, the bag travels over a rotating circular brush beneath the bag which cleans the bottom of the bag. Finally, the bag exits the device by traveling through another air lock chamber with flexible plastic stripping.

A chain conveyor is used for the entire length of the device to allow product removed from the bags during cleaning to fall into a hopper. Product collected in this hopper is then recycled back into the process normally using a screw conveyor. Once exiting the B&BCD, both the bags of product and the conveyor belt should be essentially dust-free.

During the evaluation of this specific B&BCD at two mineral processing plants, the device showed very favorable results in lowering respirable dust levels. The most relevant data from this testing was the amount of dust removed from the surface of the bags, with a 78 to 90 percent range of reduction [Cecala et al. 1997; USBM 1995].

Another common source of dust generation and liberation from bags is during the bag flattening process. In operations that use bag flatteners, this task could be performed within the B&BCD to capture the dust released during this process.

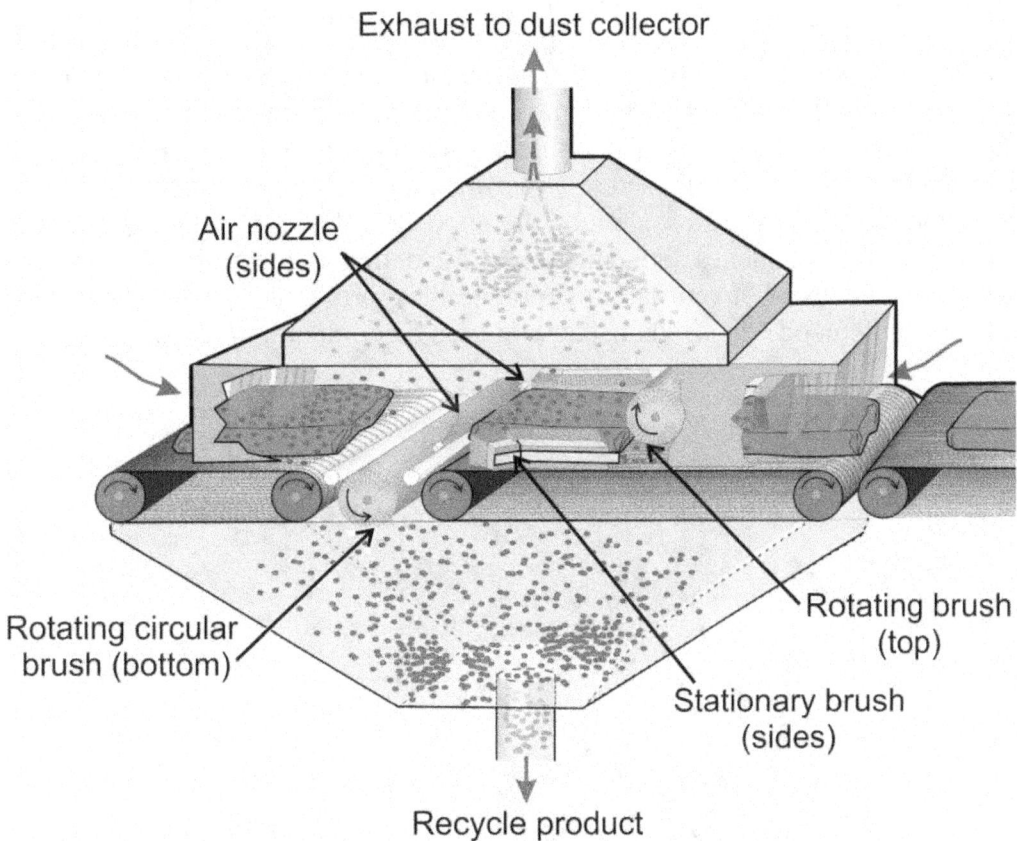

Figure 6.14. Bag and belt cleaner device. Note the use of a closed system under negative pressure.

Although this section describes one particular B&BCD system, there are numerous types that can be purchased from manufacturers, fabricated in-house by mineral processing plants, or fabricated by a local engineering firm. One key factor for a system is to be self-supporting so it can be implemented anywhere along the conveyor belt line. It is more logical to place a B&BCD in the closest possible proximity to the filling station to eliminate the dust hazards as quickly as possible. These devices normally need electrical and compressed air to power the cleaning unit and the blow-off nozzles, respectively.

One last critical aspect is that a B&BCD must be under sufficient negative pressure to ensure that any dust removed from the bags or conveyor belt does not leak from the cleaning unit into the mill. Refer to Chapter 1—Fundamentals of Dust Collection Systems for the specifics on the HVAC system.

Semi-Automated Bag Palletizing Systems

Semi-automated systems use workers in conjunction with an automated system to perform the bag stacking process. These can include a vast array of different setups and types of systems. In one case, a worker performs the bag stacking task manually, but is assisted by a hydraulic lift table. This lift table allows the height for stacking the bags to remain constant throughout the entire pallet loading cycle. The lift table is set to approximately knuckle-high for the worker, which is the most ergonomic loading height (Figure 6.15). A push-pull ventilation system is used on either side of this pallet to capture the dust liberated during the bag stacking (palletizing) process [Cecala and Covelli 1990]. This system is composed of a low-volume, high-velocity blower operating at approximately 120 cfm to direct a stream of air over the top layer of bags on the pallet. The blower system is composed of two 3-inch air jets (approximately 1,200-fpm velocity) directed towards an exhaust system on the opposite side of the pallet. As these air jets travel approximately 10 to 12 inches above the top of the pallet, they entrain the dust generated during bag stacking. The exhaust ventilation system pulls approximately 2,500 cfm of air and dust through the exhaust hood. This exhaust air can then be dumped into an LEV system.

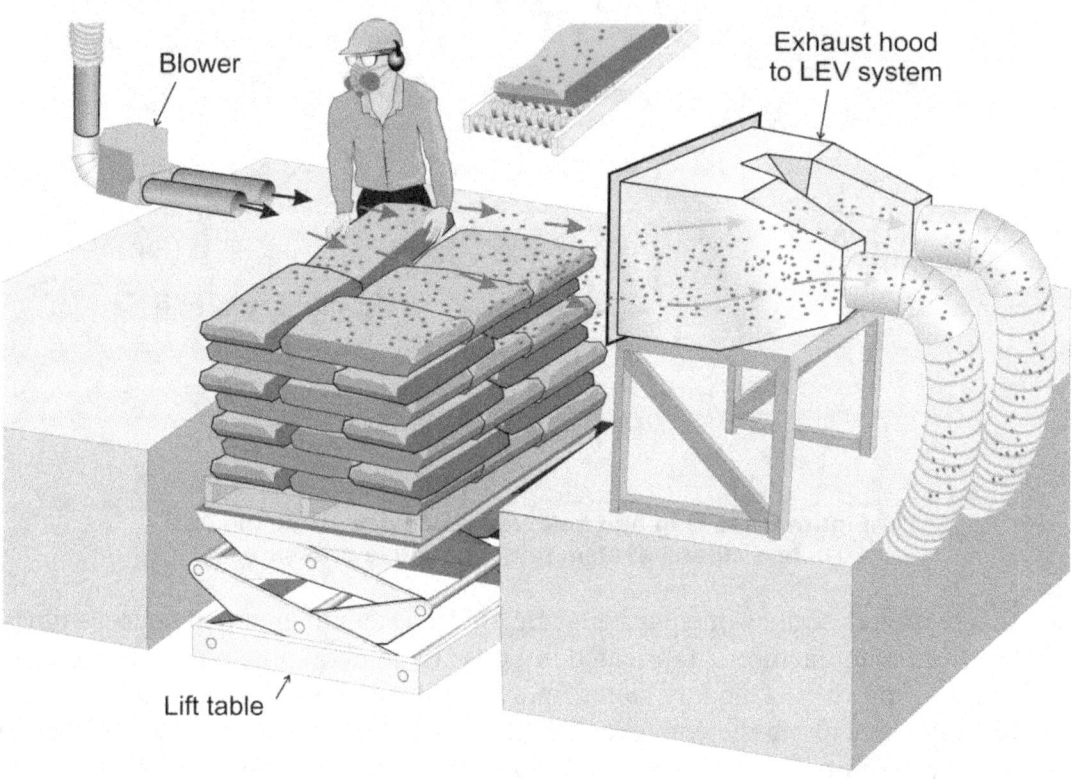

Figure 6.15. Semi-automated bag palletizing system to ergonomically improve bag stacking process with a push-pull ventilation system used to capture the dust generated.

In other cases involving palletizing systems, the worker slides the bag of product on an air table one layer at a time, but the actual stacking of the bags onto the pallet is performed automatically. Since back injuries represent such a major lost-time injury potential for bag stackers, this design significantly reduces back stress by eliminating the need to manually lift any bags of product. One problem with an air slide device is that it can cause dust to be blown from the bags of product into the worker's breathing zone. An OASIS system (see this chapter's earlier "Overhead Air Supply Island System" section) can work very effectively in conjunction with an air slide device.

In addition to the OASIS system, an exhaust hood was also used next to the air slide area to capture the dust blown from the bag (Figure 6.16) [Cecala et al. 2000]. With this setup, or any other industrial application using an exhaust hood, one area of concern for overexposure is when a worker positions himself/herself between the dust source and the exhaust hood. In this application, if the worker leans out over the air slide, he or she will be exposed to any dust being blown off the bags by the air slide as it is pulled into the exhaust hood.

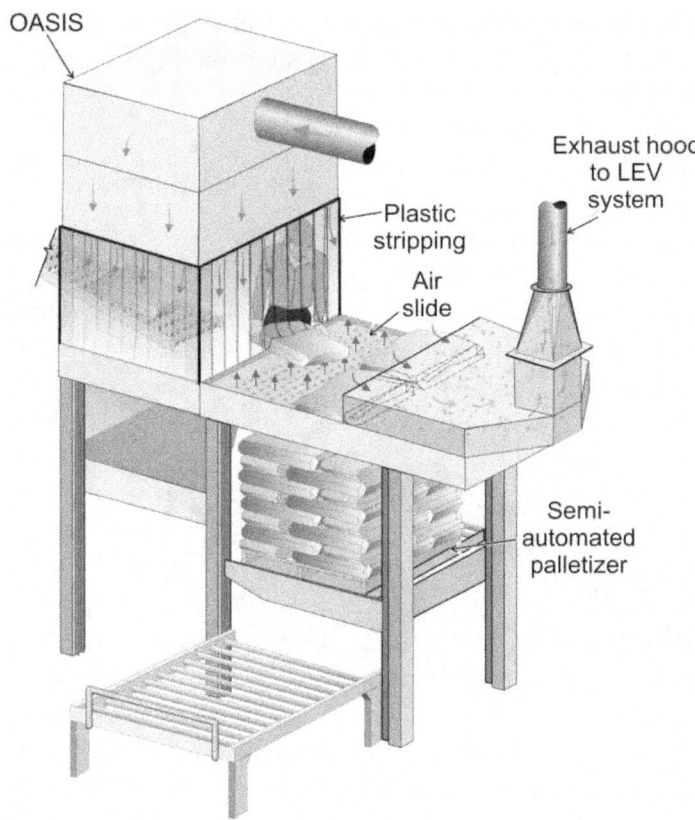

Figure 6.16. Semi-automated bag palletizing systems in which a worker removes bags from a conveyor and lifts and positions each layer on an air slide before automated loading onto the pallet. Both an OASIS system and an exhaust ventilation system are used to lower the worker's dust exposure during this bag palletizing process.

Automated Bag Palletizing Systems

Various manufacturing companies have developed advanced designs for automatic palletizing equipment that utilize either the traditional high-level pallet loading or low-level pallet loading technology. With high-level pallet loading systems, the pallets enter the automated palletizer system near the floor and are then raised to perform bag loading. When the pallet is completely loaded, the full pallet is then lowered back down to near floor level before being delivered to a plastic pallet wrapping system or loadout area.

With these high-level pallet loading systems, the bags arrive at the unit from the filling station via a conveyor belt. Once at the unit, the bags are guided onto a stripper plate that is composed of a series of rollers. Bags are arranged onto the stripper plate in a predetermined pattern as programmed into the unit by the worker overseeing the automated unit. Once a full layer of bags is positioned, the system stripper plate is automatically opened and the entire layer of bags is loaded onto the previous layer. The stripper plate then closes, allowing another layer of bags to begin loading in a prearranged pattern. Each layer of bags is loaded in an opposite pattern than the previous layer, typically with any manually or automated pallet loading system. Each time a full layer of bags is completed, the stripper plate is opened and the new layer of bags is stacked onto the pallet. When the predetermined number of layers of bags is achieved, the full pallet is lowered down near the floor level which ends the automated pallet loading process. Figure 6.17 shows a typical high-level automated bag palletizing system. There are many different manufacturers of these types of systems, with each system having its own unique design specifications.

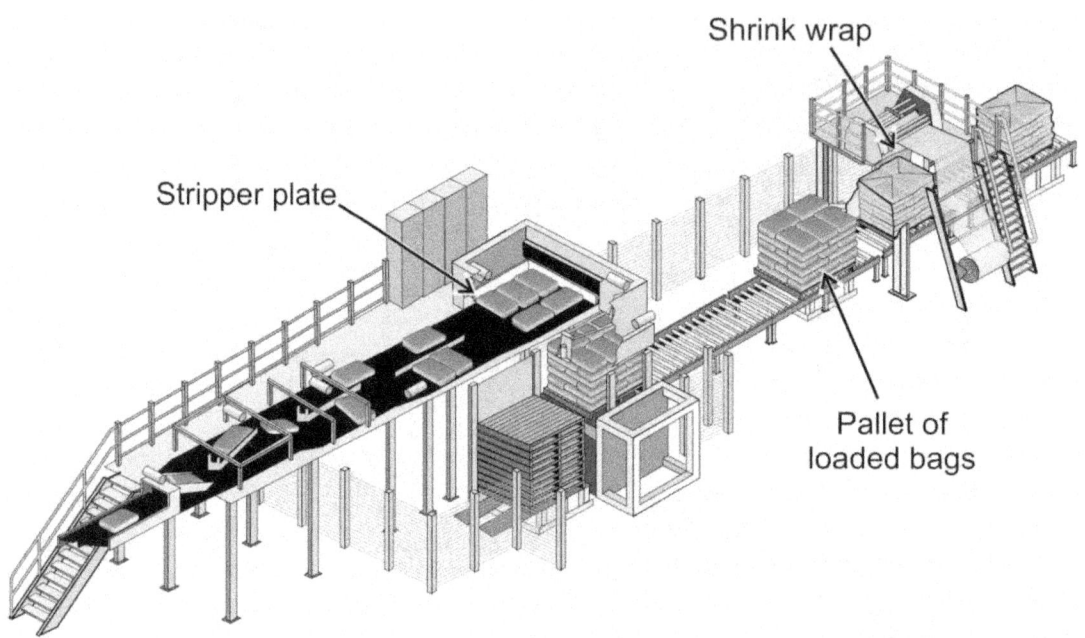

Figure 6.17. Figure of the automated high-level palletizing process.

With low-level automated pallet loading systems, the pallets are located near the floor and remain stationary throughout the entire loading process. A mechanical device takes the loaded bags and places them directly onto the pallet. The use of a robotic mechanical arm type palletizer is the most common type of automated low-level system.

The robotic arm automatic palletizer illustrated in Figure 6.18 has become very popular due to its productivity and lower cost of maintenance compared to the in-line/high-level palletizers. Filled bags travel from the bagging station to the palletizing area via a conveyor belt. At the palletizing area, a photocell detects that a bag has arrived and is ready to be loaded onto a pallet. Each bag is grasped by the fingers of the robotic arm, then taken and gently placed onto the pallet in a programmed pattern established by a worker overseeing the automated unit. The arm can be programmed to close tolerances to avoid the bag from being dropped onto the pallet, which would cause dust to be emitted from the bag valve. The robotic arm can also stack bags onto two separate pallets and at a high rate of speed to increase productivity.

The robotic arm automatic palletizers are proving themselves to be the preferred method for palletizing bags because they are much simpler, and thus more cost-effective by comparison to the high-level pallet loading systems. With these systems, operating costs, breakdowns, and maintenance are greatly minimized [Case History 2006b]. They also require much less space and minimize all the various dust generation and liberation potential when compared to high-level pallet loading systems.

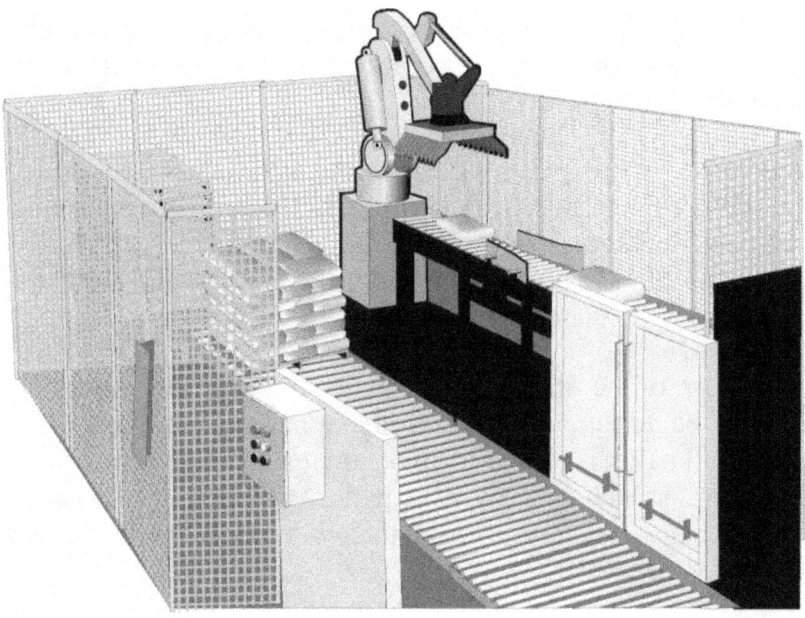

Figure 6.18. New innovation of low-level pallet loading process using a robotic arm device.

One aspect which is very important in any automated system is to provide adequate guarding to protect workers from entering automated work areas while these units are operating, or even activated. Because of the operating speeds and the mass of the unit in robotic arm systems, the swing pattern can span significant distances and the mechanical arm can cause severe injury, or even death, if it strikes a worker.

When combining these automated palletizing systems with automated bag placers, in-line bag cleaning devices, automatic bag weighing systems, printing systems, and plastic pallet wrap systems, the worker is removed from the work processes. This in turn substantially reduces the worker's dust exposure potential, and the worker has more of an oversight responsibility to ensure that the system is working properly. Another significant benefit over and above the dust exposure reduction is the elimination of MSD bag stress injuries common to performing these various tasks manually.

It should be stressed that when using automated systems, there is still a concern for dust. The amount of respirable dust generated and liberated from automated systems still needs to be evaluated and controlled to ensure that it does not flow into the surrounding mill areas and contaminate the breathing atmosphere of workers. This is normally achieved by enclosing the dust generation areas with plastic stripping and using an LEV system to exhaust the respirable dust.

Plastic Pallet Wrap

The final consideration for dust control in the palletizing of paper bags is plastic pallet wrap systems. The purpose of plastic pallet wrap is to contain product and dust leakage from bags loaded onto full pallets from becoming airborne while also providing protection from torn or ruptured bags during the pallet transportation process. Currently, there are two types of plastic pallet wrap systems being used: (1) spiral stretch wrap, and (2) shrink wrap/stretch hoods. Both systems are commonly used in the minerals processing industry and they can be used in both manual and automated applications. The size of the packaging system will normally dictate which pallet wrapping system is most practical.

A consideration when using plastic pallet wrap systems is that the bags should be completely void of entrapped air before the plastic wrap is applied. If a protective plastic coating is applied before the bags have been deflated, it will become "loose" and thus, ineffective. In operations where entrapped air is a problem, a pallet press should be used to "squeeze" the excess air from the bags immediately before the plastic wrap is applied.

Spiral stretch wrap was one of the first types of plastic wrapping techniques developed and used in the industry. This can be performed either through a manual or automated process. For manual operations, the full pallet is placed in an open area and a worker manually walks around the pallet, unwinding spiral stretch wrap over the pallet. Once completing multiple revolutions to the point that the pallet is adequately covered, the worker cuts the stretch wrap from the roll and connects the cut end to the pallet.

In the automated process, the full pallet of bags is placed on a rotating table. As this table spins the pallet round and round, a roll of stretch plastic is continually wrapped around the pallet while slightly moving up and down at a predetermined wrap tension (Figure 6.19). After enough revolutions occur that the pallet is completely sealed, a mechanical device cuts the plastic wrap from the roll and assists the unrolled end in being attached to the pallet.

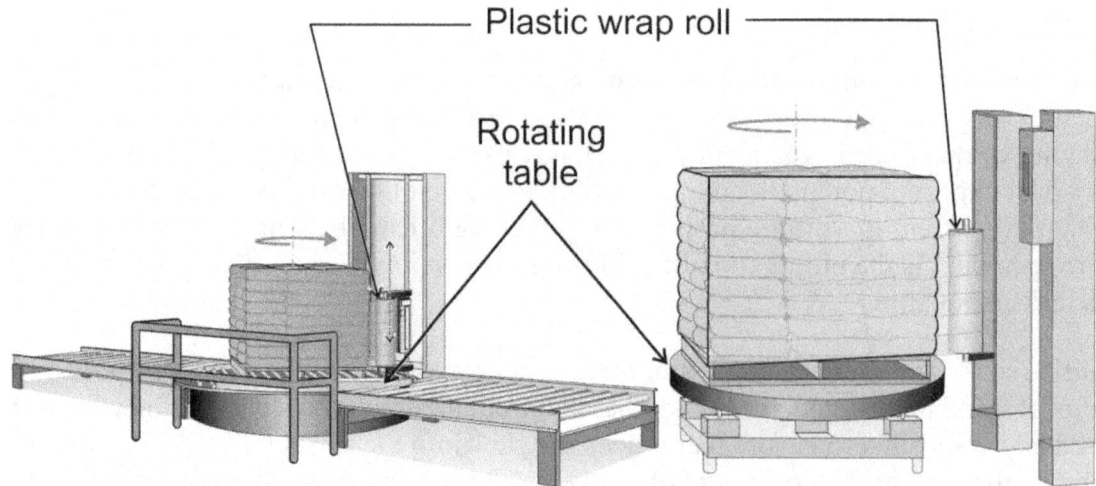

Figure 6.19. Automated spiral stretch wrap machine.

For many years, this plastic wrapping technique was the best and least expensive method of protecting loads for shipment [Packaging system 1978]. One shortcoming with the spiral stretch wrap is that the top of the pallet is not normally sealed and dust can be liberated from this top area. In addition, when the pallets are outside, rain or moisture can get down in and saturate the bags of product. To correct this shortcoming, a top sheet can first be placed over the top of the pallet before the process is started, but this normally has to be performed manually by a worker, which increases the time and the cost of performing this process.

Another method to perform this plastic wrapping technique is through the use of shrink wraps or stretch hoods. This technique uses a single sheet of plastic which is placed over the entire pallet and then heated to cause it to shrink tightly down completely around the pallet of bags. This technique can also be performed manually by a worker or through the use of an automated system. When using this technique, pallets can be stacked and stored outside without worrying about water damage during storms. Stretch-hooded pallets are safer than shrink wrap pallets because the pallets are more stable and easier to handle [New Installations 2006].

FLEXIBLE INTERMEDIATE BULK CONTAINERS

Flexible Intermediate Bulk Containers (FIBCs), also called "bulk bags," "semi-bulk bags," "mini-bulk bags," and "big bags," have become more popular over the recent years for shipment of product material. They are often more cost-effective than the 50-, 80-, or 100-pound bags for both the mineral producer and the end user. The most effective method to control the dust liberated during filling of FIBCs is by using an expandable neoprene rubber bladder in the fill spout of the bagging unit. This bladder expands against the interior of the FIBC loading spout and completely seals it, eliminating any product and dust escaping from the spout during loading (Figure 6.20). When using this expansion bladder, the feed spout must also incorporate an exhaust ventilation system to exhaust the excess pressure from the FIBC during loading (Figure 6.21).

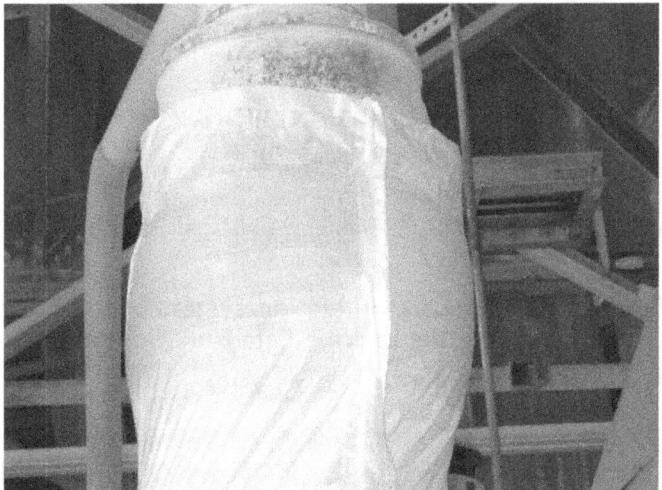

Figure 6.20. Expandable neoprene rubber bladder used in the fill spout of the bagging unit for Flexible Intermediate Bulk Containers.

If dust is still being liberated, another effective technique is to isolate the bag loading area from the rest of the plant. A worker can enter this area to manually attach the empty bag onto the loading device. The worker then leaves the area and once outside, remotely activates a start button to begin the bag filling cycle. The area is under negative pressure by virtue of being tied into an LEV system, and any dust generated will be exhausted and not allowed to contaminate the operator's work space. Once the FIBC loading is completed, the worker re-enters the area and removes the bag for shipping and begins the process again.

A cautionary note is made here regarding work habits of employees performing the sealing of the filled FIBC spout. Most FIBCs have a fabric fill-spout that requires cords to be tied or secured in some manner. In this case, it is very important that the operator "point" the position of the spout away from the body to avoid the dust-laden exhaust air exiting the bag from being directed toward the operator's breathing zone.

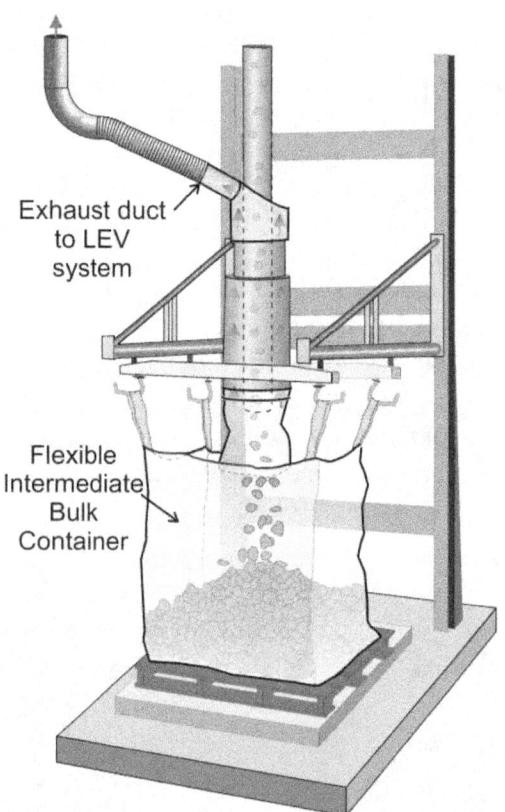

Figure 6.21. Exhaust ventilation system used during loading of Flexible Intermediate Bulk Containers.

MAINTAINING CLEAN FLOORS IN BAG LOADING, PALLETIZING, AND WAREHOUSE LOCATIONS

Pallets loaded with 50- to 100-pound bags, or bulk bags which are also loaded onto pallets, are normally moved by a forklift or tow motor directly to trailer trucks or railcars, or are stored in a warehouse location until a later time when they are then loaded into a transportation vehicle. As these forklifts and/or tow motors move around plant and warehouse locations, they can stir up significant amounts of dust if floors are not kept in a clean condition. The two most common floor cleaning techniques in mineral processing plants are washing with water and/or through the use of floor cleaning units.

To serve as an effective cleaning technique, hosing down floors with water needs to be built into the structure right from the beginning. Floors need to be sloped toward floor drains to be truly effective and to minimize the amount of standing water. Typically, hosing down floors, needs to occur on a shift-by-shift basis to be effective.

For floor cleaner units, there are many different manufacturers and companies that produce and sell a vast array of different types of units. These units vary from one that an individual rides to smaller units that an individual walks behind. One area that should be closely investigated is the disposal of the contaminated material once it has been collected. It would be counter-productive if the units were very efficient during the floor cleaning process but contaminated employees as the floor cleaning units were being cleaned or during disposal of the accumulated material. For more information on floor cleaning units, refer to "Housekeeping Practices" in Chapter 8—Controls for Secondary Sources.

ENCLOSING DUST-LADEN AREAS WITH PLASTIC STRIPPING AND USING AN LEV SYSTEM

Many of the different dust control techniques discussed in the chapter for bagging and palletizing operations can be further improved by enclosing the dust-laden air zone with clear plastic stripping and using an LEV system to capture and remove the dust. This "enclose and capture" technique has a number of benefits. The plastic stripping contains the dust within the job process and minimizes the possibility of it being liberated throughout the entire plant. It also provides a visual indication for the plant personnel of the boundary of the dust-laden air zone. Workers know not to enter the dust-laden air zone while the job processes or functions are operating.

Another benefit of the boundary area created by the plastic stripping is that it allows the LEV system to perform more effectively. Since the capture efficiency of exhaust ventilation systems in an open environment is not very efficient, the plastic stripping barrier allows the exhaust system to operate much more effectively than without one in place. From an economical standpoint, plastic stripping is relatively inexpensive compared to other types of dust control techniques.

One last area to consider when using clear plastic stripping is that it allows workers to see through the curtain for safety reasons. This technique has numerous applications not only in the bagging and palletizing area, but throughout the entire mineral processing operation. When evaluating the correct connection technique for the LEV system, as well as recommended exhaust airflow quantities, refer to Chapter 1—Fundamentals of Dust Collection Systems.

WORK PRACTICES TO MINIMIZE DUST EXPOSURES FROM BAGGING AND PALLETIZING

The following list of best practices should be considered by operations wishing to minimize respirable dust exposures to workers performing bagging and palletizing.

- Use slides or chutes whenever possible to minimize the distance that product free-falls in open areas.
- Keep floors as clean as possible to minimize dust liberation as forklifts/tow motors travel around the area transporting bag material and other items.
- Maintain the work area such that there are no sharp edges at bag drop-off locations, turning points, and/or conveyor lines that could tear bags of product material.
- Maintain material flow cut-off valves at bagging machines to minimize product dribble from fill nozzles/fill chutes.
- Determine the optimal bag construction and bag valve type for the material being processed and bagging machine being used.
- Minimize product blowback and product "rooster tail" during bag filling and ejection from the fill station. Both of these dust sources can also cause extensive dusting on the outside of bags, as well increase the overall demand on the LEV system.
- Consider using ventilated filling spouts as discussed in the "Dual-Nozzle Bagging System" section to remove product from the fill nozzle and bag valve areas after bag filling is completed.
- Slow bag filling rates when necessary to minimize the amount of product blowback.
- Carefully place loaded bags onto pallets. Throwing or dropping bags onto conveyors or pallets increases the potential for dust generation and/or bag breakage.
- Use a hydraulic lift table to minimize dust generation and MSDs to workers during bag loading onto pallets by keeping the loading height at knuckle height as described in the "Semi-Automated Bag Palletizing Systems" section. This allows for a push-pull ventilation system to more effectively capture any dust generated during bag palletizing.
- Use dust-free empty bags manufactured to Flexible Intermediate Bulk Container (FIBC) standards. Many reusable FIBC bags are returned with clean interiors, but dust is often trapped in the fabric on the bag exterior. This dust is then expelled during handling of the bag.
- Store all empty bags in a clean environment and keep them covered to prevent dust from settling on them.
- When tying or sealing the spout of a filled FIBC bag, the operator should always hold the flexible loading spout away from his or her breathing zone to minimize the escaping dust.
- Avoid positioning the bag operator or the bag palletizer between any dust source and the pickup point of an LEV system. When this is not done, dust will be drawn directly over the worker, significantly increasing the worker's dust exposure.
- Consider using a mechanically assisted bag handling system to prevent dropping filled bags onto pallets. These systems also greatly reduce the potential for workers developing MSDs.
- Consider the use of a bag cleaning device prior to bag palletizing, as discussed in the "Bag and Belt Cleaning Device" section. Mechanical brushes and air nozzles can be

used successfully to perform this cleaning. Brushes need to be adjusted or replaced often due to high wear potential. It is also critical to capture the dust removed from the bags with an LEV system.

- Develop and implement a plan for dealing with broken bags that minimizes manual handling. This can include a vacuuming system or a ventilated waste hopper device. Every effort should be made to eliminate plant personnel from manually handling any broken bags.
- Enclose bagging and palletizing areas from the general plant environment through the use of clear plastic stripping. If entry is not necessary, a solid type curtain could be used. These enclosed areas should be under negative pressure through an LEV system to prevent dust from migrating to other areas of the plant.
- Collect dust at the source through an LEV system rather than trying to redirect or to dilute with ventilation fans. The use of free-standing fans only circulates the air and dust and does not remove it from the plant.
- When using an OASIS, periodically clean and replace filters and ensure that clean filtered air is evenly distributed down over the worker. Significant airflow reductions can occur from dust-laden filters.
- Check and lubricate bearings on LEV systems, including the OASIS, as well as check power transmission belts. In addition, consider the use of a pressure differential gauge or monitor to indicate system problems or when maintenance needs to be performed.
- Ductwork to the LEV system should be examined and cleaned periodically to ensure that the system is working properly. Capture velocities at the face of shrouds and hoods should be monitored periodically to ensure proper functioning.

REFERENCES

Case History [2006a]. Dustless bag filling system puts material in, takes dust out. Powder and Bulk Eng *20*(5):82–87.

Case History [2006b]. Robotic palletizer improves company's bottom line. Powder and Bulk Eng *20*(5):88–93.

Cecala AB, Covelli A [1990]. Automation to control silica dust during pallet loading process. In: Proceedings of SME Annual Meeting. Preprint 90–28.

Cecala AB, Muldoon T [1986]. Closing the door on dust - dust exposure of bag operator and stackers compared for commercial bag valves. Pit & Quarry *78*(11):36–37.

Cecala AB, Timko RJ, Prokop AD [1997]. Bag and belt cleaner reduces employee dust exposure. Rock Products *100*(3):41–43.

Cecala AB, Zimmer JA, Smith B, Viles S [2000]. Improved dust control for bag handlers. Rock Products *103*(4):46–49.

Constance JA [2004]. Ways to control airborne contaminants in your plant. Powder and Bulk Eng *18*(5):118–119.

Godbey T [2005]. Recirculating your cleaned air: is it right for your plant? Powder and Bulk Eng *19*(10):31–37.

Kearns P [2004]. What's new in paper and plastic bags. Powder and Bulk Eng *18*(5):75–79.

New Installations [2006]. Stretch-hood fits like a glove. Powder and Bulk Eng *20*(5):76–81.

Packaging system eases mineral dust problems [1978]. Pit and Quarry *70*(10):104–106.

Robertson JL [1986]. Overhead filters reduce dust level in air supply. Rock Products *89*(7):24.

USBM [1981]. SF6 tracer gas tests of bagging-machine hood enclosures. By Vinson RP, Volkwein JC, Thimons ED. U.S. Department of the Interior, Bureau of Mines Report of Investigations 8527.

USBM [1983]. Technology News 54: Dust control hood for bag-filling machines. By Volkwein JC, Pittsburgh, Pennsylvania: U.S. Department of the Interior, Bureau of Mines.

USBM [1984a]. Technology News 207: New nozzle system reduces dust during bagging operation. By Cecala AC. Pittsburgh, Pennsylvania: U.S. Department of the Interior, Bureau of Mines.

USBM [1984b]. New bag nozzle to reduce dust from fluidized air bag machine. By Cecala AB, Volkwein JC, Thimons ED. U.S. Department of the Interior, Bureau of Mines Report of Investigations 8886.

USBM [1986a]. Impact of background sources on dust exposure of bag machine operator. By Cecala AB, Thimons ED. U.S. Department of the Interior, Bureau of Mines Information Circular 9089.

USBM [1986b]. Dust reduction capabilities of five commercial bag valves. By Cecala AB, Covelli A, Thimons ED. U.S. Department of the Interior, Bureau of Mines Information Circular 9068.

USBM [1995]. Reducing respirable dust levels during bag conveying and stacking using bag and belt cleaner device. By Cecala AB, Timko RJ, Prokop AD. U.S. Department of the Interior, Bureau of Mines Report of Investigations 9596.

Volkwein JC, Engel MR, Raether TD [1986]. Get away from dust with clean air from overhead air supply island (OASIS). International Symposium on Respirable Dust in the Minerals Industry, University Park, Pennsylvania.

White D [2006]. Is it time to automate your bagging line? Powder and Bulk Eng *20*(5):71–75.

CHAPTER 7:
BULK LOADING

CHAPTER 7: BULK LOADING

During the bulk loading of product, dust contained within the product can be liberated and emitted into the ambient air as the product falls from the loadout area to the transport container (truck, rail car, barge, or ship). This dust-laden air can expose workers to respirable dust, as well as create nuisance dust problems.

A number of factors impact the severity of dust liberation during bulk loading, including:

- the type of product and its size distribution,
- moisture content of the product,
- volume of the product being loaded and the loading rate,
- the falling distance,
- environmental factors such as wind velocity and rain, and
- physical characteristics of the receiving vessel.

The first three factors represent the characteristics of the product being loaded and may be defined by the process or by the customer. Past research has shown that larger size distributions and higher moisture contents typically result in less respirable dust liberation. As a general rule, a moisture content of 1 percent by weight is recommended as a starting point for adding moisture to a crushed product [NIOSH 2003]. Chapter 2—Wet Spray Systems provides a detailed discussion on using water to reduce dust liberation.

The last three factors represent the characteristics of the loading process. Engineering controls can be implemented to minimize the impact of these factors. For example, the falling distance can be reduced and physical barriers can be installed to protect against high winds.

The following dust controls, which have been implemented in the industry, will be discussed in this chapter:

- loading spouts,
- the Dust Suppression Hopper, and
- enclosures.

LOADING SPOUTS

The loading spout is designed to transfer product from a plant/storage container into the vehicle that will be used to transport the product. Typically, these spouts have the capability to extend down to the vehicle being loaded and then be retracted to allow the vehicle to move away from the loading station. A review of several loading spout manufacturer specifications shows that this spout travel can be as little as a few feet and can extend up to 100 feet. As shown in Figure 7.1, these loading spouts are equipped with a series of telescoping cups or pipes that extend and retract. This figure also shows that the spout typically has an outer shroud which encases the product transfer section of the spout. This outer shroud shields the product from the elements (rain, wind, etc.) and helps to contain any dust that is liberated.

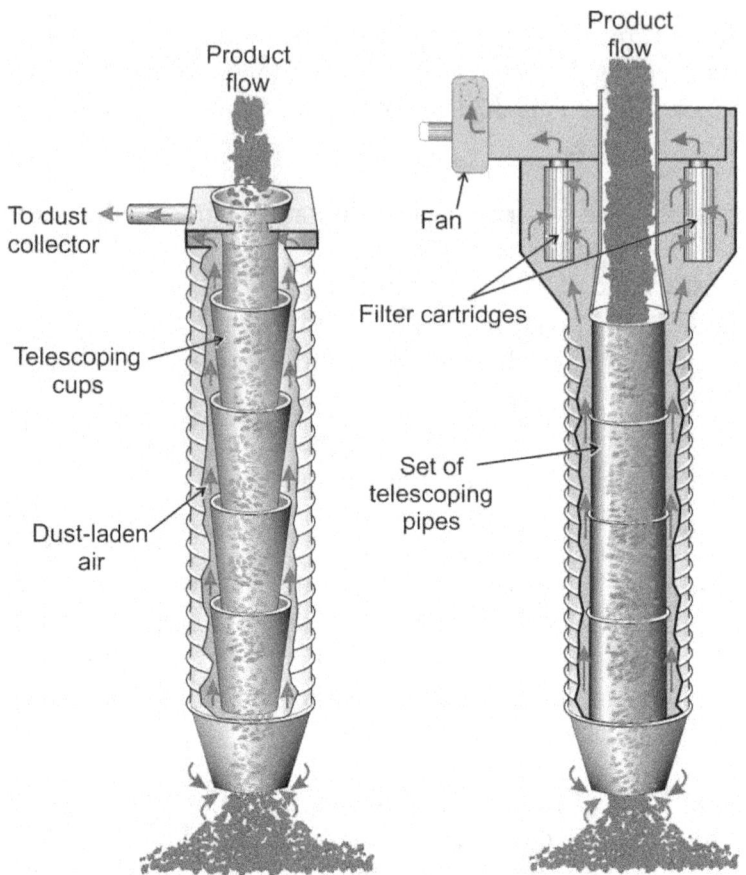

Figure 7.1. Loading spouts designed with telescoping cups or pipes and dust collection capabilities.

Depending upon the loading application, the discharge end of the spout may be specially designed to seal against the loading port of the receiving vehicle. Often times, the spout discharge is cone-shaped to partially enter and seal against the vehicle port. Figure 7.2 illustrates examples of these cone discharges designed to reduce dust emissions during loading into a closed vehicle. If the vehicle has multiple loading ports, only one port should be opened during loading to minimize dust liberation from the vehicle.

In order to realize the dust control benefits of the discharge cone, the loading spout must be properly positioned to minimize the chance of dust escaping into the ambient air during loading. This can be accomplished through the use of articulated loading systems. These systems can move the loading spout side-to-side as well as forward and back to ensure that the spout discharge is located in the optimum position above the vehicle, which minimizes dust liberation and product spillage. The operator can control these articulated loading systems from a control room or from a pendant station near the loading site. Figure 7.3 shows a schematic and photograph [Midwest International Standard Products 2010] of an articulated loading system.

Figure 7.2. Loading spout discharge cones designed to seal against the loading port of closed vehicles to minimize dust liberation.

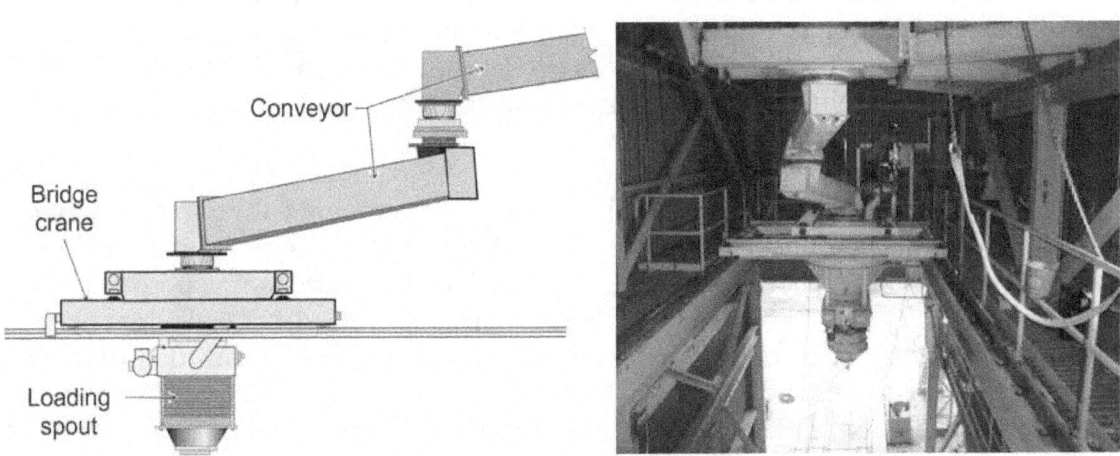

Figure 7.3. Articulated loading spout positioner.

When loading into open vehicles, the discharge end of the loading spout can be equipped with a skirt or apron which can seal against the product pile in an effort to minimize dust liberation. The skirts are typically fabricated from belting or fabric and flare out as the loaded material builds up. For the most effective dust control and to prevent clogging of the loading spout, it is important that the spout position be periodically raised as the height of the loaded material increases. To automate this procedure, the loading spouts can be equipped with tilt sensors that are suspended vertically from the spout. As the loaded material builds up, it moves these sensors from their vertical orientation (e.g. 15 degrees or more) which activates the automatic raising of the loading spout. Examples of these discharge skirts and tilt sensors are illustrated in Figure 7.4.

 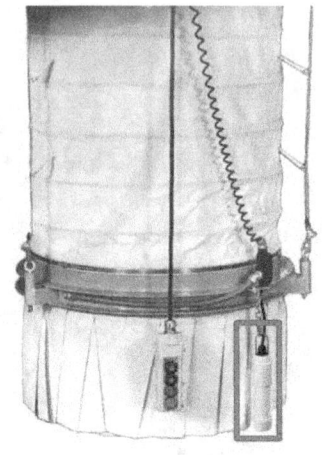 

Figure 7.4. Skirt (in yellow box) and tilt sensors (in red boxes) installed on the discharge end of loading spouts to help reduce dust emissions.

Dust Collection Systems and Loading Spouts

As material falls through the loading spout, airflow is induced which can entrain dust liberated from the product. As the falling product hits the vehicle floor or previously loaded material, the material is compressed and the induced air is driven off, creating entrained dust. As the falling distance increases, the induced air is driven off with a higher velocity, which creates larger amounts of dust [Biere et al. 2010]. When a given volume of material is loaded into a receiving vehicle, an equal volume of air in the vehicle is displaced, which can carry liberated dust with it. Chapter 1—Fundamentals of Dust Collection Systems discusses and illustrates air induction and dust generation for falling material.

To address the above dust emissions during bulk loading, a common control measure is to equip the loading spout with a dry dust collection system. These dust collection systems induce airflow in the annular region between the outer shroud and the inner loading portion of the spouts, as shown in Figure 7.1. These systems also typically utilize filter cartridges or bags to capture the airborne dust. Compressed air is periodically introduced to back-flush the filters/bags to remove captured dust from the cartridges/bags and return this dust to the product stream. The dust collector unit can be located away from the loading spout and connected to the loading spout through ductwork as shown on the left in Figure 7.1. Another option is to have the dust collector incorporated into the top of the loading spout. The loading spout shown on the right in Figure 7.1 illustrates an internal filter cartridge system and fan mounted at the top of the loading spout.

A review of various loading spout manufacturers' specifications indicates that dust collection systems are offered with available exhaust air volumes ranging from 1.8 [SLY 2010] to 4.9 [PEBCO 2010] times the rated loading volume of the spout. Equation 1.2 in Chapter 1—Fundamentals of Dust Collection Systems can be used to calculate a recommended exhaust air volume based upon material feed rate, height of free-fall, product size, and feed open area. It should also be noted that if the exhaust airflow is too high, product can be pulled into the dust collection system.

Cascading Loading Spout

Figure 7.5 illustrates a schematic and photograph [Cleveland Cascade 2010] of a loading spout designed with cascading cones that prevent the product from dropping straight down through the spout. As mentioned earlier, free-falling product can promote dust entrainment and liberation into the ambient air during the loading process. Cascading-type loading spouts are designed with a series of inclined cones that do not permit the product to free-fall through the loading spout. The goal is to minimize product velocity and induced airflow movement, by allowing the product to be transferred smoothly from one cone to the next. Originally designed to reduce degradation of the product, the cascading action was also thought to have the potential to lower dust emissions by reducing product velocity and promoting mass flow [Maxwell 1999]. Improved mass flow of the product allows the particles to maintain greater contact with one another and thus reduce the release of small dust particles into the air. Wind shrouds and discharge skirts are the primary dust control devices utilized in these designs. Typically, cascading loading spouts are not equipped with dry dust collection systems. Within the loading spout, cone size, spacing, and inclination are adjusted to accommodate different product flow rates. Also, since there is more contact between the product and the cones than found in free-falling spouts, abrasion-resistant liners are frequently utilized inside the cones.

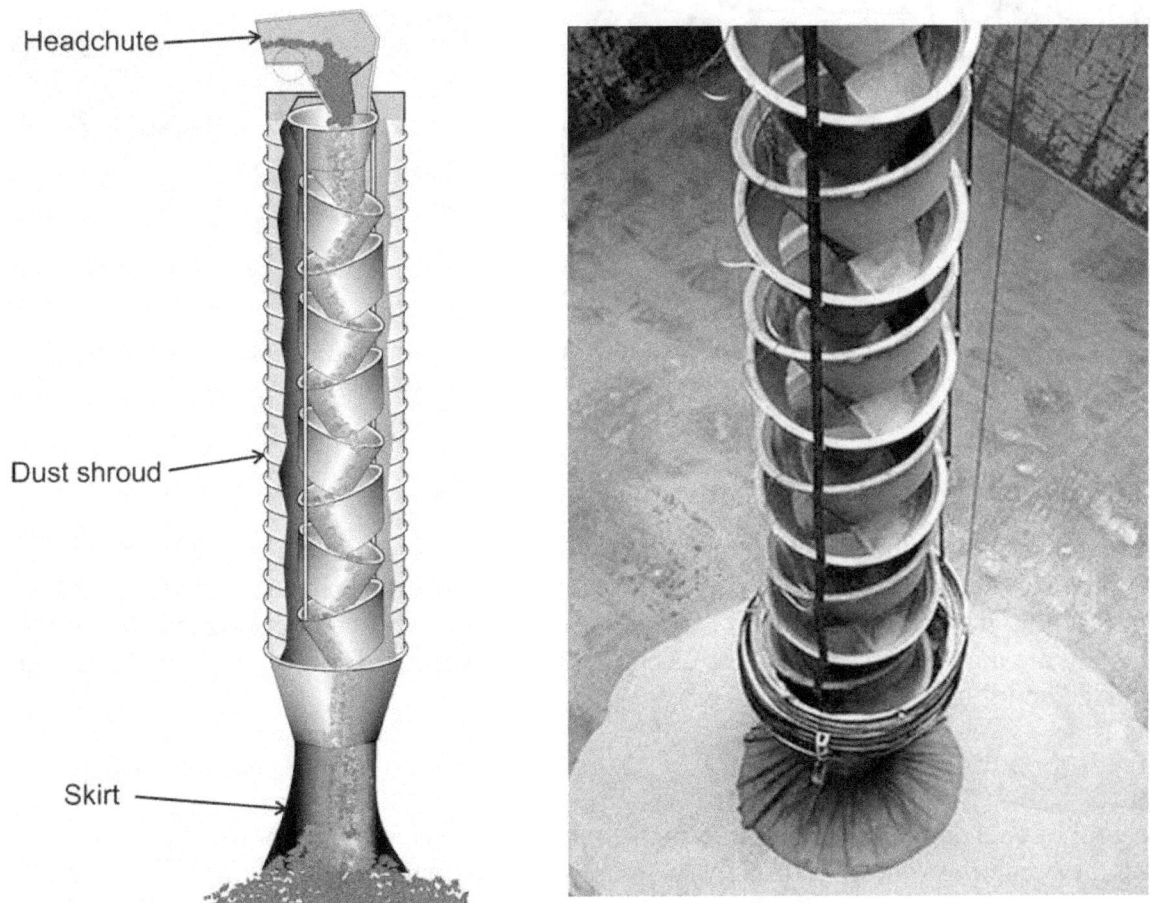

Figure 7.5. Cascade loading spout.

DUST SUPPRESSION HOPPER

Another modification that can be made in an effort to reduce dust emissions during bulk loading is the utilization of the Dust Suppression Hopper (DSH) [DSH Systems 2010]. The DSH is located at the product discharge point and is designed to help the loaded product flow more as a solid column of material with less exposed surface area to minimize entrainment into the ambient air [Wypych 2009]. The hopper is equipped with a central plug at the bottom to prevent discharge of product until a predefined quantity has accumulated in the hopper. Either springs or a programmable logic controller are used to set and control the amount of clearance between the plug and the hopper. Material is loaded into the top of the hopper and contained until sufficient weight forces clearance between the hopper and plug. Material then flows from the hopper in more of a solid column, as shown in Figure 7.6, rather than in a free-falling unconsolidated stream. This overall process results in less dust liberation [Biere et al. 2010].

Figure 7.6. Dust Suppression Hopper creating a solid stream of material during loading to reduce dust liberation.

ENCLOSURES

During bulk loading, the preferred method of dust control is to prevent dust from becoming airborne. However, if dust is released into the ambient air, efforts must be made to keep the dust from reaching the breathing zone of surrounding workers. This can be accomplished by enclosing the dust-generating process or enclosing the worker. Both of these methods will be discussed and illustrated below.

Enclosed Loadout

The goal of enclosing the loadout area is to contain and capture the airborne dust so that it cannot reach the breathing zone of nearby workers or contaminate the surrounding environment. Typically, this type of control will be used for the loading of trucks or railcars. Essentially, a physical barrier is constructed that confines the dust that is generated as loading occurs. The barrier also minimizes the impact of environmental factors such as high wind velocities. These barriers can be permanent walls with flexible plastic stripping, belting, or curtain used to provide a seal at the entry and exit points of the vehicles. Alternatively, flexible plastic stripping or curtains can be used to construct the entire enclosure as shown in Figure 7.7. In either case, the individual pieces of flexible material should overlap one another to provide a more effective barrier against the possibility of dust escaping. To further prevent dust escape, the enclosure should be supplemented with a dry dust collection system. Equation 4.1 in Chapter 4—Crushing, Milling, and Screening, can be used to calculate an initial exhaust volume for the dry dust collector. If dust is seen escaping the enclosure, it is an indication that an effective seal is not being maintained, that the exhaust air quantity is not sufficient, or both.

Figure 7.7. Rail car loadout area enclosed with plastic strips to contain dust.

The following is a list of some issues that should be addressed in designing the loading process and the loadout area so that personnel do not need to enter the enclosure.

- Vehicle hatches should be opened and closed outside of the enclosure.
- Cameras can be installed to assist the operator in properly positioning the loading spout to minimize dust generation and spillage.
- An automatic level sensing kit should be installed to signal the operator when the vehicle is fully loaded or to automatically shut off the feed. Several types of sensor technologies are available and include tilt, pressure, capacitance, or infrared. This automation allows the operator to maintain a more remote position from the area where dust might be generated.
- The loading spout can be raised automatically after the vehicle is loaded to a predefined height, which allows the vehicle to move away without the spout being fully retracted. This ensures that sufficient clearance is available and reduces the time required to begin the loading of the next vehicle. Level sensors and limit switches can be used to help accomplish this effort. Once again, this automation allows the operator to maintain a more remote position from the area where dust might be generated.
- During loading and until all liberated dust is cleared, personnel should not be allowed to enter the enclosed area. The loading process should be designed so that it is not necessary for personnel to enter the enclosure. If personnel must enter the enclosure, dust concentrations within the enclosure should be quantified in order to ensure that appropriate respiratory protection is utilized.

Operator Booths/Control Rooms

One of the most effective means of protecting a worker associated with a dust-generating operation can be the utilization of a booth/control room that provides filtered air inside the enclosure. Previous research has shown that properly operating filtration systems installed on enclosed compartments can provide over 90 percent reductions in respirable dust exposures [NIOSH 2008]. Another advantage of these booths is that they can be equipped with air-conditioning and heating systems to make the work environment more comfortable for the operator. Figure 7.8 shows an operator's booth overlooking the loadout area in a plant.

From a dust control engineering standpoint, the goal is to provide a competent operator's booth that is positively pressurized with filtered air. The positive pressure within the booth prevents dust that may be present outside of the booth from leaking in and exposing the worker. Chapter 9—Operator Booths, Control Rooms, and Enclosed Cabs provides detailed information regarding the design and operation of filtration/pressurization systems for operator compartments.

Figure 7.8. Environmentally controlled operator's booth near loadout area.

REFERENCES

Biere G, Swinderman RT, Marti AD [2010]. Conveyor upgrades increase plant availability, reduce airborne dust. Coal Power Magazine [online] Feb 16. [http://www.coalpowermag.com/plant_design/Conveyor-Upgrades-Increase-Plant-Availability-Reduce-Airborne-Dust_242.html]. Date accessed: October 18, 2010.

Cleveland Cascade [2010]. [http://www.clevelandcascades.co.uk/brochures.php]. Date accessed: November 18, 2010.

DSH Systems [2010]. Dust Suppression Hopper Systems. [http://www.dshsystems.co.nz/]. Date accessed: October 7, 2010.

Maxwell AS [1999]. Degradation, segregation and dust control of bulk materials. Solids Handling 19(3):399–403.

Midwest International Standard Products [2010]. [http://www.midwestmagic.com/products_positioners_articuloader.php]. Date accessed: October 6, 2010.

NIOSH [2003]. Handbook for dust control in mining. By Kissell FN. Pittsburgh, PA: U.S. Department of Health and Human Services, Centers for Disease Control and Prevention, National Institute for Occupational Safety and Health, NIOSH Information Circular 9465, DHHS, (NIOSH) Publication No. 2003–147.

NIOSH [2008]. Key design factors of enclosed cab dust filtration systems. By Organiscak JA, Cecala AB. U.S. Department of Health and Human Services, Centers for Disease Control and Prevention, National Institute for Occupational Safety and Health, Report of Investigations 9677, DHHS (NIOSH) Publication No. 2009–103.

PEBCO [2010]. [http://catalog.pebco.com/category/dustless-loading-spout]. Date accessed: October 8, 2010.

SLY, Inc. [2010]. [http://slyinc.com/loading-spouts/xp-bulk-material-loading-spouts/]. Date accessed: October 8, 2010.

Wypych P [2009]. Award winners can solve grain's dust problems. Australian Bulk Handling Review. [http://www.bulkhandling.com.au/news/april-8th-09/award-winners-can-solve-grain2019s-dust-problems-says-wypych]. Date accessed: October 13, 2010.

CHAPTER 8: CONTROLS FOR SECONDARY SOURCES

CHAPTER 8: CONTROLS FOR SECONDARY SOURCES

This chapter discusses various types of secondary dust sources frequently found in mineral processing facilities and that may have a significant impact on workers' respirable dust exposure. In some cases, the worker's personal respirable dust exposure can be more significant from these secondary sources than from the exposure resulting from the worker's primary job function. Because of this, these dust sources need to be recognized, identified, and controlled. Not only does this chapter identify these sources, but it provides methods and techniques that are available to minimize dust exposure from these sources.

CLEANING DUST FROM SOILED WORK CLOTHES

A significant area of respirable dust exposure to workers at mineral processing operations is from contaminated work clothing. It has been documented that the level of contamination from dust-laden work clothes was so significant in some cases that it caused a ten-fold increase in a worker's respirable dust exposure [USBM 1986]. It must be noted also that once clothing becomes contaminated, it is a continual source of dust exposure until being changed or cleaned.

One option that operations can use to deal with contaminated work clothing is to use dust-blocking coveralls that can be worn by workers. The coveralls can be removed before the worker enters a clean room/area, then discarded at the end of the shift to minimize dust being transferred to the worker's personal clothing. These types of coveralls are also very popular when dealing with different types of toxic dusts. One drawback with these coveralls is that they become uncomfortable for workers because they do not dissipate heat very well and a secondary consideration is the cost.

Currently in the United States, the only Mine Safety and Health Administration-approved (MSHA) method for cleaning work clothes requires the use of a HEPA-filter vacuuming system, which makes the clothes cleaning effort both time-consuming and unlikely to be performed effectively. It is very difficult for a worker to effectively vacuum his/her clothing, particularly in hard-to-reach areas such as the legs and back.

The use of compressed air to remove dust from work clothing is prohibited by MSHA; however, because the vacuuming method is very time-consuming and ineffective, workers may attempt to use this illegal method of a single compressed air hose to blow the dust from their clothing. Although this is a slightly more effective cleaning method than the vacuuming technique, it is also time consuming and equally as difficult to clean the same hard-to-reach areas. The primary concern with the blowing technique is that it creates a dust cloud, elevating respirable dust levels for both the worker and coworkers in the work environment [Pollock et al. 2005].

Once a worker's clothing becomes contaminated, dust is continually emitted from the material as the worker performs his/her normal work activities. Given this problem and the reality that workers can be exposed to multiple dust sources throughout the day, the most effective solution is to change the contaminated clothing. Although disposable coveralls have been in use for many years, as well as some new and improved clothing material that is less susceptible to dust retention, the vast majority of miners continue to wear standard work clothes (typically demin) that they have worn for years [Langenhove and Hertleer 2004; Hartsky et al. 2000; Salusbury 2004]. To address the above issues, a quick, safe, and effective method was developed that allows workers to clean their dust-contaminated clothing throughout the workday.

NEW TECHNOLOGY FOR CLEANING DUST FROM WORK CLOTHING

A cooperative research effort between Unimin Corporation and the National Institute for Occupational Safety and Health (NIOSH) led to the development of a new clothes cleaning system. This technology was not intended to eliminate the need to launder work clothing, but to provide an interim solution to allow workers to safely remove dust from their work clothing periodically throughout the day until laundering could be performed.

The clothes cleaning system developed under this cooperative research effort consists of four major components: (1) a cleaning booth, (2) an air reservoir, (3) an air spray manifold, and, (4) an exhaust ventilation system. Figure 8.1 represents the design of the clothes cleaning system.

In order to perform the clothes cleaning process with this new system, a worker wearing personal protective equipment (respirator, eye, and hearing protection) simply enters the cleaning booth, activates a start button and rotates in front of the air spray manifold while dust is blown from the clothing via forced air. After a short time period (18 seconds), the air spray manifold is electronically deactivated and the worker can exit the booth with significantly cleaner work clothing.

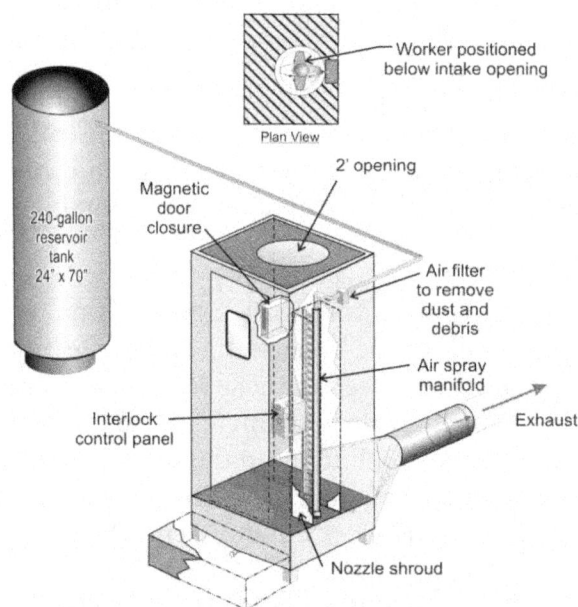

Figure 8.1. Design of clothes cleaning system.

Clothes Cleaning Booth

The clothes cleaning booth is 48 by 42 inches which provides the worker with sufficient space to rotate in front of the air nozzles to perform the cleaning process. Intake air enters the booth through a 2-foot opening in the roof, then flows down through the enclosure before exiting through an air plenum on the bottom back wall of the booth. As this intake air flows through the booth, it entrains dust removed from the worker's clothing during the cleaning process and forces it down towards the air plenum and away from the worker's breathing zone. The exhausted dust-laden air then travels from this air plenum at the base of the booth to the exhaust ventilation system.

Air Reservoir

The air reservoir is necessary to supply the required air volume to the air nozzles used in the spray manifold. The size requirement for the reservoir was calculated based upon the design of the manifold. Either a 120- or 240-gallon reservoir should be used, depending on the number of workers needing to clean their clothes in sequence. The average cleaning time required during field testing was about 18 seconds, and the 120-gallon reservoir provides approximately 22 seconds of air capacity. If multiple individuals will be using the booth one after another, then the 240-gallon reservoir should be used. This reservoir should be pressurized to at least 150 psi and it should be located close to the cleaning booth and hard-piped to the air spray manifold located in the booth. A pressure regulator must be installed immediately before the air spray manifold to regulate the nozzle pressure to a maximum of 30 psi.

Air Spray Manifold

Figure 8.2 represents the air spray manifold design. The air spray manifold is composed of 26 spray nozzles spaced 2 inches apart. The bottom nozzle is located 6 inches from the floor and is a circular designed nozzle for cleaning the worker's boots. This nozzle is used in conjunction with an adjustable ball-type fitting so that it can be directed downward. The other 25 air spray nozzles are flat fan sprays, which lab testing proved to be the most effective for cleaning at close distances. The air spray nozzles deliver slightly less than 500 cfm of air.

The original air spray manifold was designed for an individual around 5 feet 10 inches, the average height for a male worker in the United States. For shorter workers, a sliding mechanism is used to cover the top air nozzles to prevent discharged air from directly hitting the individual's face.

An in-line filter should also be incorporated between the compressed air supply and the air spray manifold. This filter substantially reduces the potential that foreign material such as a metal burr or rust particle could be blown from an air spray nozzle during cleaning.

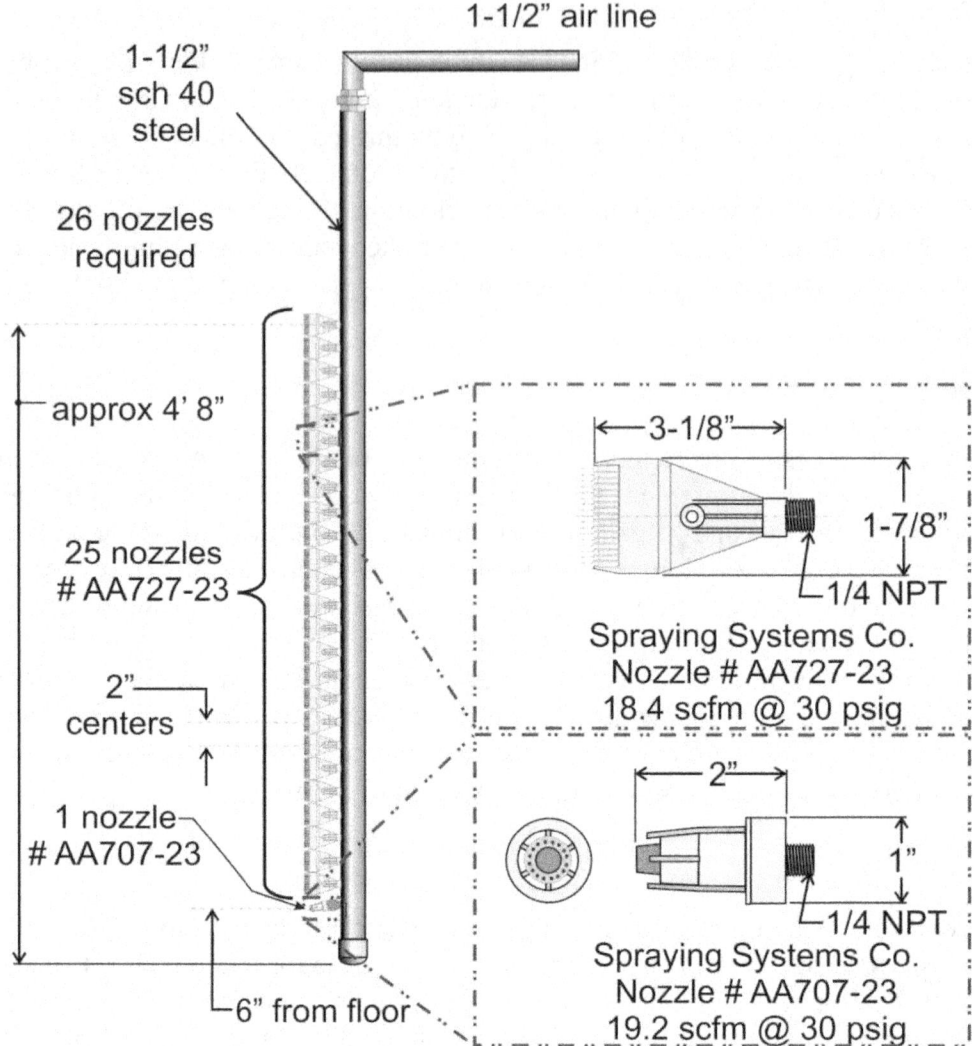

Figure 8.2. Air spray manifold design with 26 nozzles spaced 2 inches apart.

Exhaust Ventilation

For effective dust removal, it is critical that the cleaning booth be ventilated under negative pressure at all times so as to not allow any dust liberated from the clothing to escape from the booth and into the work environment. In the design stage of this technology, testing validated that an exhaust volume of 2,000 cfm was sufficient to maintain a negative pressure throughout the entire clothes cleaning cycle. A pressure control switch is also incorporated into the system which will not allow the air spray manifold to activate unless an acceptable negative pressure is maintained within the cleaning booth.

To exhaust the dust-laden air from the cleaning booth, there are two recommended methods: (1) the dust can be vented to a dust collector system (e.g., baghouse [Cecala et al. 2007]), or (2) the dust can be vented to an area outside the plant where it will not contaminate other workers nor be entrained back into the plant [Cecala et al. 2008]. There are many dust sources outside all mineral processing plants, and the amount of dust-laden air ducted outside for this application is minor in comparison to other sources.

Safety Issues

All personnel entering the cleaning booth must wear personal protective equipment, including an approved fit-tested respirator with filters that are acceptable for the material being processed, and hearing and eye protection. In addition, based upon the Occupational Safety and Health Administration (OSHA) mandated pressure regulation, the air pressure at the air spray manifold must be regulated to a maximum of 30 psi before being directed toward the individual performing the clothes cleaning process.

EFFECTIVENESS OF CLOTHES CLEANING TECHNOLOGY

During development, the new clothes cleaning technology was compared to both the MSHA-approved vacuuming approach and a single handheld compressed air hose. In this testing, several 100-percent cotton and cotton/polyester blend type coveralls were soiled with limestone dust before the worker entered the clothes cleaning booth. Table 8.1 shows the three cleaning methods, their effectiveness in cleaning dust off the two types of coveralls, and the time it took the worker to perform the cleaning task. Clearly, the use of the clothes cleaning system was much more effective at removing the dust than the other two approaches, and required only a fraction of the time, 17 to 18 seconds compared to an average of 372 seconds for vacuuming and an average of 178 seconds for the handheld compressed air hose. Table 8.1 also shows that the polyester/cotton blend coveralls were cleaned more effectively than the 100 percent cotton type. Figure 8.3 depicts a worker before and after entering the clothes cleaning booth.

Table 8.1. Amount of dust remaining on the coveralls after cleaning, and the cleaning time for 100 percent cotton and polyester/cotton blend coveralls

Cleaning method	100% Cotton		Polyester/Cotton blend	
	Dust remaining on coveralls, grams	Cleaning time, seconds	Dust remaining on coveralls, grams	Cleaning time, seconds
Vacuuming	63.1	398	45.5	346
Air hose	68.8	183	48.4	173
Clothes cleaning booth	42.3	17	21.9	18

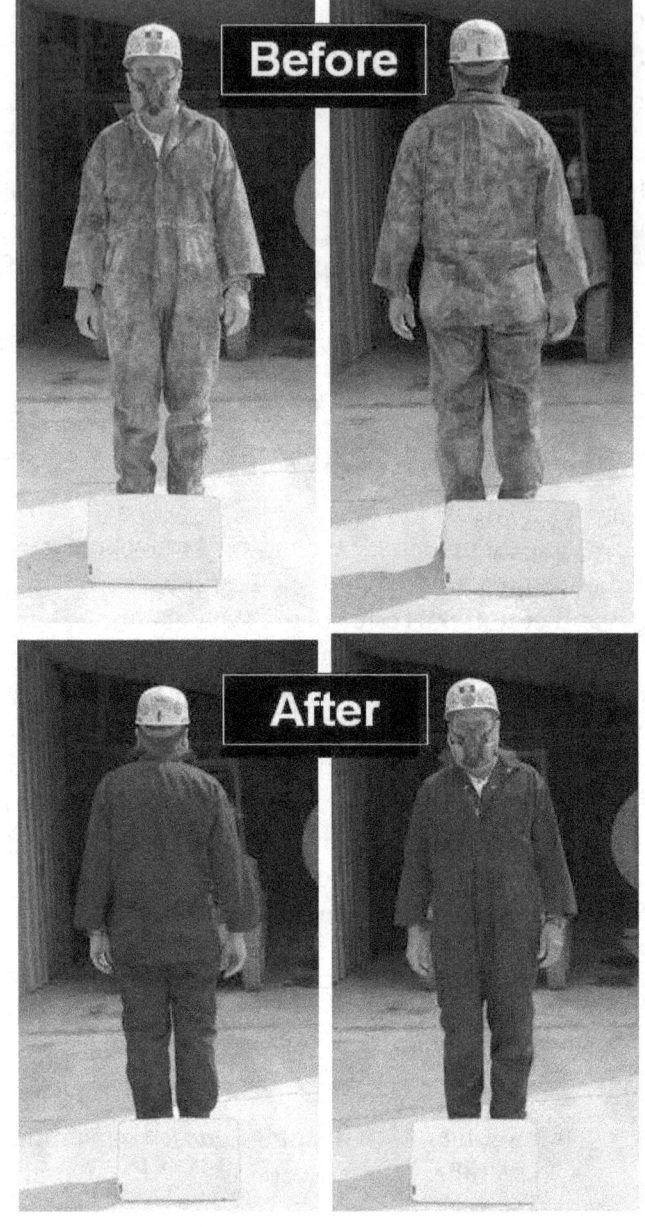

Figure 8.3. Test subject wearing polyester/cotton blend coveralls before and after using the clothes cleaning booth.

CURRENT STATUS OF CLOTHES CLEANING REGULATIONS

In the U.S., two federal regulations currently impact the clothes cleaning process. The first regulation is an MSHA standard 30 CFR Part 56.13020, which states: "At no time shall compressed air be directed toward a person. When compressed air is used, all necessary precautions shall be taken to protect persons from injury." A deflector mechanism is incorporated into the air spray manifold which allows the top air nozzles to be covered over when shorter individuals are performing the cleaning process. This only allows the nozzles to impact the worker's clothing and not the worker's face.

A second regulation is a general industry standard established by OSHA in 29 CFR 1910.242(b), which states: "Compressed air shall not be used for cleaning purposes except where reduced to less than 30 psi and then only with effective chip guard and personal protective equipment." This regulation established the 30 psi limit for the maximum air pressure for the air spray manifold.

Because of these two federal standards, operations must file a petition of modification to their MSHA District Manager to obtain approval to use this clothes cleaning system at their facility. Since this system has already been approved for use at other operations, obtaining approval for similar systems should not be difficult. Manufacturers producing the clothes cleaning system can be contacted by those interested in taking advantage of this technology.[5]

HOUSEKEEPING PRACTICES

Although good housekeeping practices might seem to be a minor or common sense issue, they can be a significant factor in a worker's respirable dust exposure at mineral processing operations. When housekeeping is performed properly and on a scheduled time frame, it can be a key factor in minimizing respirable dust exposures to workers. On the other hand, when it is not performed, or performed improperly, it can have just the opposite effect. In one study, an example of this was documented when a worker was dry sweeping the floor with a push broom. In this instance, the average exposure of a coworker located one floor up from the person sweeping the floor increased from 0.03 mg/m^3 before sweeping to 0.17 mg/m^3 during, and immediately after, the occurrence (Figure 8.4). Dry sweeping is discouraged as a viable cleaning technique because of the dust generated and liberated into the work environment [USBM 1986].

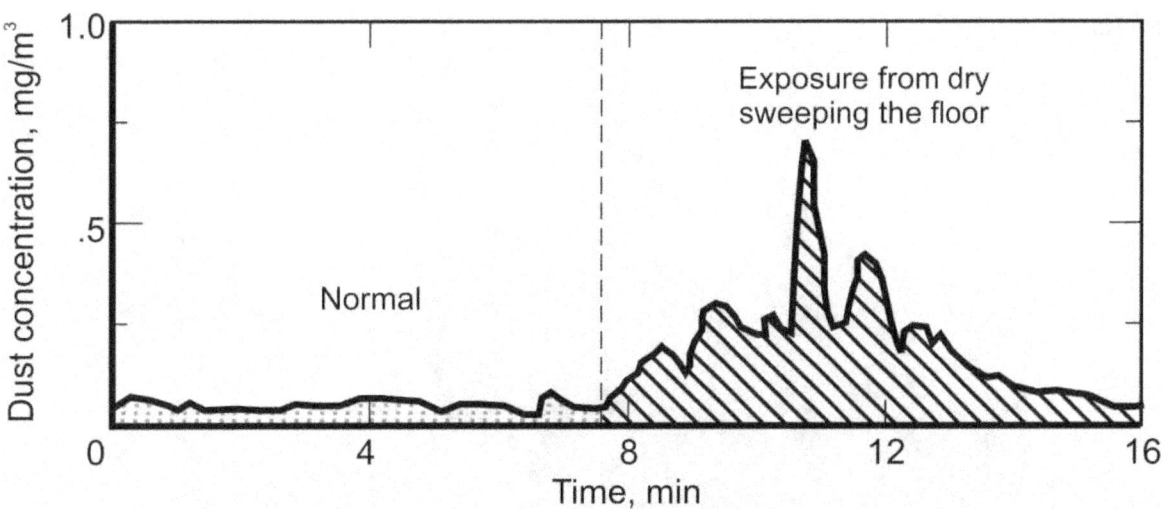

Figure 8.4. Increase in coworker's dust exposure one floor up from floor sweeping activities.

---

[5] One manufacturer of the clothes cleaning booth technology is Clothes Cleaning Systems (http://www.cleanclothbooth.com/), which is selling the system that was developed under this cooperative work effort.

Spills of product material are a common occurrence at processing operations and when they occur, the initial removal of material is most often performed with a scoop, bobcat, or through having workers using shovels. When this task is performed manually, the appropriate respiratory protection must be worn by workers.

Once the bulk of this material is removed, or for general housekeeping, it is very effective to wash down the facility with water on a regular basis. Ideally for this to work, the design should be incorporated in during construction so that drains are installed and the floor is sloped correctly to ensure that the water runs toward the drains. Although washing down is very effective, it must also be noted that this can unintentionally increase respirable dust levels during the hosing process. The problem occurs when water strikes piles of accumulated fine material, which can cause dust to be liberated into the air. When washing down is being used as a housekeeping technique, it should be done shift by shift, or at least daily, to keep product from accumulating.

In addition to washing down, the next most common housekeeping techniques are the use of floor cleaning devices and/or vacuum systems. Although both of these techniques are much more costly than washing down, neither typically generate or liberate significant levels of dust during the cleaning process, as is the case with washing down.

When considering floor cleaning options, there are many different manufacturers and companies that sell a vast array of different types of floor cleaning units. These units vary from ones that an individual rides to smaller units that an individual walks behind (Figure 8.5). One area that should be closely investigated is the disposal of the contaminated material once it has been collected. It would be counterproductive if the unit is very efficient during the floor cleaning process but contaminates employees when the unit is being cleaned or during disposal of the accumulated material.

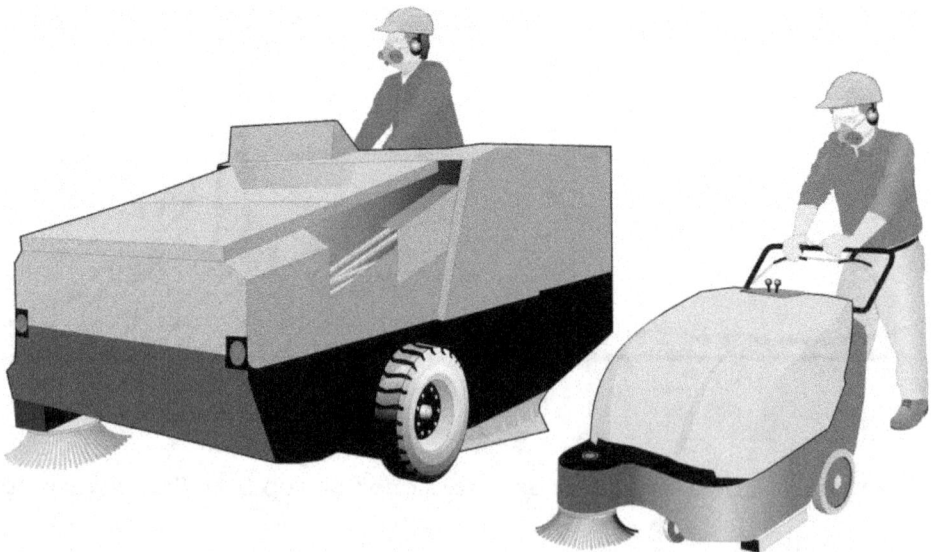

Figure 8.5. Various types of floor sweep units used to clean dust-laden floors.

In addition, there are many different types of vacuuming units. These can range from high-capacity, in-plant vacuuming systems with multiple collector pickup locations to portable tank-type units. Some operations also contract this work to independent companies who bring in high-powered exhaust trucks to perform the vacuuming process. In almost all cases, a worker uses an exhaust system collector device to manually exhaust the product and dust into the collector system (Figure 8.6). This is very time consuming and can be difficult to perform in areas that are not easily accessible in the plant. In most cases, this is not something that can be performed on a shift-by-shift or daily basis because it is so labor-intensive. One safety issue with these vacuuming devices is from static electricity charges that can be created while performing this cleaning technique. Using a nonconductive or grounded hosing material should eliminate the risk of this safety issue.

Figure 8.6. Portable system used to perform vacuuming.

In addition to hosing with water, sweeping, or using mechanical devices to clean the floors at mineral processing facilities, one last option that can be considered is to use a floor sweep compound. Although it is not a common practice to use a floor sweep compound throughout an entire facility, some operations are using it in high traffic or extremely dusty areas to assist in minimizing the dust generated. Since there are many different manufacturers selling floor sweep compounds, operations may want to consider evaluating some different types of material in some problem areas to determine the effectiveness obtained within their facilities.

Housekeeping should be performed at the end of each shift so that workers coming on to the next shift start with a clean work environment. When housekeeping is performed at the beginning of a shift, it increases the potential for worker clothing to become contaminated, which will increase the worker's personal dust exposure throughout the work shift.

One final aspect of effective housekeeping is proper upkeep and maintenance of plant equipment and processes. When product is observed building up on floors, it indicates that some function or process is causing the leakage. In some cases, visible dust can be seen leaking from holes or damaged equipment, and this must be quickly corrected to minimize dust leakage. In addition, repetitive dust leaks should not simply be, "patched," but rather, the underlying cause should be determined and corrective action taken to both fix the leak and prevent future reoccurrences.

Each process or function within mineral processing operations should have a Standard Operating Procedure (SOP) that includes routine inspections, maintenance, and cleaning schedules. When equipment is cleaned and maintained on a routine basis, it can have a significant impact on minimizing dust liberation and generation.

TOTAL STRUCTURE VENTILATION DESIGN

The first strategy to lower dust exposures in any structure is to have an effective primary dust control plan that captures major dust sources at their point of origin, before they are allowed to escape out into the plant and contaminate workers. Most mineral processing plants control the dust by using standard engineering controls such as local exhaust ventilation systems, water spray applications, scrubbers, and electrostatic precipitators (refer to Chapter 1—Fundamentals of Dust Collection Systems). These engineering controls are effective at reducing and capturing the dust generated and liberated from the primary sources at mineral processing plants; however, they do not address the continual buildup of dust from background sources, such as the following:

- product residue on walls, beams, and other equipment which becomes airborne from plant vibration or high-wind events;
- product that accumulates on walkways, steps, and access areas, and which may be released as workers move through the plant and by plant vibration;
- leakage or falling material from chutes, beltways, and dust collectors;
- lids and covers of screens that are damaged or when they are removed for inspection and cleaning; and
- product released because of imperfect housekeeping practices.

Since most mineral processing structures can be considered closed systems, the background dust sources stated above, along with numerous other unnamed sources, can cause dust concentrations to continually increase as the day or shift progresses inside these closed buildings. In order to keep dust levels at safe and acceptable levels, a method needs to be used to control these background dust sources. The most effective way to achieve this is to use a total structure ventilation system.

PRINCIPLES OF TOTAL STRUCTURE VENTILATION

The basic principle behind the total structure ventilation design is to use clean outside air to sweep up through a building to clear and remove the dust-laden air. This upward airflow is achieved by placing exhaust fans at or near the top of the structure and away from plant personnel working both inside and outside the structure. The size and number of exhaust fans is determined based upon the initial respirable dust concentration and the total volume of the structure. It must be noted that all the processing equipment within a mill generates heat and this produces a thermodynamic "chimney effect" that works in conjunction with the total structure ventilation design [Cecala and Mucha 1991; Cecala 1998; Cecala et al. 1995]. Figure 8.7 shows the concept of the total structure ventilation design. To be effective, the total structure ventilation design must meet three criteria, which are: to have a clean makeup air supply, an effective upward airflow pattern, and a competent shell structure.

Clean Makeup Air Supply

Intake air needs to be brought in at the base of the structure by strategically located wall louvers or open plant doors. It is very important that this air be free from outside dust sources such as bulk loading, high traffic areas, etc., which could cause the dust-laden air to be drawn into the structure and could increase respirable dust levels. Through the use of wall louvers or closing doors, the intake air locations could be changed based upon the outside dust conditions.

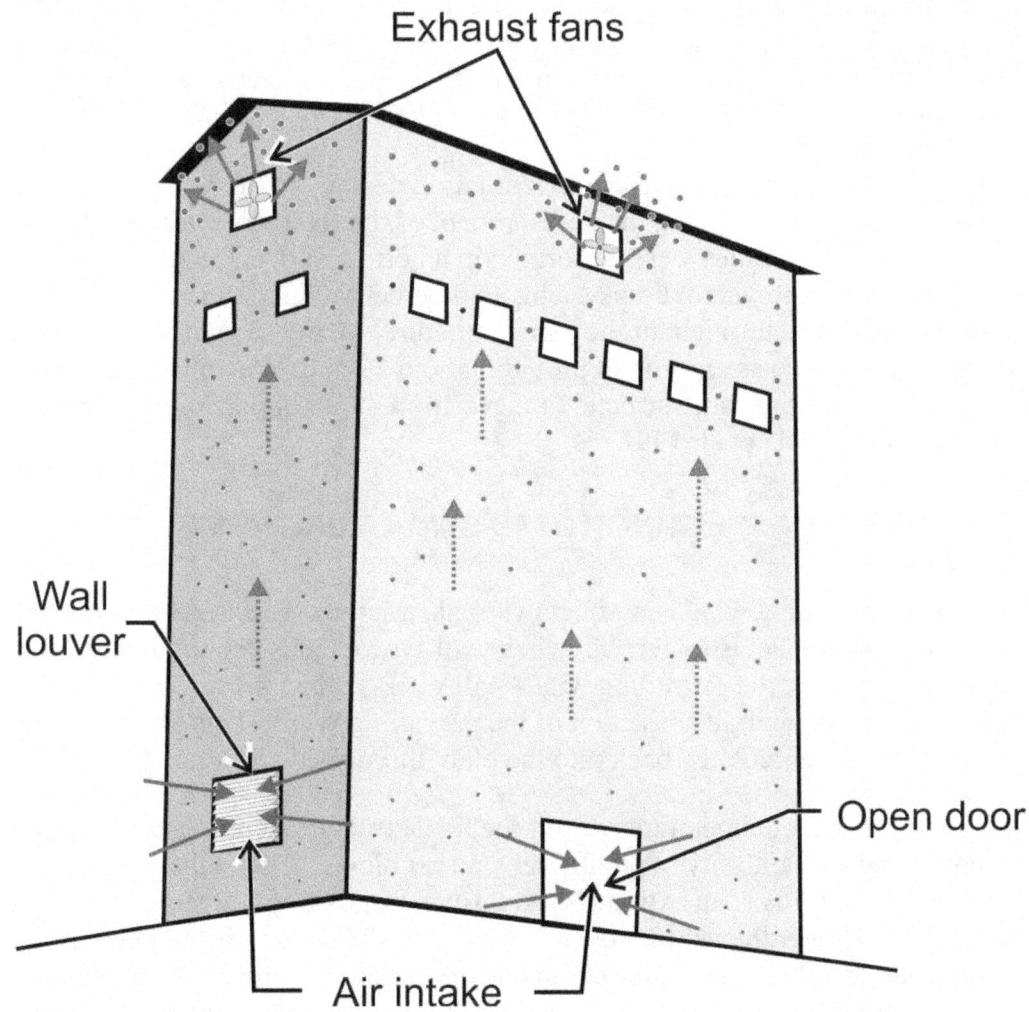

Figure 8.7. Basic design of total structure ventilation system.

Effective Upward Airflow Pattern

The system should provide an effective upward airflow pattern that ventilates the entire structure by sweeping the major dust sources and work areas. Strategic positioning of both exhaust fans (high in the walls or roof) and makeup air intakes (at the base) create the most effective airflow pattern for purging the entire building. It must be noted that the total structure ventilation is not applicable to buildings with multi-story solid floors since solid floors do not allow the air to flow up through the structure.

Competent Shell of Structure

A competent outer shell of the structure is necessary because the ventilation system draws the makeup air through the points of least resistance. Therefore, the structure's outer shell must be free of open or broken windows, holes, cracks, and openings, especially in the vicinity of the exhaust fans. The exhaust fans create a pressure, and if the structure is not competent, air will be brought in from unwanted dust sources and will not allow for an effective airflow pattern for purging the building.

A normal range of airflow for a total structure ventilation design would be in the 10–35 air changes per hour (acph) level. During the development of the total structure ventilation design concept, two different field studies were performed in an effort to document its effectiveness. In the first study, with a 10-acph ventilation system, a 40 percent reduction in respirable dust concentrations was achieved throughout the entire structure. In the second study, the total structure ventilation system was capable of providing both 17 and 34 acph. Average respirable dust reductions throughout the entire structure ranged from 47 to 74 percent for the two exhaust volumes, respectively [Cecala et al. 1995].

VARIABLES AFFECTING TOTAL STRUCTURE VENTILATION

The direction of the prevailing wind can affect the total structure ventilation design. When roof exhausters are used, the impact is minor; but when wall-type exhausters are used, fans should not be placed in locations where the prevailing winds will work against them. The ideal design is that the fan should exhaust with the direction of the prevailing wind. This also minimizes the possibility of dust being recirculated back into the plant through the makeup air.

When extremely low winter temperatures occur, the total structure ventilation system may somewhat increase the vulnerability to plant freeze-up problems. The makeup air can be heated to avoid this problem, but this is an extremely expensive proposition. A more affordable and realistic approach is to lower the ventilation volume during freeze-up air temperatures. This can be achieved by using variable-speed control fans or selectively turning off fans in a multiple fan design. When lowering the ventilation volume is not an option, operations may want to consider providing localized heating to individual work areas/stations.

Costs

The total structure ventilation design has proven to be one of the most cost-effective systems to lower respirable dust concentrations throughout an entire closed mineral processing structure. Not only is the initial cost of this technique inexpensive when compared to other engineering controls, but its operation and maintenance are also minimal. Operations can install all the components for this system in-house to reduce their costs.

OPEN STRUCTURE DESIGN

The outside ambient environment can be an effective source of ventilation to dilute and carry away dust generated and liberated within a structure. Therefore, in some cases, an open-structure design can be used as an optimal way to provide for whole structure ventilation. The open design is similar to the total structure ventilation design in that it has the potential to be a global approach to lower dust levels throughout an entire building or structure. In the open-structure design, the natural environment acts as an effective method of diluting and carrying away dust liberated during operations in mineral processing structures. No dust plume should ever be visible with an open-structure design. As with the total structure ventilation system, the first goal is to capture the major dust sources at their point of origin, before they are allowed to escape out into the plant and contaminate the work environment. The open-structure design is a secondary technique to control the residual dust not captured at the source.

In a study that compared three different building types: masonry, an open-structure design, and a steel-sided design, respirable dust concentrations were significantly lower in the open-structure building [Cecala et al. 2007; USBM 2006]. Figure 8.8 shows a conceptual drawing of a typical open-structure design with a roof in comparison with an identical sized walled processing facility. Obviously, a roof provides a little more protection from the natural elements than a totally open design. If an open-structure design is considered for an operation, a number of issues need to be addressed:

- Product residue that becomes airborne from plant vibration or high-wind events can accumulate on walls, beams, and other equipment.
- An open-structure design must be considered a secondary design. Engineering controls are needed to eliminate the major dust sources at their point of origin before being liberated into the plant and contaminating the work environment.
- A great effort must be made to provide safety railings and guards to minimize the potential for any personnel falling from the structure.
- Equipment and personnel must be protected from environmental elements such as rain, snow, sleet, and hail. One possibility to minimize this concern would be to design a structure with a roof and sufficient overhang. In addition, equipment freeze-up problems can happen when low outside air temperatures occur.

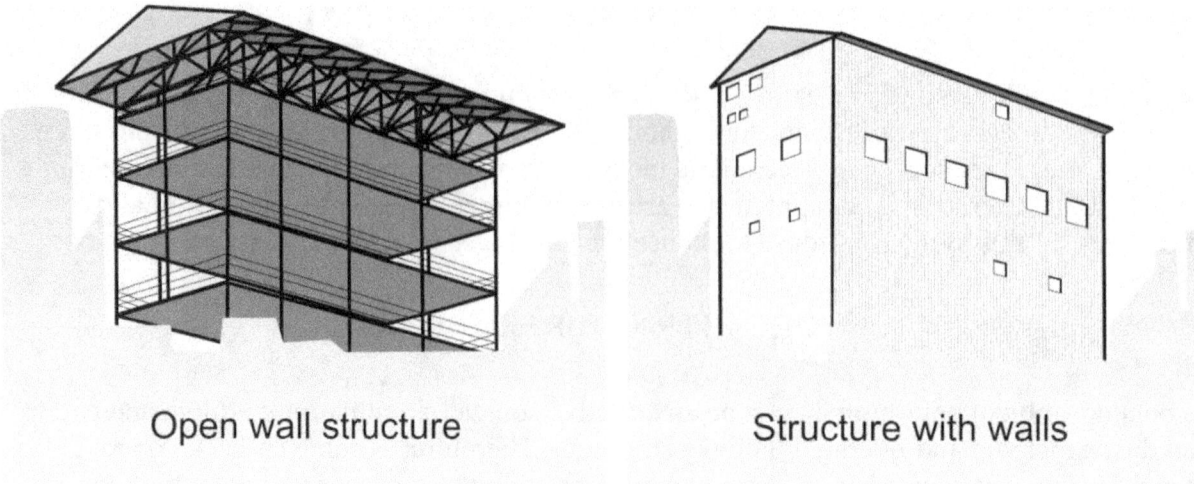

Figure 8.8. Open-structure design compared to standard walled structure.

From a federal standard basis, the only consideration would be the Environmental Protection Agency's opacity dust measurement, which is a qualitative measurement, taken by a federal regulator based on the presence of a visible dust plume. If any plumes are visible from an operation using an open-structure design, primary dust control methods need to be implemented to control the problem.

In addition to lowering dust levels throughout an entire building or structure, the open-structure design has also been determined to lower worker's noise exposure because it eliminates the reverberation effects within a closed structure. When building new facilities, obviously the open-structure design is more cost-effective because it lowers material and construction costs. Some companies may also want to consider modifying their existing structure with a more open design to further reduce dust and noise levels.

BACKGROUND DUST SOURCES

When a worker is overexposed to respirable dust, most often the assumption is that the dust exposure came from the worker's primary job function. A study was performed that documented a number of instances where this was not the case. When a worker obtains a high respirable dust measurement, the correct course of action is to evaluate the worker's job function, to determine the dust sources that contributed to this exposure, and the magnitude of the exposure from each of these sources. Sometimes a secondary or background dust source can be the major contributor to the worker's overall exposure. Controlling these less obvious dust sources can have a major impact on lowering the worker's respirable dust exposures.

EXAMPLES OF SECONDARY DUST SOURCES

The following are a few examples that were documented in actual field studies and will demonstrate the impact that secondary dust sources can have on a worker's respirable dust exposure [Cecala and Thimons 1987; USBM 1988; Cecala and Thimons 1992].

Dusts from Outside Sources Traveling Inside Structures

When dusts from outside sources travel inside structures, every worker inside the structure is impacted. Most bagging operations at mineral processing plants use an exhaust ventilation system to draw the dust generated from the bagging process down into the fill hopper and away from the breathing zone of the bagger (refer to Chapter 6—Bagging). It is important that the air being drawn into this exhaust ventilation system, commonly called makeup air, be clean air. At one operation, the makeup air was drawn directly from the bulk loading area outside the mill. The dust generated from this bulk loading process traveled through an open door into the mill, substantially contaminating the workers inside the mill. During periods when bulk loading was not performed, the bag operator's average dust exposure was 0.17 mg/m^3. As trucks were loaded at the bulk loading area, the worker's average exposure increased to 0.42 mg/m^3 due to this contaminated air (Figure 8.9). As this study demonstrates, if outside air is used as makeup air, it must be from a location where the air is not contaminated.

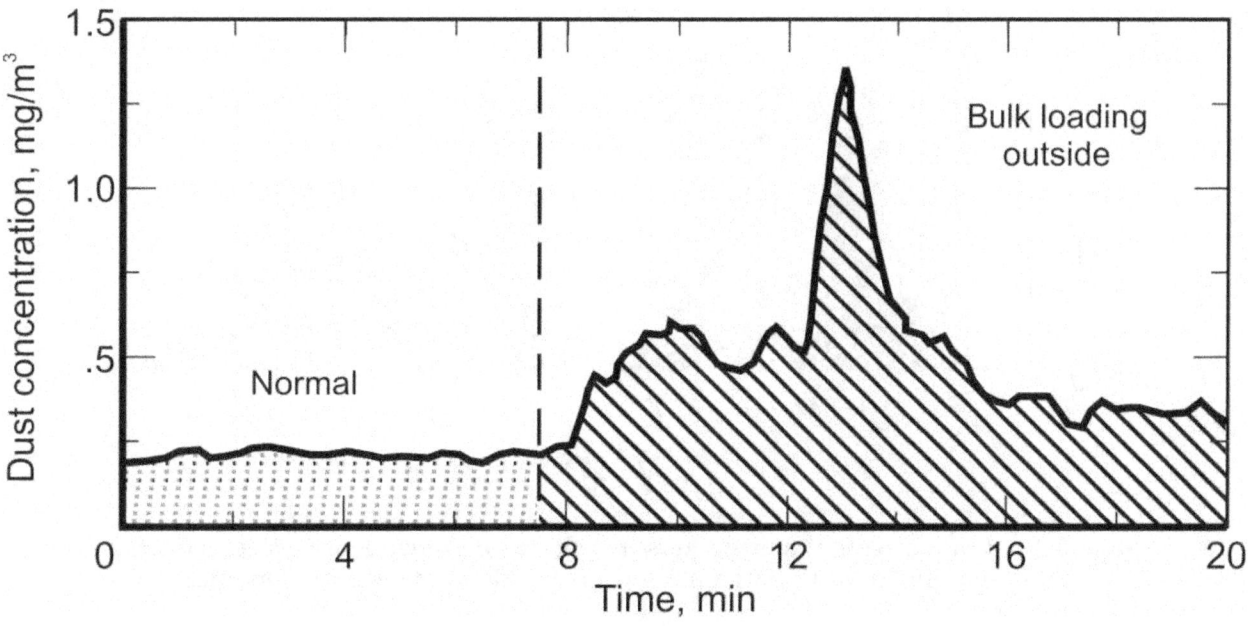

Figure 8.9. Increase in a worker's respirable dust exposure inside a mineral processing facility while bulk loading was being performed outside.

Personal Dust Exposure from Differences in Job Function Work Practices

During an evaluation of four different dust control systems at a mineral processing operation, it was determined that substantial variations occurred in the respirable dust exposure of two workers based upon their personal work practices in performing their same job function. In this case, these two workers were loading bags of product on a bag filling machine and then manually removing the bags and placing them onto a conveyor belt. Figure 8.10 shows a comparison of the respirable dust exposure of these two workers while performing the same job function when different control techniques were implemented (system type 1–4). The point to be highlighted from this testing was that regardless of the system being evaluated, worker #2's respirable dust exposure was always significantly less than that of his coworker (worker #1). The main reason for these differences was worker #1 performed his job function in a rough manner relative to worker #2, who performed the exact same tasks in a very conscientious or careful manner. Ultimately, this resulted in worker #2's respirable dust exposure being approximately 70 percent less than that of his coworker.

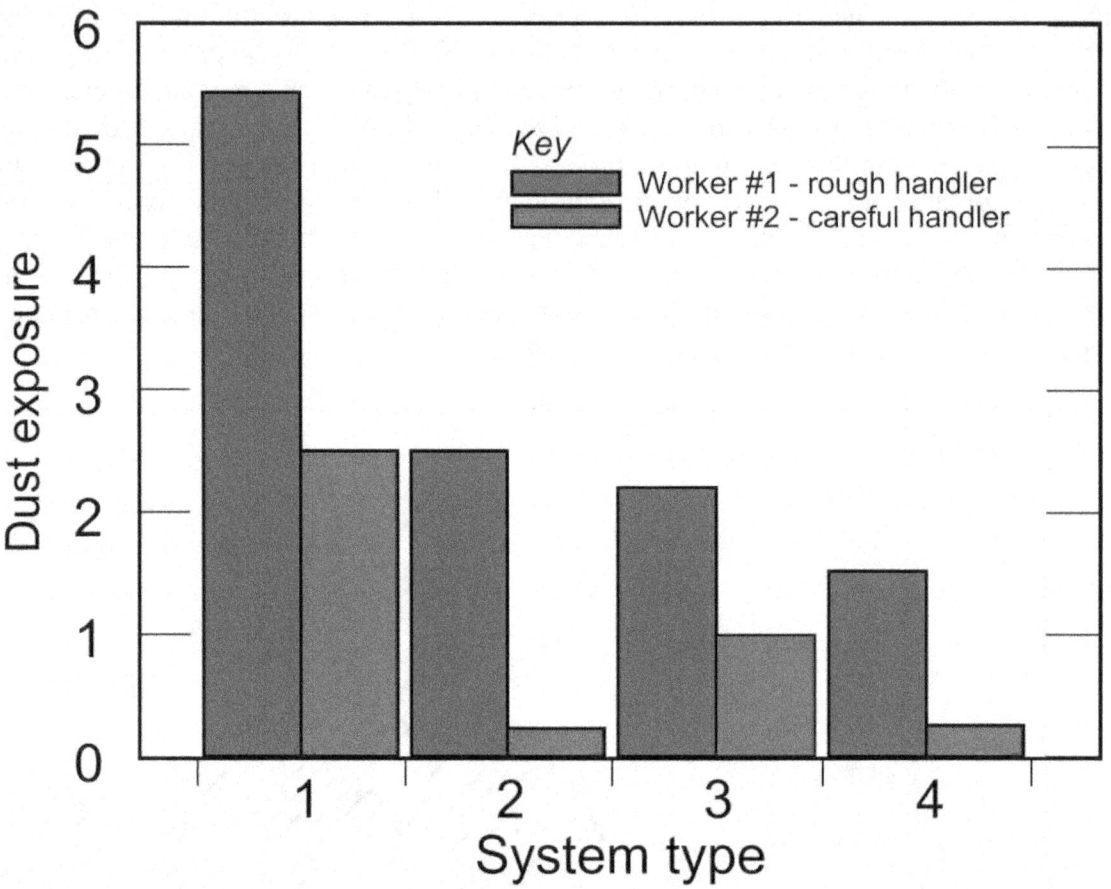

Figure 8.10. Comparison of two workers' respirable dust exposure while performing their job function with different control techniques being implemented.

One key element in how worker #2 performed his tasks was the way he manually closed the bag valve with his hand as he removed the bag from the fill station and transferred it to the conveyor (Figure 8.11). This technique was believed to be instrumental in lowering the worker's respirable dust exposure. Comparing the difference between these two workers indicates the impact that individuals can have on their dust exposure based upon how they perform their job function.

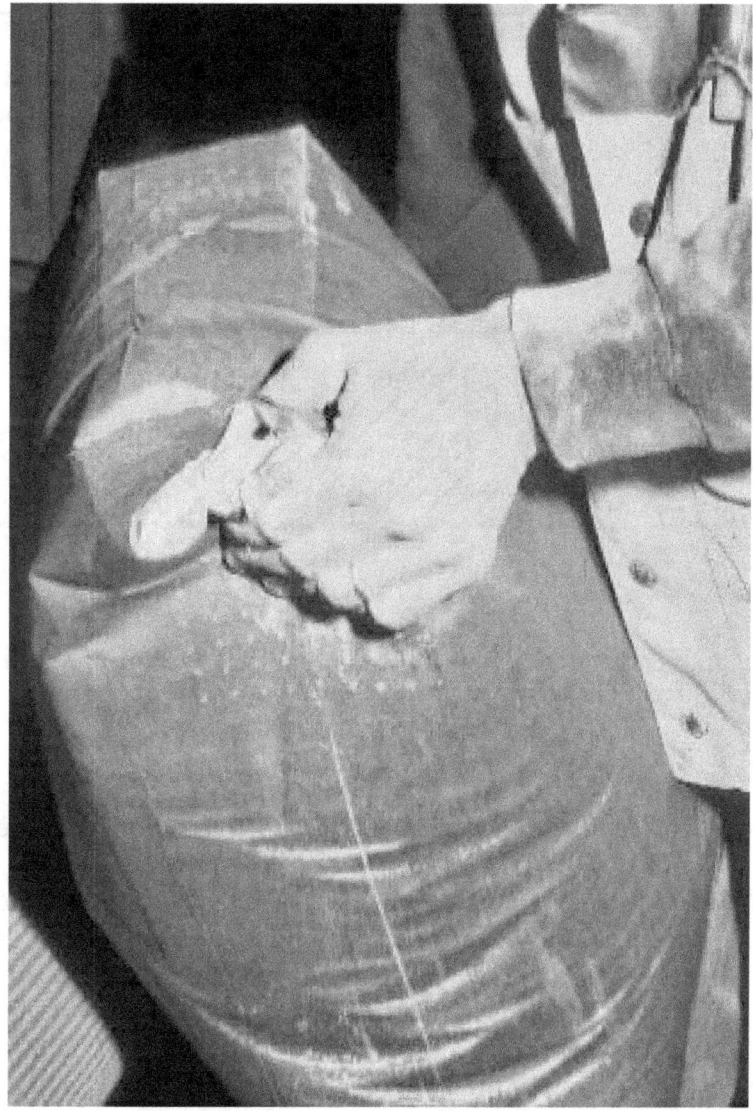

Figure 8.11. Worker closing bag valve with his hand as he removes the bag from the fill nozzle, significantly reducing his dust exposure.

Broken Bags of Product

In most cases, bag breakage occurs because of flaws in the bags delivered from the bag manufacturer. At one particular operation, a bag operator's average dust exposure went from 0.07 mg/m^3 before the bag break to 0.48 mg/m^3 afterwards. Since the bag broke during the conveying process and not directly in front of the worker, the dust substantially contaminated the surrounding mill air, which in turn, flowed over the bag operator (Figure 8.12). This occurred because the exhaust ventilation system in the bag loading area creates a negative pressure that draws background air from the contaminated mill air.

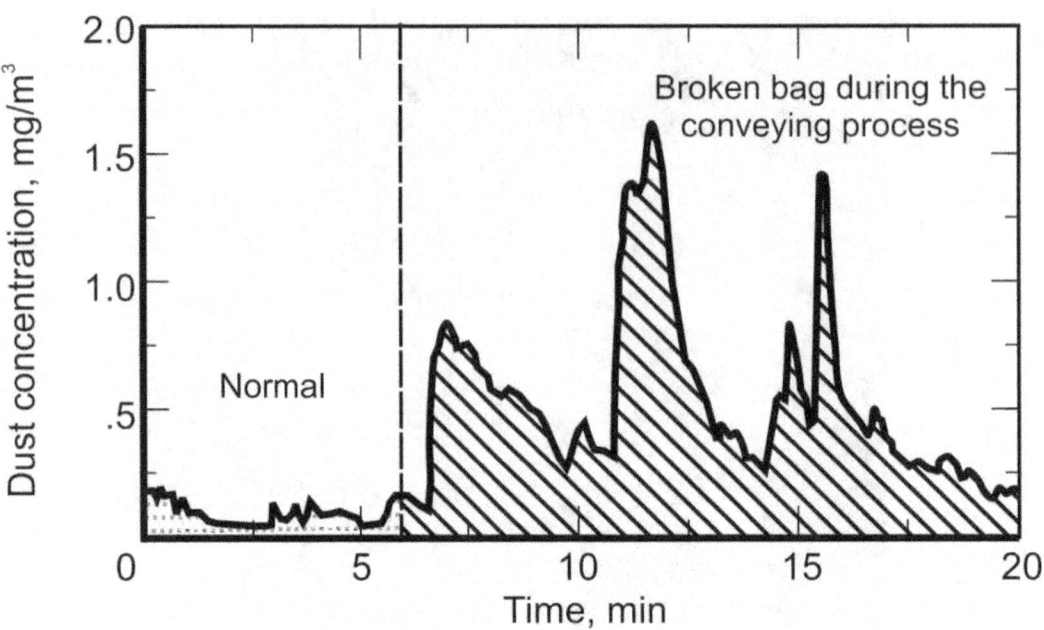

Figure 8.12. Increase in worker's respirable dust exposure from broken bag during the conveying process.

Cloth Seat Material/Worn Acoustical Material

At many locations throughout mineral processing operations, workers are required to sit in chairs to perform their job functions. This includes all mobile equipment operators, control room operators, crusher operators, some bag loading operators, bulk loading operators, fork lift operators, lab workers, as well as other site-specific applications. In addition to these job functions where workers are required to sit, there are many chairs located throughout facilities such as in break rooms, lunch rooms, and other locations, which allow workers an opportunity to get off their feet for a break. When cloth is used for the seat material, it can retain a substantial amount of dust which is liberated each time a worker sits or moves on the material. Because of this, cloth material should not be used and should be replaced with some type of vinyl material. It should also be noted that even with vinyl seats, if the vinyl material has deteriorated to a point where the foam padding is being exposed, this also need to be changed because the foam will also hold and liberate dust.

Material selection can also be a problem when acoustical material is used in equipment, enclosed cabs, controls rooms, operator's compartments, etc. to minimize the effects of noise. As this material gets old and starts to deteriorate, if the foam material becomes exposed, it also becomes a source of dust accumulation and exposure to the worker, similar to cloth or exposed foam in seats or chairs.

Dirty Work Boots

As with dust-laden work clothing, dirty work boots can also be a source of background dust exposure to workers. This occurs when a worker walks through wet product, overburden, and/or mud and it adheres to his/her boots. Obviously, wet conditions are common at mineral processing operations from rain and/or any wet plant processes. Once a worker's boots are soiled with product, it is then tracked everywhere the worker goes until it either all falls off or the boots are cleaned. Figure 8.13 shows product tracked onto the floor of a piece of mobile equipment. It is easy to understand the amount of dust that can be generated when this level of material is accumulated on the floor.

Figure 8.13. Product tracked onto the floor of a piece of mobile equipment from dirty work boots.

In cases where a substantial amount of dust and product gets tracked into a work area, housekeeping should be performed on a daily or shift basis. In other cases where the contamination is not as great, housekeeping could be done on a more infrequent basis, but in any case, it is critical that this contamination is dealt with to minimize the source of dust to workers.

One strategy that was tried and shown to be effective in reducing the amount of dust generated from dirt and product on the floor was through the use of sweeping compounds. A study was conducted to determine the effectiveness of using a gritless, nonpetroleum based sweep compound in a number of enclosed cabs [NIOSH 2001]. Figure 8.14 shows the average respirable dust concentrations measured inside two enclosed cabs for three to five shifts for both test conditions (baseline and sweep compound). The respirable dust levels were reduced from 0.18 and 0.68 mg/m^3 to 0.07 and 0.12 mg/m^3, respectively, for these two drills. These results indicate that the sweeping compound had a positive effect on suppressing dust when it was able to make direct contact with and bind up the dirt on the steel and rubber floor mats.

It should be noted that before any sweeping compound is used, companies should examine the Material Safety Data Sheet (MSDS) to ensure there are no ingredients that would cause problems in an enclosed area. The sweeping compound should also be changed and disposed of properly once it has reached a state where it is no longer containing the dust.

In other cases, an operation may want to consider installing a boot wash or boot cleaning device to minimize the negative effects of dirty work boots. These boot cleaning stations are relatively inexpensive and can be placed at various locations around an operation.

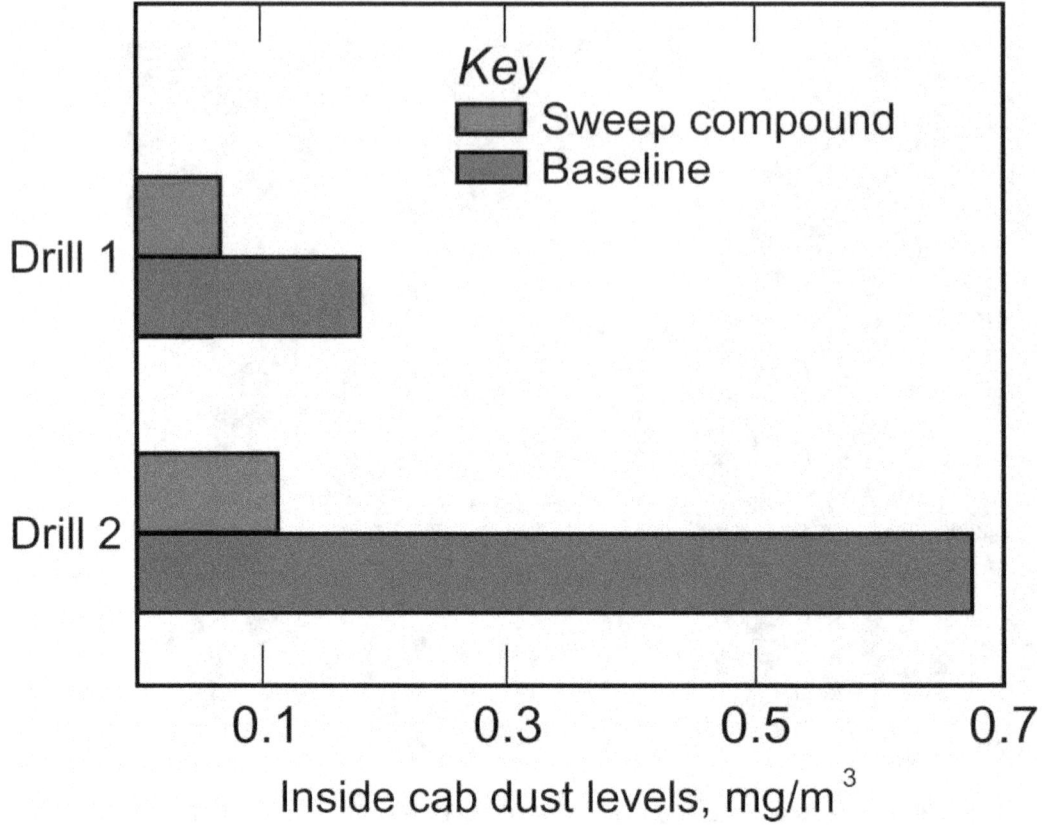

Figure 8.14. Respirable dust concentrations inside enclosed cab of two different pieces of equipment with and without the use of a gritless, nonpetroleum sweep compound.

MAINTENANCE

Maintenance of dust control systems is critical to ensure continuing worker protection [USBM 1974]. Preventive maintenance schedules should be established based on observed component wear, system performance measures, and exposure sampling results.

Work Practices to Reduce Dust Exposure during Maintenance Activities

Workers performing maintenance can be exposed to higher than normal dust concentrations because the maintenance task may involve working in areas where dust controls are not functioning effectively, working on highly contaminated equipment, or the work may have to be performed in unusual nonroutine operating conditions.

- Clean equipment and work areas when needed by washing or vacuuming before and during maintenance work. Working in dusty work areas results in increased dust exposure to maintenance workers.

- Maximize the use of remote access fittings such as extension grease lines. This practice lessens the potential for dust exposure to maintenance workers from having to remove guards, covers, and other items of equipment that may have surface dust or exposed dusty areas.
- Keep outside intake air and recirculating air dust filters clean and replace when necessary in mobile equipment, cabs, control rooms, and operator booths. Proper attention must be given to the filter efficiency when filter changes are necessary. Refer to Chapter 9—Operator Booths, Control Rooms, and Enclosed Cabs for more information.
- Provide travelways to routine maintenance sites. This eliminates dust exposure from climbing in and around equipment to access maintenance sites.
- Provide lifting and mechanical handling equipment to reduce handling of dust-laden equipment.
- Clean dust-laden work clothing immediately following the completion of maintenance work activity using a clothes cleaning booth system or another approved technique. (Refer to the "Cleaning Dust from Soiled Work Clothes" section, earlier in this chapter.) In addition, disposable coveralls made of dust-blocking material, such as Tyvek® or Kleenguard™, can be worn to greatly reduce dust contamination of work clothing and the transfer of dust to clean work areas or to the worker's home.
- When gloves are worn, a nonporous material that does not retain dust should be used. Using an open weave fabric material that holds the dust is discouraged because of its ability to increase the workers dust exposure.
- Where feasible, wash down mobile equipment prior to entering maintenance bays for conducting maintenance work.
- Avoid the use of compressed air to clean surfaces.
- Periodically clean the inside of mobile equipment, control rooms, and operator booths to remove accumulated product, dust, and debris.

REFERENCES

Cecala AB [1998]. Supplementing your dust control equipment with whole-plant ventilation. Powder & Bulk Engineering *12*(1):19–32.

Cecala AB, Mucha R [1991]. General ventilation reduces mill dust concentrations. Pit & Quarry *84*(1):48–53.

Cecala AB, Thimons ED [1987]. Significant dust exposures from background sources. Pit & Quarry *79*(12):46–51.

Cecala AB, Thimons ED [1992]. Some factors impacting bag operator's dust exposure. Pit & Quarry *85*(5):38–40.

Cecala AB, Klinowski GW, Thimons ED [1995]. Reducing respirable dust concentrations at mineral processing facilities using total mill ventilation system. Min Eng *47*(6):575–576.

Cecala AB, O'Brien AD, Pollock DE, Zimmer JA, Howell JL, McWilliams LJ [2007]. Reducing respirable dust exposure of workers using an improved clothes cleaning process. Inter J of Min Res Eng *12*(2):73–94.

Cecala AB, Rider JP, Zimmer JA, Timko RJ [2007]. Dial down dust and noise exposure. Aggregates Manager *12*(7):50–53.

Cecala AB, Pollock DE, Zimmer JA, O'Brien AD, Fox WF [2008]. Reducing dust exposure from contaminated work clothing with a stand-alone cleaning system. In: Wallace, ed. Proceedings of 12th U.S./North American Mine Ventilation Symposium. Omnipress. ISMN 978–0–615–20009–5, pp. 637–643.

Hartsky MA, Reed KL, Warheit DB [2000]. Assessments of the barrier effectiveness of protective clothing fabrics to aerosols of chrysotile asbestos fibers. Performance of Protective Clothing: Issues and Priorities for the 21st Century: Seventh Volume, ASTM STP 1386, pp. 141–154.

Langenhove LV, Hertleer C [2004]. Smart clothing: a new life. Inter J of Clothing Science and Tech. *16*(1&2):63–72.

NIOSH [2001]. Technology news 487: Sweeping compound application reduces dust from soiled floor within enclosed operator cabs. By Organiscak JA, Page SJ, Cecala AB. Pittsburgh, PA: U.S. Department of Health and Human Services, Centers for Disease Control and Prevention, National Institute for Occupational Safety and Health.

Pollock DE, Cecala AB, O'Brien AD, Zimmer JA, Howell JL [2005]. Dusting off. Rock Products. March. pp. 30–34.

Salusbury I [2004]. Tailor-made-self-cleaning clothing, chemical warfare suits that trap toxins. Materials World. August. pp. 18–20.

USBM [1974]. Survey of past and present methods used to control respirable dust in noncoal mines and ore processing mills–final report. U.S. Department of the Interior, Bureau of Mines Contract No. H0220030. NTIS No. PB 240 662.

USBM [1986]. Impact of background sources on dust exposure of bag machine operator. By Cecala AB, Thimons ED. U.S. Department of the Interior, Bureau of Mines Information Circular 9089.

USBM [1988]. Technology news 299: Reduce bagging machine operator's dust exposure by controlling background dust sources. U.S. Department of the Interior, Bureau of Mines.

USBM [2006]. Lower respirable dust and noise exposure with an open structure design. By Cecala AB, Rider JP, Zimmer JA, Timko RJ. U.S. Department of Health and Human Services, Centers for Disease Control and Prevention, National Institute for Occupational Safety and Health, NIOSH Report of Investigations 9670.

CHAPTER 9: OPERATOR BOOTHS, CONTROL ROOMS, AND ENCLOSED CABS

CHAPTER 9: OPERATOR BOOTHS, CONTROL ROOMS, AND ENCLOSED CABS

Operator booths, control rooms, and enclosed cabs have all been used for many years to isolate workers from dust sources in mineral processing operations. When they are properly designed, installed, and maintained, they can provide a safe work environment that supplies clean and acceptable air quality to the worker. There has been a substantial amount of research performed in this area over the past decade identifying and correcting problematic issues and determining the critical components that are necessary for an effective system.

Workers in operator booths, control rooms, and enclosed cabs at mining operations are surrounded by dynamic working conditions that have highly variable dust sources. These enclosures create a microenvironment for the workers where they can be either more protected or more vulnerable to respirable dusts. Workers can be more vulnerable to in-cab dust sources (floor heaters, dirt on floors/walls, or on operator's clothing, etc.) that are trapped within the enclosure. Enclosed cabs on mobile equipment are actually harder to control and maintain than enclosed stationary areas (operator booths and control rooms) since the moving of the equipment constantly stresses and can compromise the integrity of the enclosure.

Figure 9.1 shows a general design of a filtration and pressurization system. At mineral processing operations, filtration and pressurization systems are the primary engineering control to reduce worker exposure to airborne dust in operator booths, control rooms, and enclosed cabs. These enclosures should have an effective filtration system that cleans both outside and inside air that is brought into the enclosure and sufficient enclosure integrity in order to achieve pressurization. The most effective filtration and pressurization systems are integrated into the heating, ventilating, and air-conditioning (HVAC) units. A substantial amount of research has been performed to improve the air quality in enclosed cabs of surface mining equipment, and this research is directly applicable to operator booths and control room dust control systems.

EFFECTIVENESS TERMINOLOGY

When evaluating a pressurization and filtration system—comparing outside with inside respirable dust concentrations—a number of different descriptors can be used to provide a numerical value that ranks the system's effectiveness. The following three descriptors are commonly used for this purpose:

Protection Factor (PF) = $\dfrac{C_o}{C_i}$; (ratio)

Efficiency (η) = $\dfrac{C_o - C_i}{C_o}$; (fraction, or multiplied by 100 for percent value)

Penetration (Pen) = $1 - \eta$; (fraction)

where C_o = outside respirable dust concentration, and
C_i = inside respirable dust concentration.

A comparison of these descriptors can be provided by the following:

$$\text{PF} = \frac{C_o}{C_i} = \frac{1}{1-\eta} = \frac{1}{\text{Pen}} \tag{9.1}$$

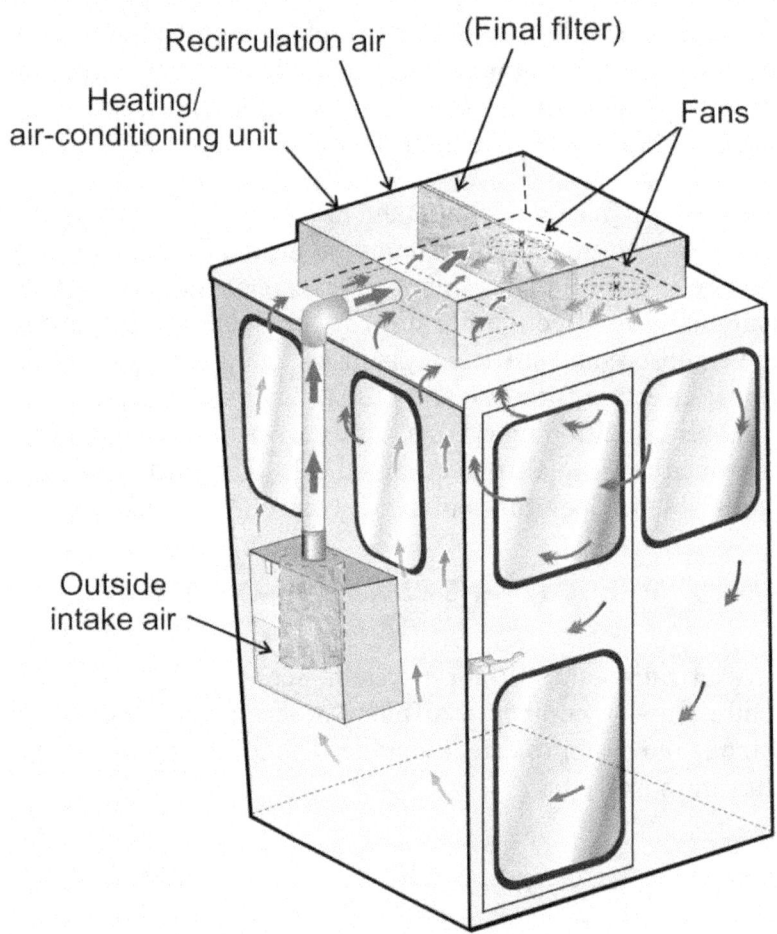

Figure 9.1. General design of an effective filtration and pressurization system.

Obviously, the higher the value for both *protection factor* and *efficiency*, and the lower the value for *penetration* that can be achieved, the better the air quality inside the operator booth/control room/enclosed cab. Table 9.1 shows a comparison of the three different descriptors.

Table 9.1. Comparison of three different descriptors for the effectiveness at providing clean air to an enclosed cab, operator booth, or control room

Protection Factor	Efficiency, %	Penetration
2	50	0.50
5	80	0.20
10	90	0.10
100	99	0.01
1,000	99.9	0.001

For this chapter, the term *protection factor* (PF) will be used when discussing enclosed cab, operator booth, and control room effectiveness.

FILTRATION AND PRESSURIZATION SYSTEMS

Field Studies

A health screening study performed during the mid-1990s in central Pennsylvania identified a significant number of silicosis cases attributed to operators of mobile mining equipment in enclosed cabs that were not providing an acceptable level of protection [CDC 2000]. Because of this, a number of organizations began investigating enclosed cabs to better understand the problem and to determine methods and solutions to improve the air quality and protect workers. This led to a number of studies in which new filtration and pressurizations systems were installed on older pieces of mining equipment in an attempt to improve the air quality inside these enclosed cabs. The results of a few of these studies can be seen in Table 9.2, listed in ascending order of effectiveness [Organiscak et al. 2004; Chekan and Colinet 2003; Cecala et al. 2005; Cecala et al. 2004].

Table 9.2. Summary of field studies evaluating upgraded cabs

Cab evaluation	Cab pressure, inches wg	Average inside cab dust level, mg/m^3	Average outside cab dust level, mg/m^3	Protection factor, out/in
1. Rotary drill	None detected	0.08	0.22	2.8
2. Haul truck	0.01	0.32	1.01	3.2
3. Front-end loader	0.015	0.03	0.30	10.0
4. Rotary drill	0.20 to 0.40	0.05	2.80	56.0
5. Rotary drill	0.07 to 0.12	0.07	6.25	89.3

These studies highlighted some very important factors relevant to improving the air quality in enclosed cabs and ultimately protecting the workers. Cab integrity, and the related ability to achieve positive pressurization, was found to be a critical component. As seen in the first two studies listed in Table 9.2, when there was very little to no cab pressure detected, this resulted in minimal improvement in the cab's air quality. In fact, similar filtration and pressurization systems were installed on the rotary drill and front-end loader listed as items 1 and 3 in Table 9.2, with the PF varying from 2.8 to 10. One notable difference between these two systems was that a small amount of pressurization was achieved in the front-end loader, whereas it was not possible to achieve any pressurization in the rotary drill. It should also be noted that low outside dust concentrations may have a negative impact on the PF.

Another critical factor was the quality and effectiveness of the filtration system. The various studies presented in Table 9.2 indicated substantial improvement in the interior air quality from effectively removing the dust particles from the outside air and delivering this clean filtered air into the enclosed cab. When sufficient pressurization was achieved along with an effective filtration system, very good air quality was obtained in these cabs as indicated by significant PFs.

Laboratory Study

From these various field studies, a number of different factors emerged that were relevant to the effectiveness of filtration and pressurization systems. In an effort to evaluate this area, a controlled laboratory experiment was performed to systematically examine multiple cab designs. Figure 9.2 shows the cab filtration system setup used for this controlled laboratory study and the various parameters evaluated [NIOSH 2008b; Organiscak and Cecala 2008].

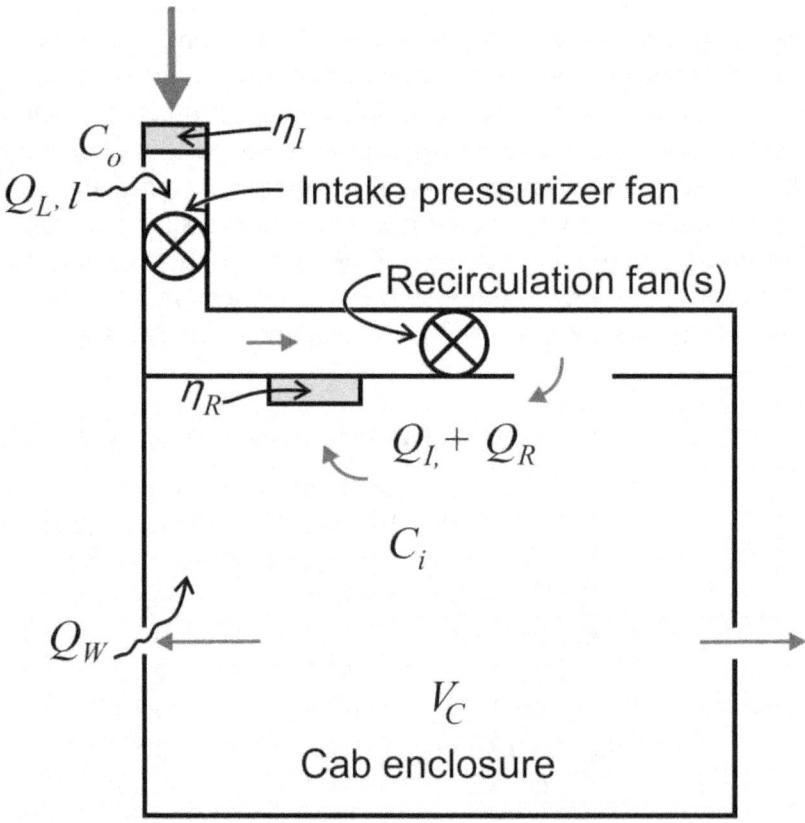

Figure 9.2. Laboratory test setup to evaluate various operational parameters on a filtration and pressurization system for an enclosed cab.

In Figure 9.2, the parameter values are as follows:

PF = protection factor, C_o/C_i;
C_o = outside cab concentration;
C_i = inside cab concentration;
η_I = intake filter efficiency, fractional;
Q_I = intake air quantity;
Q_L = leakage air quantity;
l = intake air leakage, Q_L/Q_I;
η_R = recirculation filter efficiency, fractional;
Q_R = recirculation air quantity;
Q_W = wind quantity infiltration; and
V_C = cab volume.

The results of this laboratory study indicate that intake filter efficiency and the use of a recirculation filter had the greatest impact on improving the air quality. When considering the use of an intake air filter, the addition of the recirculation component significantly improved the air quality due to the repeated filtration of the cab's interior air. The addition of an intake pressurizer fan to the filtration system increased both intake airflow and cab pressure significantly. The cab air quality was also affected by intake filter loading and air leakage.

Mathematical Model to Determine Enclosure Protection Factor

In the course of the laboratory study, the significance of the filtration system parameters was evaluated and the following mathematical model was developed. Equation 9.2[6] was formulated from a basic time-dependent mass balance model of airborne substances within a control volume with steady state conditions. The equation determines the PF in terms of intake air filter efficiency, intake air quantity, intake air leakage, recirculation filter efficiency, recirculation filter quantity, and outside wind quantity infiltration into the cab.

$$\text{PF} = \frac{C_o}{C_i} = \frac{Q_I + Q_R \eta_R}{Q_I(1 - \eta_I + l\eta_I) + Q_W} \tag{9.2}$$

Equation 9.2 allows for a comparison of how changes in the various parameters and components in the system impact the PF. The wind quantity infiltration (Q_w) can be assumed to be zero if the cab pressure exceeds the wind velocity (Figure 9.3). By using Equation 9.2, operations have the ability to determine the desired parameters necessary to systematically achieve a desired PF in an operator's booth, control room, or enclosed cab to improve the air quality to safe levels and to ultimately protect their workers.

---

[6] This equation is dimensionless; therefore, air quantities used must be in equivalent units. Also, filter efficiencies and intake air leakage must be fractional values (not percentage values).

RECOMMENDATIONS FOR FILTRATION/PRESSURIZATION SYSTEMS

Based on the knowledge gained from both laboratory and numerous field studies, the practices recommended in this section need to be incorporated in order to achieve an effective filtration and pressurization system.

Ensuring Enclosure Integrity by Achieving Positive Pressurization against Wind Penetration

Enclosure integrity is necessary in order to achieve pressurization, which is critical for an effective system. Testing has shown that the installation of new door gaskets and plugging and sealing cracks and holes in the shell of the enclosure have a major impact on increasing the enclosure pressurization. To prevent dust-laden air from infiltrating into the enclosure, the enclosure's static pressure must be higher than the wind's velocity pressure [Heitbrink et al. 2000]. Equation 9.3 is used to determine the wind velocity equivalent for an enclosure (the wind velocity at which the cab is protected from outside infiltration as determined by the static pressure):

$$\text{Wind velocity equivalent} = (4000\sqrt{\Delta p_{cab}}) \text{ fpm} \times 0.011364 \text{ mph/fpm} \tag{9.3}$$

@ standard air temperature and pressure

where Δp = cab static pressure in inches wg.

Figure 9.3 provides a graphical display of this wind velocity equivalent. This shows that an enclosure pressure of 0.05, 0.1, 0.25, 0.5, and 1.0 inches wg would be able to withstand wind velocities of 10.2, 14.4, 22.7, 32.1, and 45.5 mph from penetrating dust into the enclosure. Although minimum pressurization has been shown to have positive results from field studies, a good rule of thumb is to have at least 0.05 inches wg of positive pressure in enclosures. A reasonable range of enclosure pressure is between 0.05 to 0.25 inches wg.

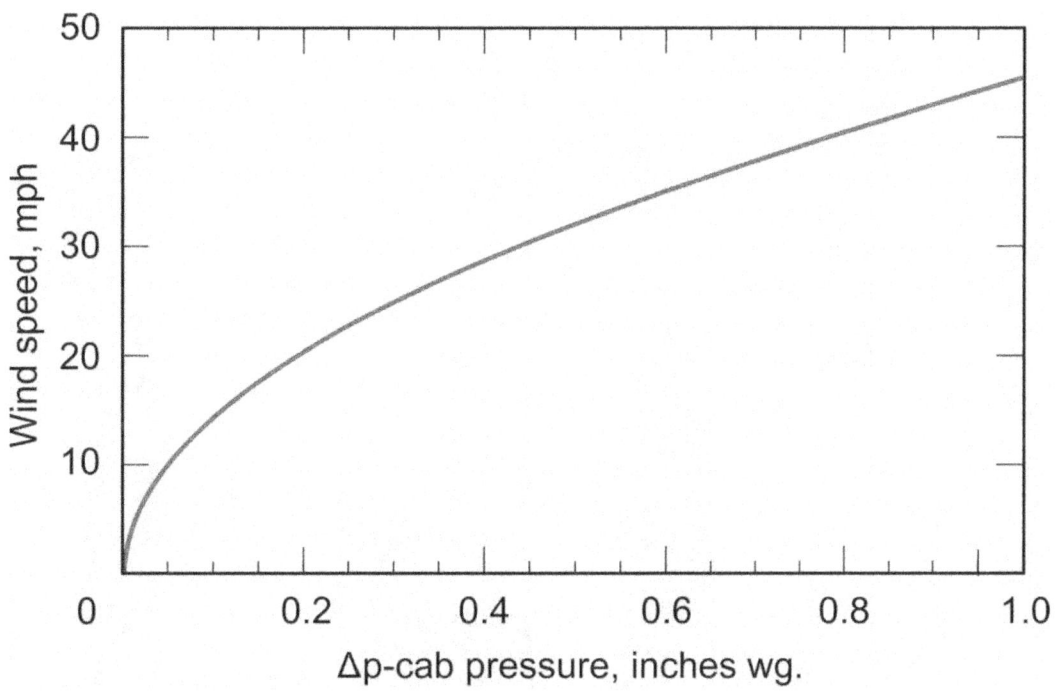

Figure 9.3. Positive cab pressure necessary to prevent dust-laden air from infiltrating the enclosed cab at various wind velocities.

Keeping Doors and Windows Closed

In order to achieve and maintain enclosure pressurization, doors and windows must be closed at all times except while the operator is entering or exiting the enclosure. This problem was noted during a field study on a surface drill when the operator repeatedly opened the cab door to manually guide the drill steel into place each time an additional section was needed [Cecala et al. 2007; NIOSH 2008a]. The cab door was usually open somewhere between 20 and 45 seconds each time this process took place before being closed again. Because no drilling was occurring and no dust cloud was visible as the cab door was opened, the impact to the drill operator's respirable dust exposure was initially thought to be insignificant. However, when dust data from inside the enclosed cab were analyzed, a substantial increase in respirable dust concentrations was noted during the periods when the door was open. This significant increase was unexpected when one considers that drilling had ceased approximately 2 minutes before the door was opened. Figure 9.4 shows average concentrations for each of the 3 days of testing for the time period when the cab door was closed and open. The average concentration was 0.09 mg/m^3 with the cab door closed and 0.81 mg/m^3 with the door open. Despite no visible dust cloud during the time when the cab door was open, respirable dust concentrations inside the cab were nine times higher than when the door was closed and drilling was being performed.

The results of this testing clearly stress the importance of keeping doors and windows closed at all times in an effort to keep the compartment pressurized and working properly. Again, the only exception to keeping the door closed should be when the equipment operator enters or exits the cab. It also needs to be stressed that even when dust clouds are not visible outside, respirable dust levels can be significantly higher than filtered levels inside cabs.

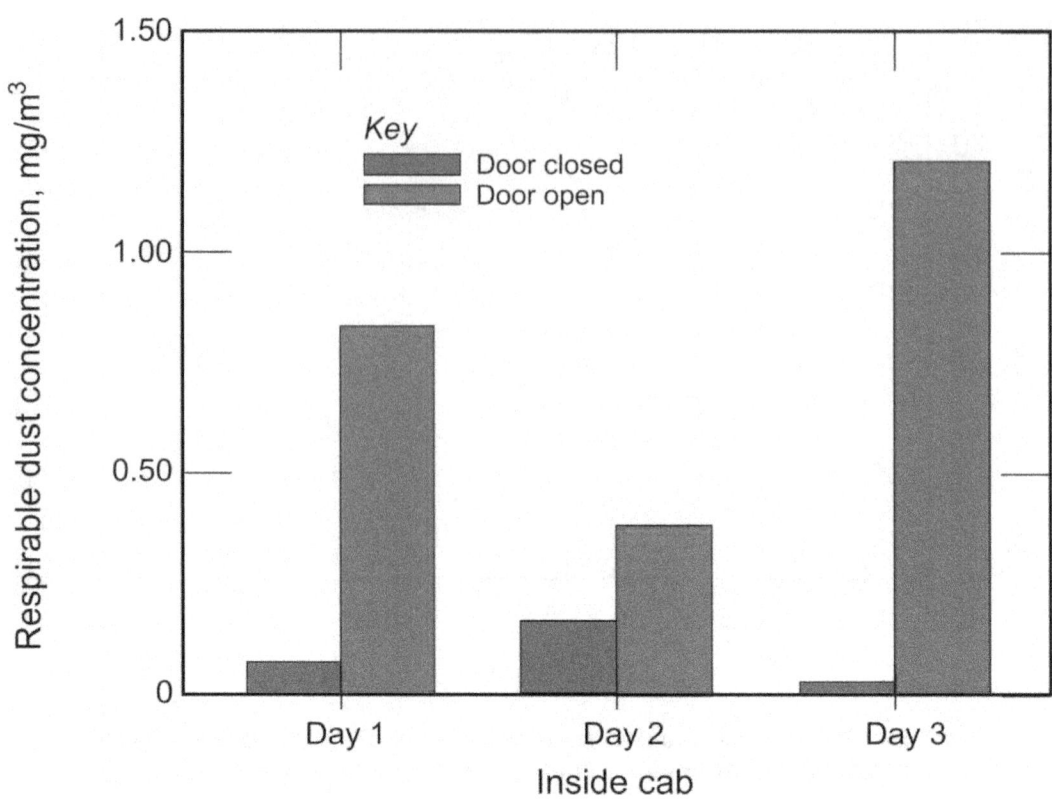

Figure 9.4. Respirable dust concentrations inside enclosed cab for three days of testing with cab door closed and open.

Effective Filtration

In order to achieve effective filtering and provide clean air to an operator's enclosure, the system needs to be composed of both an effective outside intake air and a recirculation filtration component.

Effective Outside Air Filtration and Enclosure Pressurization

A minimum quantity of at least 25 cfm of intake/outside air per person is necessary to dilute CO_2 quantities exhaled by each worker [ASABE 2003]. A typical range for most enclosed cabs on surface equipment is 40 to 70 cfm. The outside air delivered into an enclosure is the only component that creates pressurization, which is critical for an effective system. The amount of intake air delivered to create this pressurization must be carefully controlled and optimized. Once the optimal point is determined, increasing the air quantity degrades the system by increasing particle penetration and decreasing filter efficiency by allowing more contaminants to flow through the filter media.

With outside air, the optimal design is to provide enough air to achieve an acceptable level of pressurization, but further increasing airflow only decreases the air quality in the enclosure. The mathematical model derived from the laboratory study to determine the PF can be used to emphasize this point. The following is a hypothetical situation: a recirculation system that provides 300 cfm (Q_R), a recirculation filter efficiency of 0.7 (70 percent) (η_R), an intake filter efficiency of 0.95 (95 percent) (η_I), and assuming no air leakage ($l=0$)—increasing the intake air quantity from 50 to 70 cfm (Q_I) will cause the PF to decrease from 104 to 80, respectively, just by increasing the intake air quantity by 20 cfm.

High efficiency filters are a necessity for the outside intake air. Laboratory experiments showed a 10-fold increase in protection factors when using a 99-percent efficient filter versus a 38-percent efficient filter on respirable sized particles. The balance is to have a sufficient quantity of intake air to create a positive enclosure pressurization to eliminate the wind from blowing dust into the enclosure, but not so much airflow as to burden the filtration system beyond what is necessary. In addition, all intake air must be conditioned for temperature, either via heating in the winter or air-conditioning in the summer, and thus, the greater the intake air quantities, the greater the cost for providing for this temperature component.

Outside Air Inlet Location

Another key consideration is the location of the outside inlet for the intake air. Locating the cab air inlet near major dust sources causes high dust loading on the air filtration system. This high dust loading burdens the filtration system and reduces its effectiveness by increasing the pressure drop across the loaded filter and decreasing the quantity of air and cab pressurization. In addition, the increased pressure drop across the loaded filter also increases the potential for dust leakage around the filter cartridge. This requires that the filter cartridge be cleaned or changed more frequently which also increases the filter cost. Finally, air filtration is based on relative dust capture efficiency, so filtering higher outside dust levels creates higher inside cab dust concentrations.

In an effort to minimize these effects, it is recommended to place the enclosure's air inlet location strategically away from dust sources to reduce dust loading of the filter cartridge [NIOSH 2001a]. This can usually be accomplished by locating the outside air intake inlet at higher levels away from the ground and on the opposite side of the enclosed cab, operator booth, or control room away from dust sources. This location also enables the enclosure to shield some of the dust from the inlet.

PreCleaner

It is recommended that the outside air filtration system use a precleaning and final-filtering arrangement. The precleaning should be performed to remove oversized dust particles and to increase the life cycle and the filtering capacity of the final intake filter. This could be performed by using some type of centrifugal design to spin out oversized particles or a back-flushing compressed air system, similar to those in a local exhaust ventilation (LEV) system. These back-flushing systems can be set to clean the filter cartridge on a predetermined time basis or when the pressure builds to a predetermined level.

Effective Recirculation Filtration

The use of the recirculation component with effective filtration is critical for an optimal design. Laboratory experiments showed a 10-fold increase in protection factors when using an 85 to 94.9 percent efficient filter on respirable sized dusts as compared to no recirculation filter. Laboratory testing also showed that the time for the interior to stabilize after the door was closed (decay time) was reduced by more than 50 percent when using the recirculation filter. The average decay times were between 16 and 29 minutes without the recirculation filter and between 6 and 11 minutes with the recirculation filter. Thus, the use of a recirculation filter greatly improved the air quality and reduced the exposure time after the cab door was closed [NIOSH 2007]. One additional benefit of using a recirculation filter is that it allows cleaner air to be circulated through the HVAC system, thus providing better thermal efficiency.

Floor Heaters

Any type of floor heater or fan located low in the enclosure that will stir up dust should be eliminated. During a field study, it was found that a floor heater fan used during the winter months to provide heat to an operator of a surface drill greatly increased the respirable dust concentrations inside the enclosed cab (Figure 9.5) [Cecala et al. 2001; NIOSH 2001b]. The floor heater can be a serious problem because the floor is the dirtiest part of the cab from the operator bringing dirt in on his or her work boots. Then as the operator moves his or her feet around, dust is created, which is then blown throughout the cab by the fan on the floor heater. This fan also tends to stir up dust that may be on the drill operator's clothes.

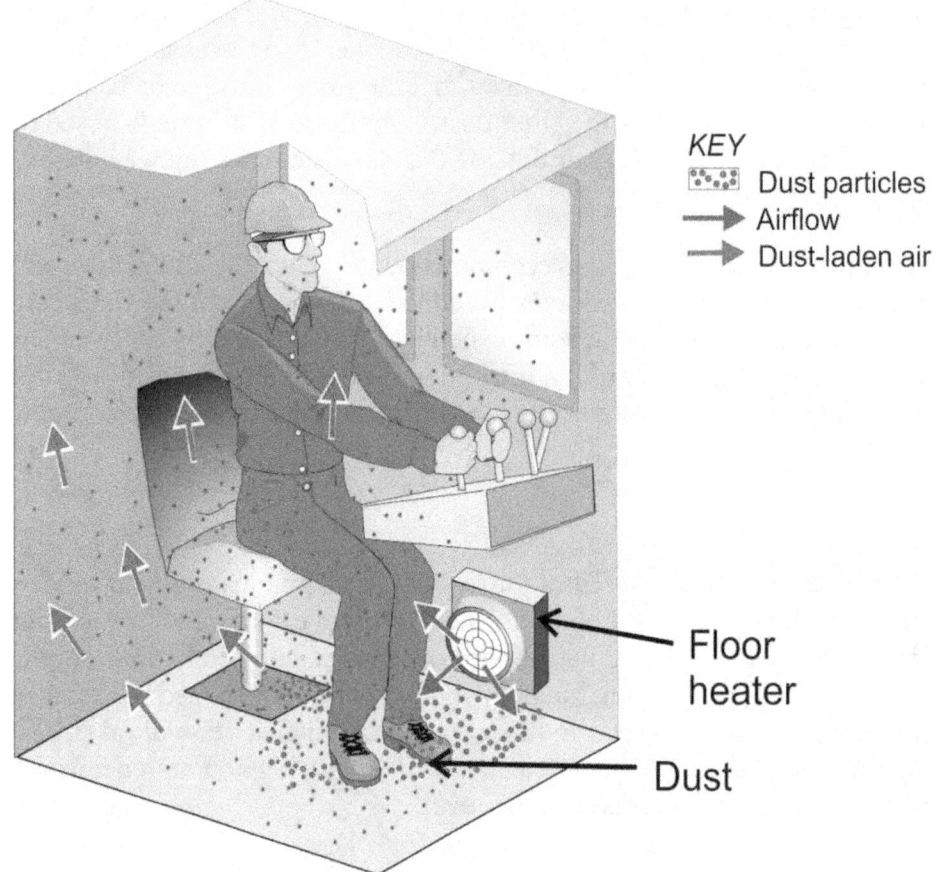

Figure 9.5. Problem created by heater stirring up dust from the floor and blowing dust off of the worker's clothing.

Because of the significant increase in dust levels with floor heaters, it is recommended that they not be used. If removal is not an option, they should be repositioned to a higher area in the enclosure where they are less prone to pick up dust from the floor and operator's clothing. Also, no type of fan should be used low in the cab because of the potential to stir up in-cab dust sources. Ideally, the heater unit should be tied into the filtration and pressurization unit to deliver the heated air at the roof of the cab.

Good Housekeeping (Enclosure Cleanliness)

To maintain pressurization and filtration systems, good housekeeping practices are essential in that systems need to be cleaned periodically and filters need to be changed when necessary. In addition, the integrity of the enclosure must also be inspected to ensure that pressurization is maintained by replacing gaskets and seals when wear appears, and by plugging and sealing holes and cracks in the walls. It must be understood that a system that is not properly maintained will deteriorate over time to a point where it is no longer providing an acceptable level of protection, thus causing workers to be exposed to respirable dusts.

During the field studies, a number of filtration units were found in all forms of disarray and had deteriorated to a condition where they were no longer providing acceptable levels of protection to the worker (Figure 9.6, left). In some cases, it appeared that the air quality or the protection provided to the worker was not a priority as long as the operator remained comfortable in regard to temperature controls. Although many cabs used standard heaters and air conditioners to control temperatures, in some instances workers resorted to just opening windows in an effort to be comfortable, thus bypassing the protection provided by the enclosed cab. With a little time, effort, and finances, effective maintenance can be performed on filtration and pressurization systems to transition them from poor systems to ones that will again provide clean and acceptable air quality to workers (Figure 9.6, right).

Enclosure floors are commonly soiled from workers tracking dirt and product inside the enclosure upon entering from the mine site. In cases when a substantial amount of dust and product gets tracked into the compartment, housekeeping should be performed on a daily or shift basis. In other cases where the contamination is not as great, housekeeping could be done on a more infrequent basis. It is critical that the inside of an enclosed cab, operator's compartment, or control room be maintained in a manner which minimizes the worker's respirable dust exposure.

Mechanical Filter Media and Visual Indicator for Filter Changing

It is highly recommended that both the outside air and recirculation filters be a mechanical type filter media, as compared to an electrostatic media. Mechanical filters become more effective as they load with product. This occurs because as the filter loads with dust, a dust cake forms on the filter media and captures additional dust particles which further improves the filter efficiency. As the size of this filter cake continues to increase, the efficiency continues to increase. This causes the pressure differential across the filter to increase, which in turn causes the airflow to decrease. As the pressure increases further, it will become so restrictive that the filter will need to be cleaned or replaced. The ideal filtration and pressurization system includes a pressure differential visual indicator that notifies the operator when the filter needs to be changed.

Damaged pre-filter canister   Unit cleaned and intake filtration unit added (inside cylinder)

Figure 9.6. Filtration unit showing lack of maintenance and care (left side) and after cleaning and the addition of a new intake air filtration unit (right side).

Ease of Filter Change

When designing filtration and pressurization systems, one key component is the ease with which filters can be replaced when necessary. It defeats the purpose of a good system if a filter to be changed is so difficult to access that the operator or maintenance workers do not want to take the time to perform the task. Another consideration is dust contamination during the filter change. The easier a filter is to change, the less contamination should occur to the worker performing the task and to the work area. When changing a canister filter, a common and effective technique is to remove the new filter from the cardboard box and then insert the old dust-laden filter into the box, tape it closed, and dispose of it.

Uni-Directional Design

The use of a uni-directional airflow pattern, as shown in Figure 9.7, should be considered whenever possible to maximize the air quality at the breathing zone of the operator inside the enclosure. In most systems, both the intake and discharge for the recirculation air are located in the roof. Unfortunately, this location causes the dust-laden air within the enclosure to be pulled directly over the worker as it is drawn into the ventilation system. Further, in many designs, the contaminated return air and clean filtered air are ducted within inches of each other at the ceiling. This poor design allows for recirculated air to be short-circuited and allows dust-laden return air to be pulled directly back into the ventilation system and over the operator's breathing zone. A more effective design is to draw the recirculated air from the bottom of the enclosure, away from the worker's breathing zone.

Generalized Parameters for an Effective System

The key for an effective retrofit enclosed cab filtration and pressurization system is to balance the various components and factors. Obviously, if one is dealing with hazardous dust, the design should be directed to the higher efficiency range. On the other hand, if one is dealing with more of a nuisance dust, a lower range value can be chosen.

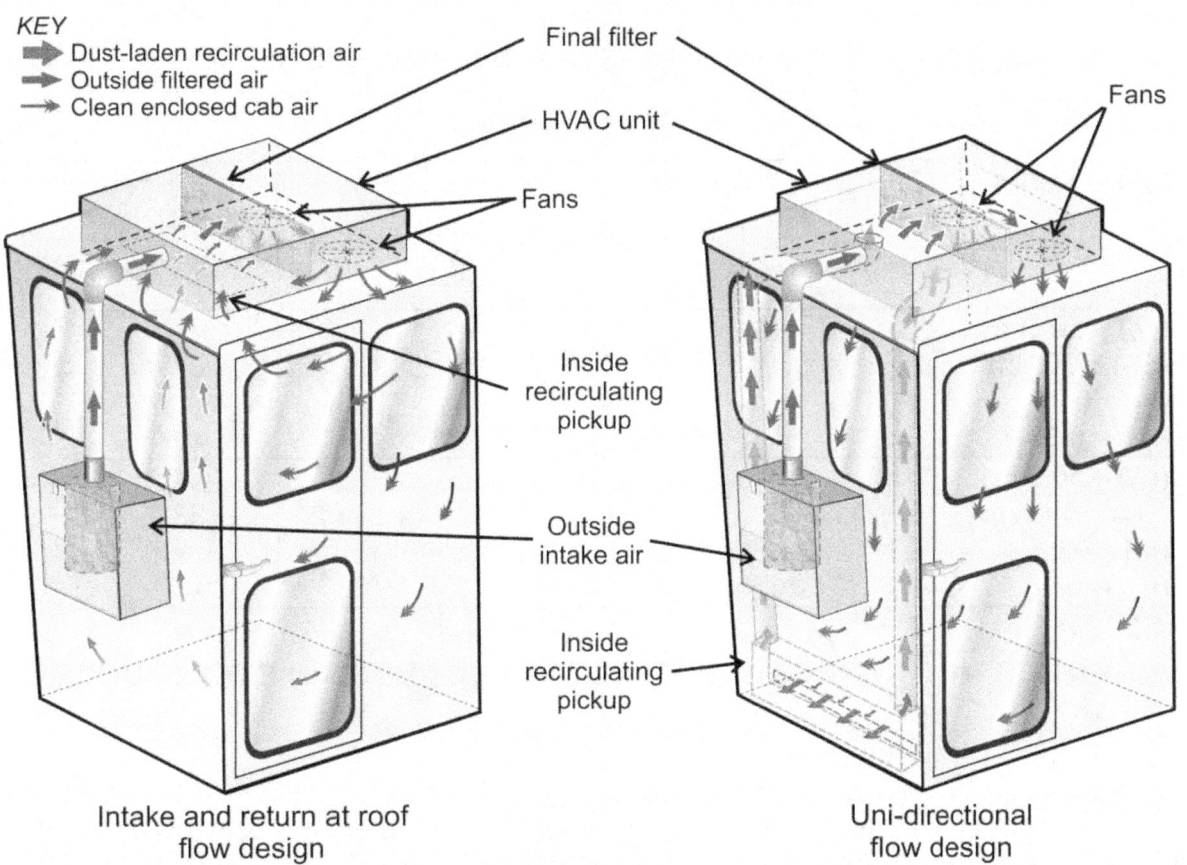

Figure 9.7. Airflow pattern for intake and return at roof of cab and uni-directional airflow design.

Table 9.3 shows a summary of the impact of the various critical components to an effective filtration and pressurization system. This table presents the calculated PF using the mathematical model derived from the laboratory study (Equation 9.2). In this table, the same component parameters have been chosen from the previous examples, being: intake air quantity (Q_I) of 50 cfm, recirculation system (Q_R) of 300 cfm, a recirculation filter efficiency (η_R) of 0.7 (70 percent), an intake filter efficiency (η_I) of 0.95 (95 percent), and leakage (l) of 0, 5 percent, 10 percent, and 15 percent (l = 0, 0.05, 0.10, 0.15).

This table highlights a number of critical factors. First, it shows how the effectiveness of each system deteriorates as the leakage into the enclosure increases. Leakage occurs from either a lack of cab integrity or an insufficient intake component. The second critical factor is the need for a recirculation system. When one evaluates an enclosure with no leakage, the PF is increased from 20 to 104 by the addition of a 300 cfm recirculation system with a 70 percent efficient filter. This highlights the critical need of the recirculation component. A high efficiency recirculation filter is not critical as long as a filter within reasonable filter efficiency range (70–95 percent) is used and maintained. This is demonstrated by viewing the first column in Table 9.3 where $l = 0$ (leakage equals zero) and the PF only increases by 24 points when going from a 70 to a 90 percent efficiency recirculation filter. Since the air is constantly being recirculated, any respirable dust inside the cab will be removed within a few passes through the recirculation system.

Table 9.3. Calculated PF derived from mathematical model

	$l = 0$	$l = 0.05$	$l = 0.10$	$l = 0.15$
(Q_I) = 50 cfm (Q_R) = 300 cfm (η_I) = 0.95 (η_R) = 0	20.0	10.3	6.9	5.2
(Q_I) = 50 cfm (Q_R) = 300 cfm (η_I) = 0.95 (η_R) = 0.70	104.0	53.3	35.9	27.0
(Q_I) = 50 cfm (Q_R) = 300 cfm (η_I) = 0.95 (η_R) = 0.90	128.0	65.6	44.1	33.2

OTHER METHODS TO LOWER RESPIRABLE DUST IN ENCLOSURES

Small In-Unit HEPA Filters

In some operator booths and control rooms, stand-alone air purifier units containing a HEPA filter have been installed. These stand-alone systems are small and portable units that are available at a fraction of the cost of permanent systems. However, these systems can only be effective if they are sized to handle the volumetric capacity of the booth or control room, and if the filters are replaced when necessary [Logson 1998/1999]. An obvious and significant shortcoming with these types of systems is that they do not provide any pressurization to keep dust from leaking into the booth or room. A few studies have shown that in many instances, they do not provide sufficient protection and companies would be better served to invest their money and time into improving their existing roof-mounted unit [Cecala and Zimmer 2004a,b]. Therefore, stand-alone air purifier units containing a HEPA filter are not recommended in most cases and should be replaced with more comprehensive and properly sized recirculation systems.

REFERENCES

ASABE [2003]. Agricultural Cabs—Engineering Control of Environmental Air Quality, Part 1: Definitions, Test Methods, and Safety Practices [Standard 5525–1.1]. St. Joseph, Michigan: American Society of Agricultural and Biological Engineers.

CDC [2000]. Notice to readers: NIOSH Alert; silicosis screening in surface coal miners—Pennsylvania, 1996–1997. Cincinnati, OH: U.S. Department of Health and Human Services, Centers for Disease Control and Prevention, National Institute for Occupational Safety and Health.MMWR Weekly *49*(27):612–615.

Cecala AB, Zimmer JA [2004a]. Clearing the air. Aggregates Manager J *9*(4):12–14.

Cecala AB, Zimmer JA [2004b]. Filtered recirculation—a critical component to maintaining acceptable air quality in enclosed cabs for surface mining equipment. Proceedings of 10th U.S./N.A. Mine Ventilation Symposium. Anchorage, Alaska. May 16–19, pp. 377–387.

Cecala AB, Organiscak JA, Heitbrink WA [2001]. Dust underfoot—enclosed cab floor heaters can significantly increase operator's respirable dust exposure. Rock Products *104*(4):39–44.

Cecala AB, Organiscak JA, Heitbrink WA, Zimmer JA, Fisher T, Gresh RE, Ashley JD II [2004]. Reducing enclosed cab drill operator's respirable dust exposure at surface coal operations with a retrofitted filtration and pressurization system. SME Transactions 2003, Littleton, Colorado: Society for Mining, Metallurgy and Exploration, Inc, *314*:31–36.

Cecala AB, Organiscak JA, Zimmer JA, Heitbrink WA, Moyer ES, Schmitz M, Ahrenholtz E, Coppock CC, Andrews EH [2005]. Reducing enclosed cab drill operator's respirable dust exposure with effective filtration and pressurization techniques. J of Occ and Env Hyg *2*:54–63.

Cecala AB, Organiscak JA, Zimmer JA, Moredock D, Hillis M [2007]. Closing the door to dust when adding drill steels. Rock Products, October, pp. 29–32.

Chekan GJ, Colinet JF [2003]. Retrofit options for better dust control—cab filtration, pressurization systems prove effective in reducing silica dust exposures in older trucks. Aggregates Manager, December, *8*(9):9–12.

Heitbrink WA, Thimons ED, Organiscak JA, Cecala AB, Schmitz M, Ahrenhottz E [2000]. Static pressure requirements for ventilated enclosures. Proceedings of the Sixth International Symposium on Ventilation for Contaminant Control, Helsinki, Finland, June 4–7.

Logson R [1998/1999]. Controlling respirable dust in plant control rooms. Stone Review, The Bimonthly Publication of the National Stone Association. December 1998/January 1999, *14*(6):43–44.

NIOSH [2001a]. Technology news 485: Improved cab air inlet location reduces dust levels and air filter loading rates. By Organiscak JA, Page SJ. Pittsburgh, PA: U.S. Department of Health and Human Services, Centers for Disease Control and Prevention, National Institute for Occupational Safety and Health.

NIOSH [2001b]. Technology news 486: Floor heaters can increase operator's dust exposure in enclosed cabs. By Cecala AB, Organiscak JA. Pittsburgh, PA: U.S. Department of Health and Human Services, Centers for Disease Control and Prevention, National Institute for Occupational Safety and Health.

NIOSH [2007]. Technology news 528: Recirculation filter is key to improving dust control in enclosed cabs. By Organiscak JA, Cecala AB. Pittsburgh, PA: U.S. Department of Health and Human Services, Centers for Disease Control and Prevention, National Institute for Occupational Safety and Health, DHHS (NIOSH) Publication No. 2008–100.

NIOSH [2008a]. Technology news 533: Minimizing respirable dust exposure in enclosed cabs by maintaining cab integrity. By Cecala AB, Organiscak JA. Pittsburgh, PA: U.S. Department of Health and Human Services, Centers for Disease Control and Prevention, National Institute for Occupational Safety and Health, DHHS (NIOSH) Publication No. 2008–147.

NIOSH [2008b]. Key design factors of enclosed cab dust filtration systems. By Organiscak JA, Cecala AB. U.S. Department of Health and Human Services, Centers for Disease Control and Prevention, National Institute for Occupational Safety and Health: NIOSH Report of Investigations 9677.

Organiscak JA, Cecala AB [2008]. Laboratory investigation of enclosed cab filtration system performance factors. Min Eng *60*(12):74–80.

Organiscak JA, Cecala AB, Thimons ED, Heitbrink WA, Schmitz M, Ahrenholtz E [2004]. NIOSH/Industry collaborative efforts show improved mining equipment cab dust protection. SME Transactions 2003 Littleton, Colorado: Society for Mining, Metallurgy and Exploration, Inc., *314*:145–152.

CHAPTER 10: HAUL ROADS, STOCKPILES, AND OPEN AREAS

CHAPTER 10: HAUL ROADS, STOCKPILES, AND OPEN AREAS

This chapter examines dust control for emissions from haul roads and wind erosion of stockpiles and open areas. Generally classified as a fugitive dust, the source material for these emissions is typically soil and mineral particles disturbed during the mining or stockpiling processes.

The main focus of this chapter is on haul road dust control as it is a significant issue at surface mine sites (Figure 10.1). Past research has shown that haul trucks generate the majority of dust emissions at these sites, with their contribution being 78–97 percent of total dust emissions for particulate matter less than 10 micrometers (µm) [Cole and Zapert, 1995; Amponsah-Dacosta and Annegarn 1998; Reed et al. 2001]. The emissions are determined using the United States Environmental Protection Agency's (U.S. EPA) published emission factors, which is the accepted method for calculating emissions from the various operations at industrial sites (including mining). The high contribution of dust emissions from haul trucks has the potential to expose personnel working nearby (within 100 feet or downwind) to significant amounts of respirable dust, creating potential overexposures. Further, high concentrations of respirable dust (up to 21.50 mg/m^3) have been documented in areas near the vehicles [Reed and Organiscak 2005], and elevated percentages of silica can be associated with haul road dust [Organiscak and Reed 2004].

Figure 10.1. Example of haul road dust from a typical mine haul truck.

Fugitive dust from haul roads, stockpiles, and open areas can contain particle sizes ranging from 10 to 1x10^{-5} mm, with the majority of the material ranging from 10 to 1x10^{-3} mm [Bagnold 1960]. The amount of larger material residing within this dust is dependent upon the wind velocity (or disturbance) at the source of emissions. Generally, the respirable dust fraction is only a small part of the total mass of total dust emissions. For example, an evaluation of dust particle size from haul truck emissions demonstrated that 85.5 percent of the fugitive dust is >10 µm in diameter, with 14.5 percent <10 µm [Organiscak and Reed 2004; Reed and Organiscak 2005]. Albeit a small amount, these smaller particle sizes are a concern from a health and safety standpoint, especially if silica dust is present.

Reduced visibility can be a safety problem with the generated fugitive dust from haul roads, stockpiles and open areas. Although there is no known documented correlation between dusty conditions and vehicle accidents at mine sites, observations of extreme dusty conditions make it obvious that there could be a connection. Coarse materials contribute to obstructing visibility by obstructing light, whereas fine materials obstruct visibility by scattering the light. Light scattering generally occurs with fine particles <2.5 µm [Cowherd and Grelinger 1997; Moosmüller et al. 2005]. Visibility reduction occurs with dust concentrations ranging from 100–400 mg/m^3. However, of greatest concern here is the ongoing health problem associated with high dust concentrations being inhaled by workers at mine operations, especially respirable silica dust.

HAUL ROADS

Haul roads and access roads are used extensively in mining operations by mobile equipment to move material in and out of the mining areas. The road network at a surface mine site can be quite extensive over the entire property and the potential for dust generation is dependent upon the traffic patterns at the site. In areas where the traffic pattern is heavy, the possibility for dust generation is much higher than areas where the traffic pattern is light, creating areas of potentially high dust exposures. Overexposures to respirable dust can occur to both equipment operators and workers in the vicinity of the road. The majority of the fugitive dust is generated through the forces of the wheels on the road surface and by the turbulence created by the vehicles [Moosmüller et al. 2005]. Therefore, the first step in haul road dust control is proper road construction.

Access roads are similar in nature to haul roads, except that vehicle traffic is less frequent and is generally smaller in weight, although this may not always be the case. The following principles that are discussed for haul road dust control can also be applied to access roads. In particular, the road construction discussion pertains to haul roads for large haul vehicles, but it can easily be adapted to be applied to access roads. It is generally accepted to overdesign access roads; because of the transient nature of mining, there may be times when an access road may be upgraded to a haul road. Additional information for the proper design of the smaller access roads can be found in *Gravel Roads, Maintenance and Design Manual* [Skorseth and Selim 2000].

Basics of Road Construction

The proper construction of haul roads is an important consideration for effective dust control. A properly constructed road will have higher initial costs, but will require less road maintenance, reduce equipment maintenance cost, and will aid in effective dust control and increased tire life. In particular, dust control will be augmented from a reduction of fine material generated and through an increase in the longevity of dust suppressants.

When a vehicle travels over a haul road, the wheels exert forces on the road surface. Of all the forces, the normal shear stresses created by the vehicle are the most critical. At the location of the wheel, the road is put into compression. Once the wheel passes this location, the road rebounds to its initial position through tension. On a properly designed road, this compression and tension cycle will occur within the elastic limits of the road structure.

Permanent deformation of the road can occur if it is not properly designed. Roads constructed of weak materials will readily degrade, which can produce fine material that can be entrained by mobile equipment traveling the road. Roads constructed of the proper materials will degrade less rapidly, which in turn lessens the dust emission potential of the road over the same period of use. Additionally, road failure will cause any dust suppressant to exceed its bonding strength, thereby reducing the effectiveness of the suppressant. Generally, dust suppressants work by forming a layer, or crust, over the top of the road. If the bonding strength of the dust suppressant is exceeded, the suppressant will break up, deteriorate, and wear away.

Materials selected for road construction must possess certain physical properties. They are: *resistance to wear*, *soundness*, *maximum size*, *particle shape*, and *gradation* [Midwest Research Institute 1981].

- A high *resistance to wear* is desired for the road material. It should be hard and tough such that it will not be easily disintegrated under traffic load. The resistance to wear can be measured by conducting abrasion tests such as the Los Angeles Abrasion test.[7] Generally, the most desirable materials include granite or limestone. Soft and unsound materials such as shale, coal, mica, or vermiculite should be avoided as the use of such material will lower the road's strength and durability.
- *Soundness* represents the ability of the material to withstand climatic conditions. A material is desirable if it can resist weathering. Materials that are weak, extremely absorptive, easily cleavable, swell when saturated, or are susceptible to breakdown through natural weathering processes are not recommended for road construction.

---

[7] The Los Angeles Abrasion test provides information on aggregate resistance to crushing, degradation, and disintegration from activities such as manufacturing, stockpiling, production, placing, and compaction. Testing is conducted by placing a coarse aggregate sample in a rotating drum with steel spheres. As the drum rotates the material degrades by abrasion and impaction, with itself and with the spheres. Once the test is complete, the calculated mass of aggregates that has broken apart to smaller sizes is expressed as a percentage of the total mass of aggregate. Lower Los Angeles Abrasion values indicate that the aggregate is tougher and more wear resistant to abrasion. Specifications for the test can be found in AASHTO T96 or ASTM C 131 "Resistance to Degradation of Small-Size Coarse Aggregate by Abrasion and Impact in the Los Angeles Machine."

- *Maximum size* is the largest particle size allowable in the road material. Generally it is undesirable to use aggregate sizes larger than 1 inch for the surface course, in order to maintain ease of maintenance with the road grader.
- *Particle shape* of the road material affects the stability, density, and durability of the road surface. Shapes which contain angular and rough-surfaced material result in good interlocking and produce a desirable road surface.
- *Gradation* is the distribution of particle sizes throughout the road material. A well-graded aggregate material contains a good representation of particle size fractions from large to small sizes of material. An acceptable aggregate particle size distribution of a material for structural purposes is generally represented by Equation 10.1 [National Stone Association 1991]:

$$P = 100 \left(\frac{d}{D} \right)^n \tag{10.1}$$

where P = percent by weight finer than the sieve;
d = sieve opening dimensions (inches) where common sieve sizes are 2-inch, 1-inch, 3/4-inch, 3/8-inch, No. 4 (0.187 inches), No. 40 (0.0165 inches), and No. 200 (0.0029 inches);
D = maximum size of aggregate, inches; and
n = an empirical gradation exponent ranging from 0.33 to 0.5. Generally 0.5 represents the maximum density of material.

In addition to possessing the appropriate material properties, a properly constructed road will consist of three components: *subgrade*, *subbase*, and *wearing surface*. These features are displayed in Figure 10.2. The American Society for Testing and Materials (ATSM) and the American Association of State Highway and Transportation Officials (AASHTO) have standards which define the minimum properties required.

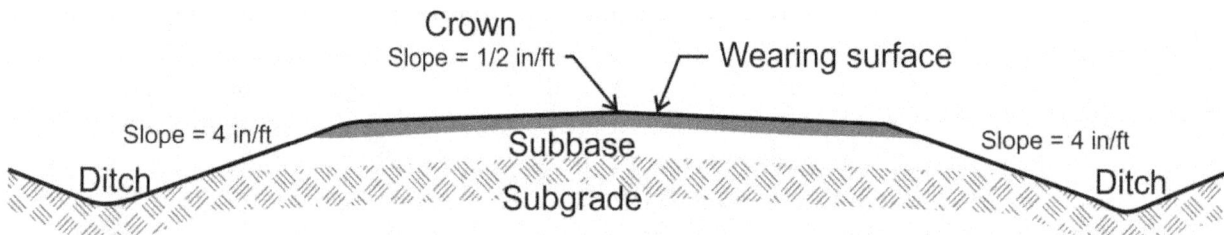

Figure 10.2. Cross section of haul road.

The *subgrade* is the underlying soil or rock that serves as the foundation of the road. It is important because it supports the entire load of the vehicles traveling the road. There are two techniques for maximizing subgrade strength: compaction at optimum moisture and adequate drainage. Most subgrade soils will compact to their maximum density at a given moisture. To achieve this, the subgrade area should be scarified prior to adding water to the location. Then compaction is generally accomplished using a vibratory compactor with a smooth drum roller [National Stone Association 1991]. Additionally, the subgrade should be designed with a crown in the center to promote water drainage away from the road. Proper drainage ditches should be designed and built to carry the drainage away from the road surface. Another consideration is that any areas of the subgrade consisting of poor quality material should be upgraded/repaired prior to adding additional bases; otherwise the location of these areas will be problematic in the future, requiring frequent maintenance [Midwest Research Institute 1981].

The *subbase* is the layer between the subgrade and the wearing surface. The material for this layer is a compacted angular aggregate. This layer has less stringent requirements for strength, aggregate type, and gradation. If the subgrade meets the requirements for the subbase, the subbase can generally be omitted from the road design [Merritt et al. 1996].

The road subbase is well-graded and consists of material that has a broad particle size range, which does not bias toward any particular size. Generally, the maximum size of the material is 3 inch, with a 1 1/2-inch maximum size material being more common for light duty traffic [Midwest Research Institute 1981]. It is best to spread the material for the base using a spreader box similar to those used in paving applications (Figure 10.3). The material is placed in layers to a uniform loose depth of 8 to 10 inches, then compacted with a vibratory roller using a minimum amount of blading to avoid material particle size segregation. End-dumping (also called "tailgating"), which is commonly used, is undesirable in that segregation of the uniform layer into areas of all fines or all coarse can occur easily with this method [National Stone Association 1991]. However, end-dumping is commonly used. To minimize the segregation of the fine and coarse material from end-dumping, it is important that a grader be used to blade the material to a uniform surface (Figure 10.4). A vibratory roller can then be used to compact the road material.

Figure 10.3. Use of a spreader box to lay gravel onto road surface.

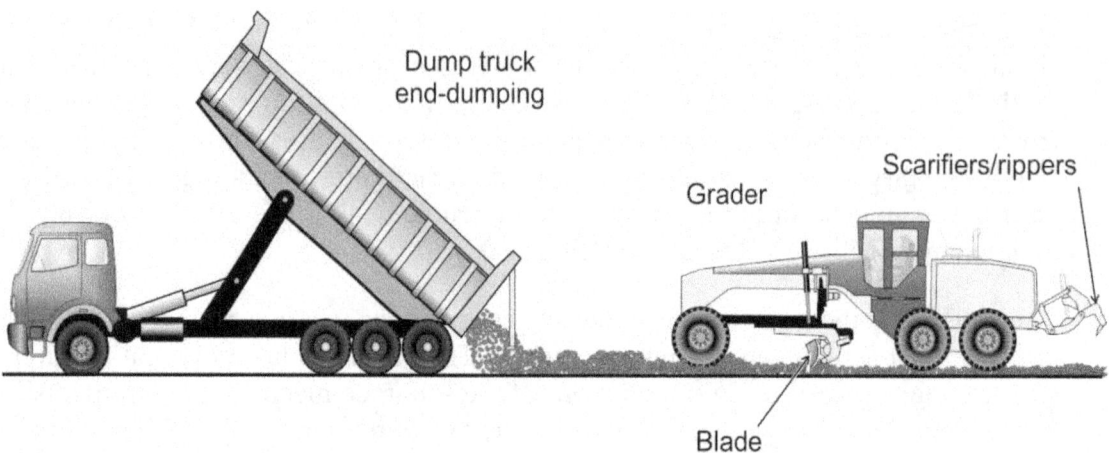

Figure 10.4. Use of end-dumping with grader to lay gravel onto road surface.

The *wearing surface* should be constructed of material that is resistant to grinding, weathering, and large particle displacement from the road. The material should contain sufficient fines to fill the voids between the larger particles. Sufficient fines will minimize large particle loss. The wearing surface should be compacted at optimum moisture to achieve maximum strength. Additionally, the crown should be maintained throughout the construction process to remove water away from the road surface.

Thicknesses of these three layers are important considerations in road designs. Sufficient thickness of these layers enables the road to support the load for which it was designed. A road that is not able to support the vehicle loading will deteriorate rapidly, which can in turn result in higher dust generation. The thickness of the subbase is an important design consideration to ensure that the surface load will be properly distributed over the subgrade area. As an example, a doubling of the thickness of the subbase can result in a reduction of more than twice the pressure on the subgrade. This design consideration can improve the life of the road by preventing the rapid deterioration of the road that can occur due to heavy loading [Midwest Research Institute 1981].

Mine haul roads have unique construction requirements due to the higher wheel loads from off-road mine haul trucks that they are required to support. Ordinary design guidelines for dirt and gravel roads for normal highway traffic are not adequate for mine roads. To determine the proper thickness for the subbase of a mine haul road, Figure 10.5, which was developed by the U.S. Bureau of Mines, should be used. This graph uses the California Bearing Ratio [8] (CBR) and the wheel load of the haul truck—which is calculated using haul truck specifications provided by the manufacturer—to determine the

---

[8] The California Bearing Ratio test provides results as a ratio from the comparison of the bearing capacity of a material to the bearing capacity of a well-graded crushed stone. It is used for evaluating the strength of cohesive materials having maximum particle sizes of 0.75 inch. The test involves applying a load to a small diameter penetration piston, which is applied to the sample material at a rate of 0.05 inches per minute, recording total load at penetration distances ranging from 0.025 inches up to 0.300 inches. Specifications for the test can be found in AASHTO T193 "The California Bearing Ratio," or ASTM D 1883 "Bearing Ratio of Laboratory Compacted Soils."

thickness of the subbase, wearing surface, and any intermediate layer. These same specifications can be used as design guidelines for mine haul roads.

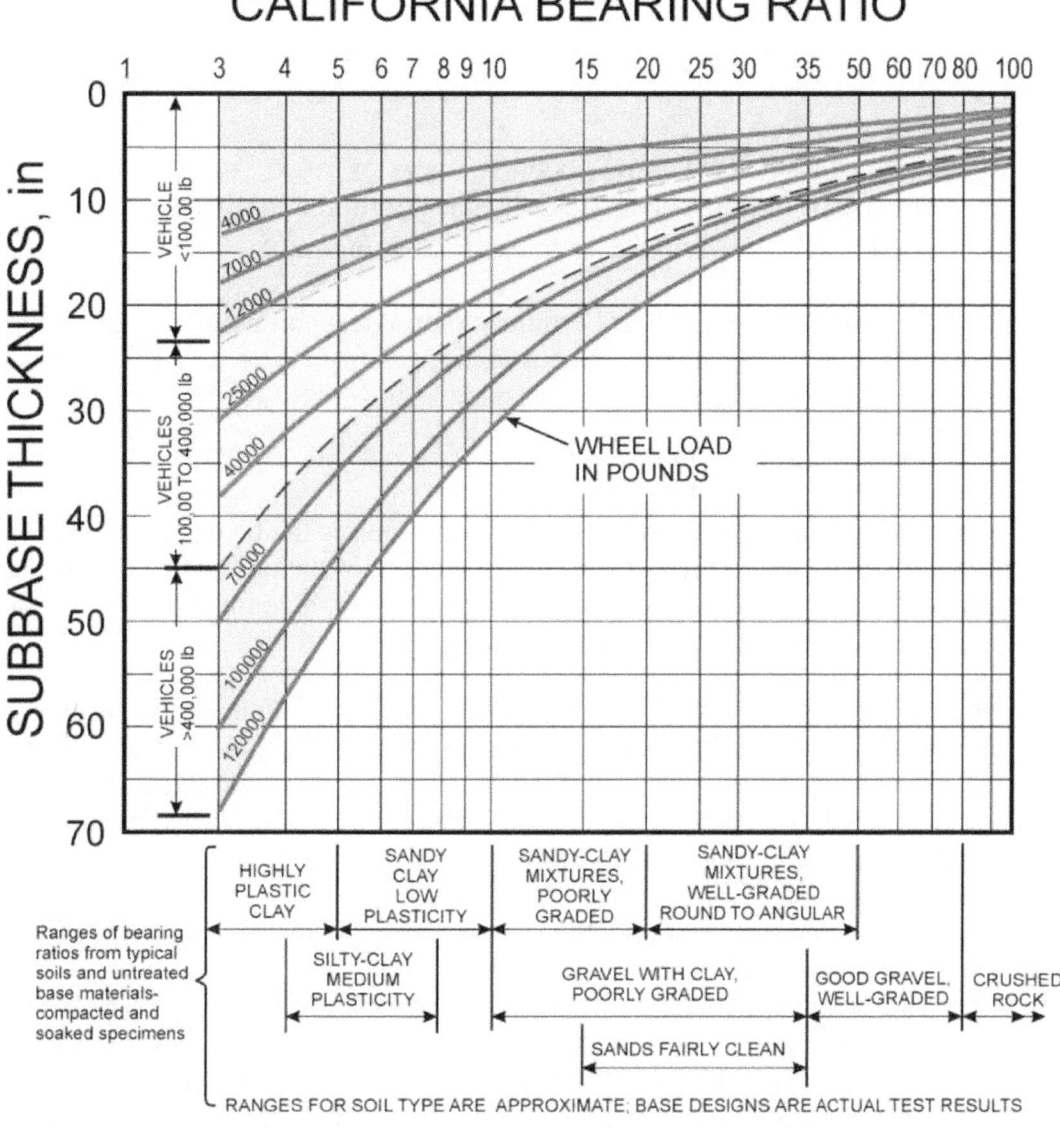

Figure 10.5. Subbase thickness from California Bearing Ratio tests on road material [after USBM 1977].

Haul Road Dust Control Measures

The most common method of dust control on haul roads is using dust suppressants, with water being the most common. Other suppressants are surfactants, salts, petroleum emulsions, polymers, and adhesives. Although haul road construction is important to the reduction of dust emissions from trucks, there is still the need to implement dust control measures for haul roads, as even the best constructed haul roads will provide material for dust entrainment. Selection of the proper agent for use is dependent upon mine site conditions. Performance of each of the above dust control agents is evaluated through the use of control efficiency. Control efficiency compares the dust concentration of a treated road surface to that of an untreated or uncontrolled road surface and is defined in Equation 10.2:

$$CE = 1 - \left(\frac{T}{U}\right) \times 100 \qquad (10.2)$$

where CE = control efficiency, percent;
U = untreated or uncontrolled road dust concentration, mg/m³; and
T = treated or controlled road dust concentration, mg/m³.

Generally, the higher the control efficiency for a dust control agent, the better its performance, which results in less dust emissions from the road surface. The control efficiencies presented are for total suspended solids (TSP).

Road Preparation

Application of many of the dust suppressants requires the haul road to be properly prepared. The equipment used for the application of the dust suppressants includes: a road grader equipped with scarifiers (refer to Figure 10.4), a compactor, and a water truck for distribution of the suppressant to the road surface.

Proper planning is necessary when applying many dust suppressants, as many require a significant amount of time to cure before traffic can return to the road. Additionally, depending upon the amount of road that requires application, several days of dry weather will be needed to properly apply the suppressant. The road should be fully prepared before applying the suppressant, as follows:

- any blade work required to eliminate road corrugations and potholes should be completed before dust suppression is initiated;
- any large material that would not make for a good road surface should be bladed off the road;
- the road should have a good crown to eliminate the potential for any standing water because it can have a detrimental effect on the road surface, creating potholes; and
- the dust suppressant is applied as the final step.

Accomplishing the proper road preparation during application will allow the dust suppressant to work at its maximum efficiency. For the application of the dust suppressant, the road surface should be scarified to loosen the top 1 to 2 inches of road surface. There are two variations to applying the dust suppressant at this point, as discussed below.

- The first variation, generally used with salt suppressants, is to ensure that a 1- to 2-inch loose layer is spread evenly across the road surface. Next, the dust suppressant is applied to the loose layer at the recommended rate. Finally, the surface is compacted, preferably with a pneumatic roller, although other types can be used. Traffic should be kept off the road until the dust suppressant cures [Skorseth and Selim 2000]. Curing time is dependent upon the dust suppressant applied (generally 24 hours is sufficient).

- The second variation is to blade most of the loose material into windrows on both sides of the road. These windrows are used to ensure that application of the dust suppressant does not run off the road surface. One-third of the dust suppressant is applied to the road surface at the recommended rate between the windrows. Next, the windrows are bladed to the center, spreading evenly across the road surface. Next, another 1/3 of the dust suppressant is applied to the road surface. The road is regraded to mix the dust suppressant and the aggregate. The final 1/3 of dust suppressant is applied to the road surface as a topcoat, after regrading is complete. Excess runoff of the dust suppressant should be avoided when applying this final topcoat. Finally, the road surface is compacted, preferably with a pneumatic roller, completing compaction before the dust suppressant dries. The road should be allowed to cure (generally 24 hours, but dependent upon the dust suppressant used) before traffic returns to it [Midwest Research Institute 1981].

Keep in mind that these are general guidelines for dust suppressant application. Specific guidelines for dust suppressant application should be obtained from the manufacturer of the dust suppressant chosen for use on the road.

Water

Watering roads is the most common method used for haul road dust control. Its application is the simplest and the easiest of all dust control measures, as it does not require any road preparation prior to application. Application is generally achieved through the use of a water truck which sprays water onto the road (Figure 10.6).

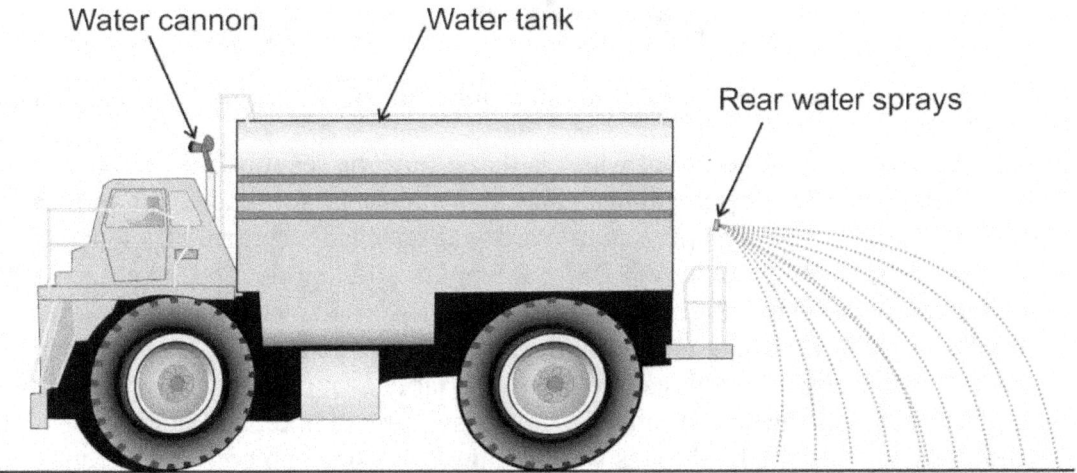

Figure 10.6. Water truck equipped with a front water cannon and rear water sprays.

Haul Roads, Stockpiles, and Open Areas 251

A water truck, used to apply water to the haul roads, consists of a tank, a pump, and the plumbing associated to send water through nozzles which are located at the rear of the truck. Additional side sprays may be available with some spray configurations but are optional. The tank capacity can range up to 30,000 gallons depending upon the size of the truck, with the larger capacities used with large off-road truck chassis. Some water trucks may have a water cannon mounted (see Figure 10.6) that can be used for wetting down muck piles for dust control during loading operations. The water cannon is controlled by the operator and can be rotated and elevated to send water to the desired location.

The spray nozzles for road watering are generally fan sprays and are mounted on the truck at a stationary position. These nozzles can be elaborate fan spray nozzles fabricated by a manufacturer (Figure 10.7) or they can be simple nozzles. Simple nozzles are constructed by drilling holes along the bottom of a water pipe along the back of the water truck to allow water to wet the road or by cutting horizontal slots in water stand-pipes (or endcaps) at the corners of the water truck. It is best to orient the nozzles in a manner to minimize overlap of the water spray from other nozzles [James and Piechota, 2008] in order to achieve the best spray coverage. The operator has the ability to start and stop the water spraying by turning the pump on or off through a switch located in the cab. Some water trucks allow the operator to turn on or off each individual water spray on the truck, allowing flexibility in water application.

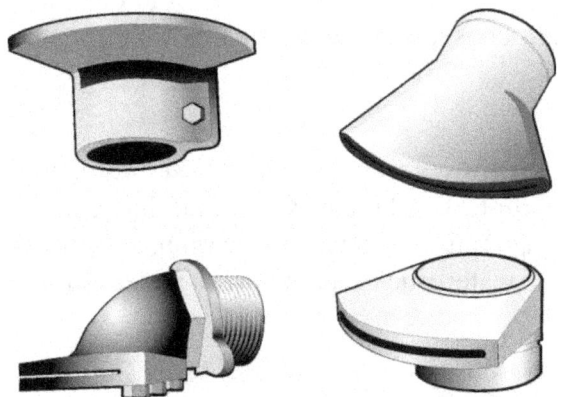

Figure 10.7. Various types of manufactured fan spray nozzles for use on a water truck.

The disadvantage of the use of water is that it must be reapplied on a consistent basis. The U.S. EPA has reported several test results of watering haul roads. An application of water at the rate of 0.13 gallons per square yard (gal/yd^2) had a control efficiency of 95 percent for total suspended particles (TSP) for 0.5 hours after application. Another test showed that an application of water at a rate of 0.46 gal/yd^2 had a control efficiency of 74 percent for TSP for the 3-4 hours following the application of water [EPA 1998]. However, the control efficiency for water can be highly variable as it is highly dependent upon road material type, traffic, and weather conditions. For example, another independent study showed that watering haul roads with a water truck once an hour had a control efficiency of 40 percent for total suspended particulates (TSP). When watering increased to once every half hour, the control efficiency for TSP increased to 55 percent [USBM 1983].

During a field study by the National Institute for Occupational Safety and Health (NIOSH), the application of water was shown to prevent high dust emissions from the haul road for several hours after application during a hot dry day when the skies were clear of any clouds [Organiscak and Reed 2004]. The application rate of the water was not measured, but was observed to be very liberal as observations of the water truck application noted water runoff from the haul road. During the day the temperature ranged from a low of 77°F increasing to 90°F and the relative humidity ranged from a high of 64 percent decreasing to 37 percent throughout the day. Figure 10.8 shows the dust concentrations measured throughout the day with one water application at approximately 10:00 a.m.

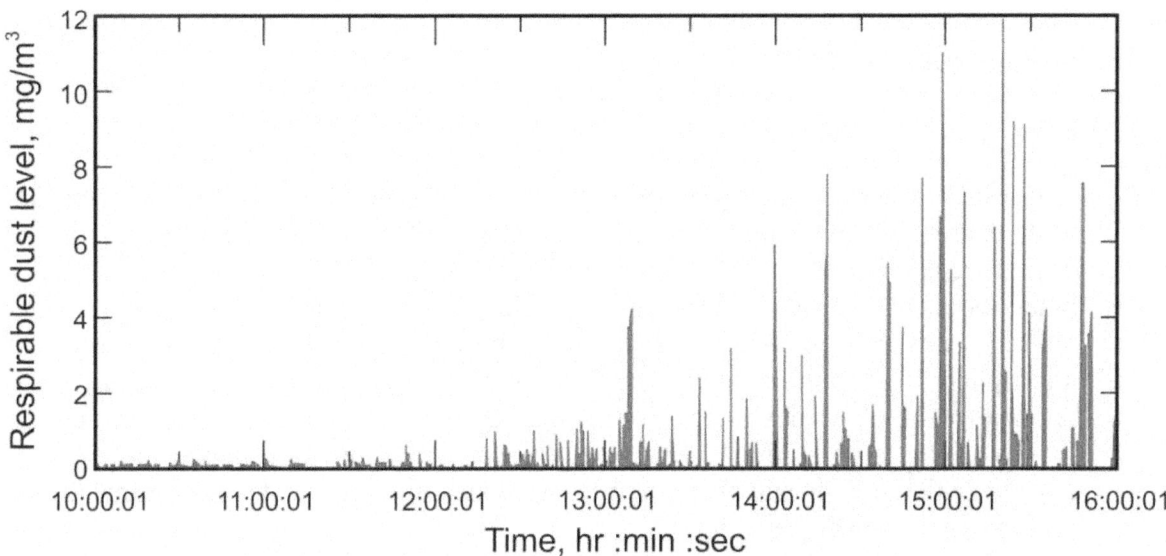

Figure 10.8. Respirable dust concentrations measured from haul road after water application occurred at 10:00 a.m. [Organiscak and Reed 2004]

There are no published guidelines for the amount of water to use for dust control on haul roads. In the past, an old U.S. EPA emissions factor had stated that days having as little as 0.01 inches of precipitation resulted in no dust emissions from haul trucks [EPA 1998]. However, it was not until recently that the optimal haul road moisture content for best reduction of respirable dust was found to be approximately 2 percent in a study completed in Kansas [Mulesky and Cowherd 2001]. Additionally, except for the application intervals previously mentioned, there are no published guidelines for determining optimum haul road watering intervals. Determining optimum haul road watering intervals would be dependent upon the amount of water applied to the road, the time between water applications, the traffic volume of the road, and the prevailing meteorological conditions [Cowherd et al. 1988].

Surfactants

Surfactants or wetting agents can be added to the water to extend its life as a dust control agent. Like water, surfactants generally do not require any special application procedures, except to remove any excess loose material from the road surface.

Surfactants work by reducing the surface tension of water. This allows the particles from the road surface to better penetrate the water droplets, thereby increasing the potential surface area coverage of the droplet (i.e., the penetration of the water droplet lets the water wet a larger surface area). The reduced surface tension also permits easier penetration of the treated water into a few inches of the sub-surface. This action allows the wetted sub-grade to act as a reservoir of water such that, once the surface moisture evaporates, the sub-grade can provide moisture for dust control through capillary action, thus extending the life of the water [Midwest Research Institute 1981].

There is very little information about the use of surfactants to extend the effective life of watering haul roads. However, observations have noted that the time between watering roads can be extended 33–50 percent when surfactants are used [Midwest Research Institute 1981].

Salts

Salt solutions are commonly mixed with water and used for haul road dust control. Magnesium chloride is one of the most common salt-based dust control agents. Other agents are calcium chloride, hydrated lime, and sodium silicates, with calcium chloride being the most common and having properties similar to magnesium chloride. Calcium chloride is hygroscopic and deliquescent, meaning that it absorbs moisture from the atmosphere to keep the road in a moist condition. In very dry climates, periodic application of water may be required to maintain road performance at acceptable levels [Midwest Research Institute 1981]. Application procedures for magnesium chloride/calcium chloride are similar to those described for dust suppressants in the section titled "Road Preparation."

After application to control haul-truck-generated dust, magnesium chloride control efficiencies were shown to average 95 percent for up to 22 days after application [USBM 1987]. For calcium chloride, two weeks after the initial application, the control efficiency was measured as 82 percent and decreased over time to 14 percent at seven weeks after application [USBM 1983]. This decline has been attributed to road surface wear due to traffic volume and precipitation, in that chlorides are water soluble and can be washed away.

Advantages of using chlorides for dust control are that they absorb moisture from the atmosphere to maintain the moisture content of the road at a higher level than normal. They can also act as a de-icer to thaw ice and snow on roads in freezing conditions. Also, chlorides generally do not require time to cure after application.

Disadvantages are that chlorides may cause corrosion on equipment using the roads sooner than normal. They can also be harmful to vegetation and to personnel if skin or eye contact occurs. Chlorides are also water soluble and may leach from the road surface during precipitation, thereby degrading in performance over time [Midwest Research Institute 1981].

Petroleum Emulsions

Petroleum resins are generally engineered products or byproducts of lubrication oil manufacturing. They generally consist of stable emulsions of petroleum residuals, solvent extracts, and acid sludge. If the road surface is in good condition, then application is relatively simple. Some products require the steps of scarifying to a 1-inch depth, applying the resin, and compacting the road surface; others can be applied by direct application of the resin to the road surface without any conditioning. Most petroleum resins do require a 24-hour time period allowing the road surface to cure. Therefore, it is important to keep traffic off the road for the 24-hour time period after application.

The practice of using petroleum resins for dust control on access roads has reduced dust potential considerably, with reports that the access roads have been kept dust free for periods of 6 months to 1 year. Haul roads have been reported to be relatively dust free for a period of 3 to 4 weeks.

For petroleum emulsions, documented control efficiencies are highly variable. For haul-truck-generated dust at a sand and gravel operation, control efficiencies can be up to 70 percent for petroleum derivatives for up to 21 days after application [USBM 1987]. Other testing has shown control efficiencies ranging from 4 to 38 percent during a 4-week period after application at several surface coal mines [USBM 1983]. Other differences in the variations of control efficiency could be due to the type of petroleum emulsion used, method of application, type of vehicle traffic, and measurement method.

A primary advantage to using petroleum emulsions is that they are not corrosive. They are also not water soluble; therefore, rain will not displace them (nor wash them away unless improperly applied) and they will not evaporate. They are relatively nontoxic and nonflammable, and do not have adverse effects on plant growth (for revegetation needs). Disadvantages are that they must cure for 24 hours after application. Therefore, traffic should not be allowed during this time period. Traffic should be limited to wheeled equipment only, because tracked equipment running on the road surface will break down the treated surface, resulting in poor dust control of the road surface. Finally, storage temperatures of the emulsion products prior to application must be controlled as they cannot endure freezing or boiling conditions [Midwest Research Institute 1981].

Polymers

Polymers include acrylics and vinyls, which are chemical additives, mixed with water to form a diluted solution, then applied to the road surface topically. Application is dependent upon the brand of polymer used, but it is generally a simplified procedure of preparing the road surface prior to application, then spraying the solution topically onto the surface in a series of applications.

Although many brands of polymers are used in haul road dust control, only a few studies have documented their control efficiency. One study conducted on mine haul roads reported that a specific polymer had control efficiencies of 74 to 81 percent within 4 weeks of application, reducing to 3 to 14 percent after 5 weeks of application [USBM 1983]. The variability in the control efficiency was stated to be caused by precipitation, which differed for the various sites. Another study, conducted on a public dirt road, demonstrated control efficiencies of 94 to 100 percent within a week of application of the same specific polymer, reducing to control efficiencies of 37 to 65 percent at approximately 11 months after application [Gillies et al. 1999].

The advantage to using polymers is that they are generally noncorrosive and nontoxic. As a dust control agent, they are also generally long-lived, although it has been shown that precipitation can affect their longevity. Application procedures can vary by product from the simple procedure of initial road preparation and spraying the solution onto the road surface with a water truck to complex procedures requiring initial road preparation and additional blading and regrading during polymer solution application. Additionally, polymers can be utilized for soil stabilization.

Adhesives

Adhesives are compounds, solutions, formulas, etc. that are mixed with the soil surface to form a new road surface. One of the most common adhesives is lignin sulfonate, which is basically a waste product from the paper/pulp industry and a well-established dust control product. It is created when the wood chips are placed in a sulfonate solution. The solution absorbs the lignin or binder from the wood chips, creating the lignin sulfonate solution used for dust control [Midwest Research Institute 1981].

Application of lignin sulfonate involves procedures mentioned previously in the section titled "Road Preparation." When properly applied, lignin sulfonate has been observed to keep access roads dust free for periods ranging from 6 months to two years with periodic application of the solution [Midwest Research Institute 1981]. Heavily traveled haul roads have been observed to be kept dust free for periods of 3 to 4 weeks, after which time another application of the solution should be applied. The reapplication can be a simple process by spraying the solution onto the haul road, unless the road is in poor condition, in which case the procedures previously mentioned in "Road Preparation" would need to be followed [Midwest Research Institute 1981].

Field testing of lignin sulfonate demonstrated control efficiencies ranging from 50 to 63 percent for up to 4 weeks after application, with the highest control efficiencies occurring directly after application [USBM 1983]. Other test sites had lower control efficiencies of 31 to 45 percent, but this could be due to different road material types, equipment types, and precipitation events. Precipitation events can have a significant influence on control efficiency, especially since lignin sulfonate is water soluble and can be washed away with large rainfall events [USBM 1983].

Other advantages to using lignin sulfonate as a haul road dust control agent are that it is noncorrosive and is easily obtainable due to the large size of the paper/pulp mill industry. Disadvantages are that it can interfere with some mineral processing processes, such as flotation, and since it is water soluble it can be washed away from the road surface, requiring reapplication of the lignin sulfonate to maintain proper dust control [Midwest Research Institute 1981].

Other Dust Control Methods

Speed Control

Reducing speed of the vehicles traveling on haul roads can be an effective method for dust control. However, this method conflicts with maximizing production. The haul trucks are designed to operate at an optimum speed; to lower the speed of the trucks would lower the production rate of the mine, which may not be desirable. Nevertheless, reducing the speed of vehicles has been shown to reduce the potential generation of dust particles <10 µm by approximately 58 percent when speeds were reduced from 25 to 10 miles per hour (mph), and 42 percent when speeds were reduced from 25 to 15 mph [Watson et al. 1996]. In another study, limiting speeds on unpaved roads to 25 mph demonstrated a control efficiency of 44 percent [Countess Environmental 2006].

Traffic Control

In addition to the mine haul trucks on-site, many mine sites have on-road trucks that enter the site to ship material to their customers. These trucks can often times come in batches of 2-3 trucks or more at a time. If the trucks travel on unpaved roads, maintaining a following distance between trucks by a time interval of 20 seconds has been shown result in up to a 52 percent reduction of respirable dust exposure to the trailing truck driver [Reed and Organiscak 2005]. Additionally, this 20-second time interval allows for the dust cloud generated from the lead truck, which can impair the visibility of the trailing driver, to dissipate. This can also reduce the possibility of an accident due to impaired visibility from the dust generated by trucks.

Load Covers

Covering the load of a haul truck using tarps can prevent the loaded material from becoming airborne during transportation. This method is not commonly used for off-road mine trucks as the speed of loaded trucks rarely exceeds speeds where entrainment of material occurs (neglecting ambient wind effects). Entrainment of material can occur when airflows coming in contact with the material exceed 13 mph for small diameter material (0.10 mm); higher velocities are required for larger material [Chepil 1958]. However, loaded highway trucks can exceed speeds where airflows contacting the loaded material exceed this threshold, resulting in dust emissions as the material becomes airborne. Therefore, covering with tarps (Figure 10.9) or surface wetting of the loaded material should be done to prevent it from being windblown. Additionally, the truck should not be overloaded. Material should be centered and a freeboard—which is the distance from the top of the dump truck bed down to the top of the material loaded into the bed—of three inches or more should be maintained. In cases where the material in the dump bed is mounded, the crest of the mound should not exceed the height of the dump bed [Blue Skies Campaign, Arizona 2010]. This will minimize spillage and allow the tarp to effectively trap the material in the truck bed.

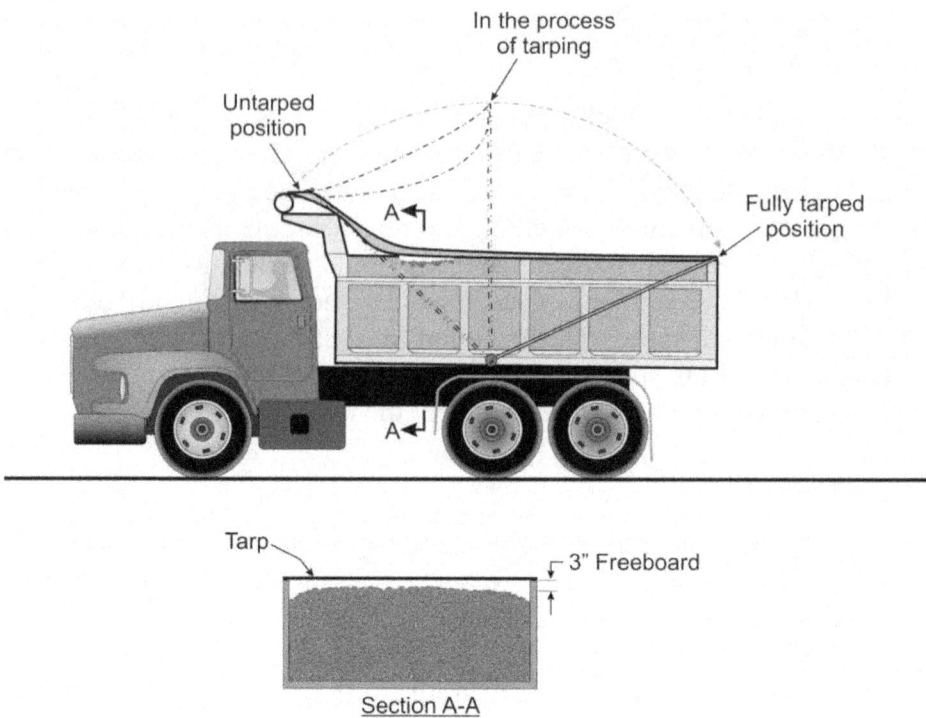

Figure 10.9. Self-tarping dump truck. The blue arrow indicates the direction of movement when tarping. Section A-A shows the recommended freeboard.

Cab Maintenance

Maintaining equipment cabs in good operating condition also reduces operator exposure to respirable dust. It has been demonstrated that properly maintained cabs can attain dust reductions of 90 percent or more for drills [Chekan et al. 2003] and between 44 and 100 percent for dozers, with the variations of the dust reductions for dozers attributable to re-entrainment of internal cab dust [Organiscak and Page 1999]. For haul truck cabs that were retrofitted with a filtration/pressure air conditioning system to produce positive pressure in the cab, it was demonstrated that properly maintained cabs can also reduce respirable dust exposures by 59 to 84 percent [Chekan and Colinet 2003; Chekan et al. 2003]. It should be noted that the retrofitted trucks tested were old model (20-30 years old) haul trucks, and that newer haul trucks would have better cab integrity. The proper use and maintenance of cabs is discussed in more detail in Chapter 9—Operator Booths, Control Rooms, and Enclosed Cabs.

STOCKPILES AND OPEN AREAS

Stockpiles and open areas are common at surface mine sites. There are generally two types of stockpiles: long-term and short-term. A long-term stockpile undergoes no disturbance to the pile over a long time period (several months or more). An example would be a topsoil stockpile where topsoil is stored until reclamation of the mine site is performed. Short-term stockpiles could consist of product or material that is being stored temporarily until it is used in a process. Surge piles are common short-term stockpiles. Open areas are self-explanatory as they are large barren areas.

Dust can be generated from stockpiles and open areas by wind velocities that entrain material into the air. The United States Department of Agriculture states that erosion will take place when "the surface soil is finely divided, loose, and dry; the surface is smooth and bare; and the wind is strong."[9] As stated previously, wind speeds as low as 13 mph can result in entrainment of material for small diameter material (0.10 mm) [Chepil 1958]. Conversely, there are several factors that can reduce wind erosion:

- soil that contains stable aggregates or clods that are large and dense to resist wind erosion;
- soil that is compacted, roughened, or kept moist;
- vegetated soil or soil covered by vegetative residue (for example, straw); and
- when wind velocity near the ground surface is reduced or eliminated [Chepil 1958].

These four factors translate to the four dust control measures that can be applied to both stockpiles and open areas: surface roughness, water, coatings, and wind barriers. Water, coatings, and to some extent wind barriers are generally applicable to stockpiles, while all four measures can be applied to open areas.

Surface Roughness

Surface roughness is important in wind erosion from open areas, but less important for stockpiles. Generally, the rougher the surface the less wind erosion affects it. This is true for both land plots containing vegetative stubble and bare surface plots. The roughness for each type of plot is created by the amount and orientation of the stubble for the stubble plots, and the amount of surface irregularities and large clods for the bare surface plot. The mechanisms for controlling wind erosion are to decrease the force of the wind on the surface or to trap the eroding material with larger particles. Increasing the magnitude of the roughness for each plot can result in a decrease in soil erosion [Zingg 1951]. The best dust control for open areas is to plant vegetation to cover the entire area. Typically the vegetation is a grass, clover, alfalfa, etc.

One control measure for open areas is to roughen the surface by using chisels, such as that provided by a chisel-pointed cultivator (Figure 10.10) used in agriculture [Chepil and Woodruff 1955]. The depth of tilling should be 3 to 6 inches. If higher wind speeds are expected then the depth should be increased to 8 to 14 inches. The objective of this tilling is to expose the rough large cloddy soil particles to the surface (Figure 10.11). The large clods protect the smaller particle material from wind erosion [Woodruff et al. 1977]. In order for the clods to be truly effective, they must cover the majority of the surface. This tillage also produces ridges and furrows which have the potential to reduce erosion 50–90

---

[9] Chepil, W.S., "Soil Conditions that Influence Wind Erosion." *United States Department of Agriculture Technical Bulletin No. 1195.* June 1958. pg. 4.

percent [Fryrear and Skidmore 1985]. However, this method is considered to be a temporary emergency measure to be used as a last resort in controlling wind erosion in agricultural settings [Chepil 1958]. Additionally, for this method to be most effective, the tilling must be accomplished at right angles to the wind direction and completed slowly to prevent the large clods from disintegrating [Woodruff 1966].

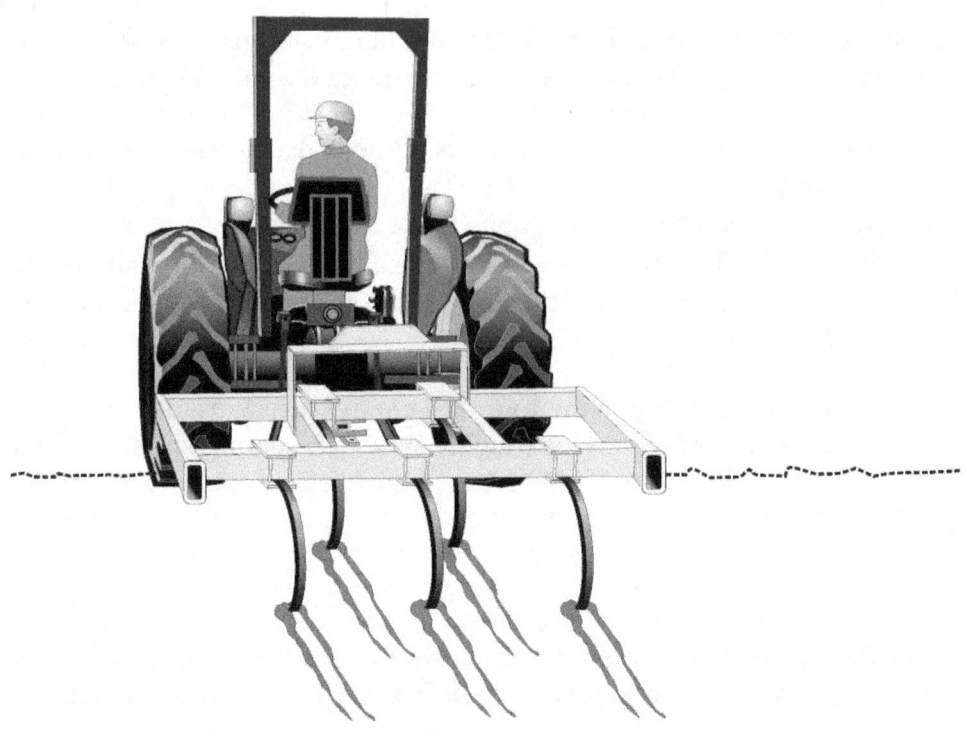

Figure 10.10. Chisel plow used for creating large clods in open areas to reduce wind erosion.

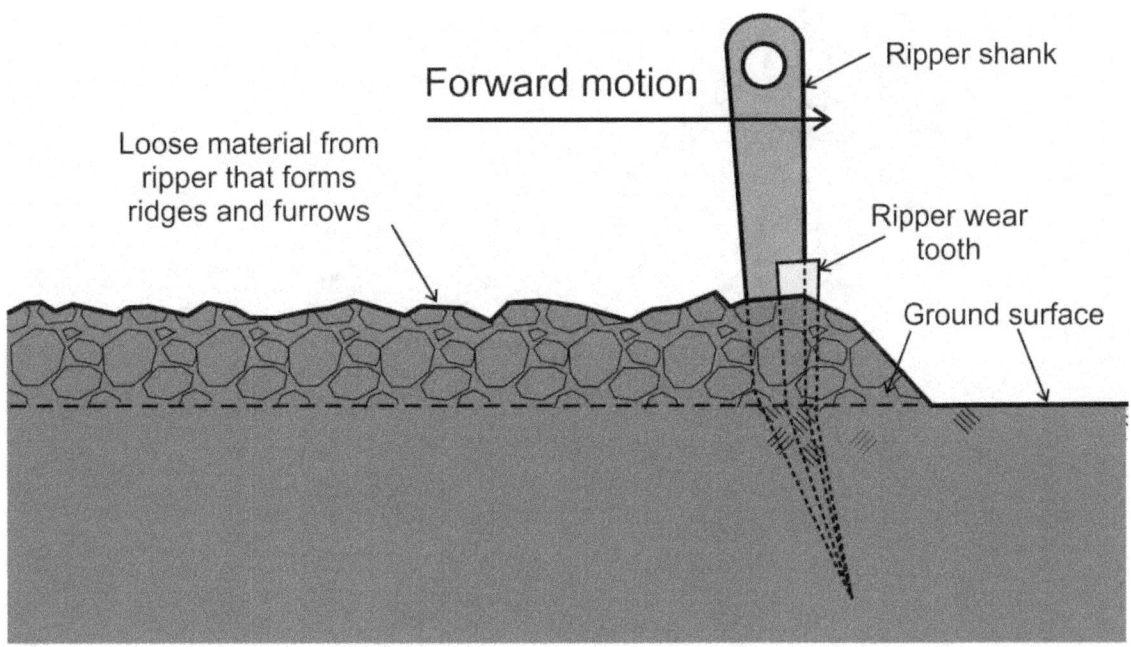

Figure 10.11. Example of tillage (from chisel plowing, for example) exposing large clods.

Tillage ridges prevent wind erosion by increasing the surface roughness so that the velocity required for material entrainment for a tilled surface is higher than that for a smooth surface. The ridge tops generally consist of loose erodible soils (unlike the un-ridged surfaces) that become armored through a process where the larger particles are left behind when the smaller loose particles erode away [Hagen and Armbrust 1992; Lyles et al. 1974]. After this armoring occurs, the wind velocities required for entrainment for the ridge tops become higher than those for smooth surfaces, thus helping to prevent wind erosion. For a series of tillage rows, this same armoring process also occurs at the upwind side of the ridges, which results in any loose particles leaving the armored upwind side to have to cross over many ridges, and eventually many of the loose particles become trapped in the troughs and on the armored faces (Figure 10.12). These processes can result in the tilled or ridged surfaces that reduce emissions up to 50 percent as compared to smooth untilled surfaces [Hagen and Armbrust 1992].

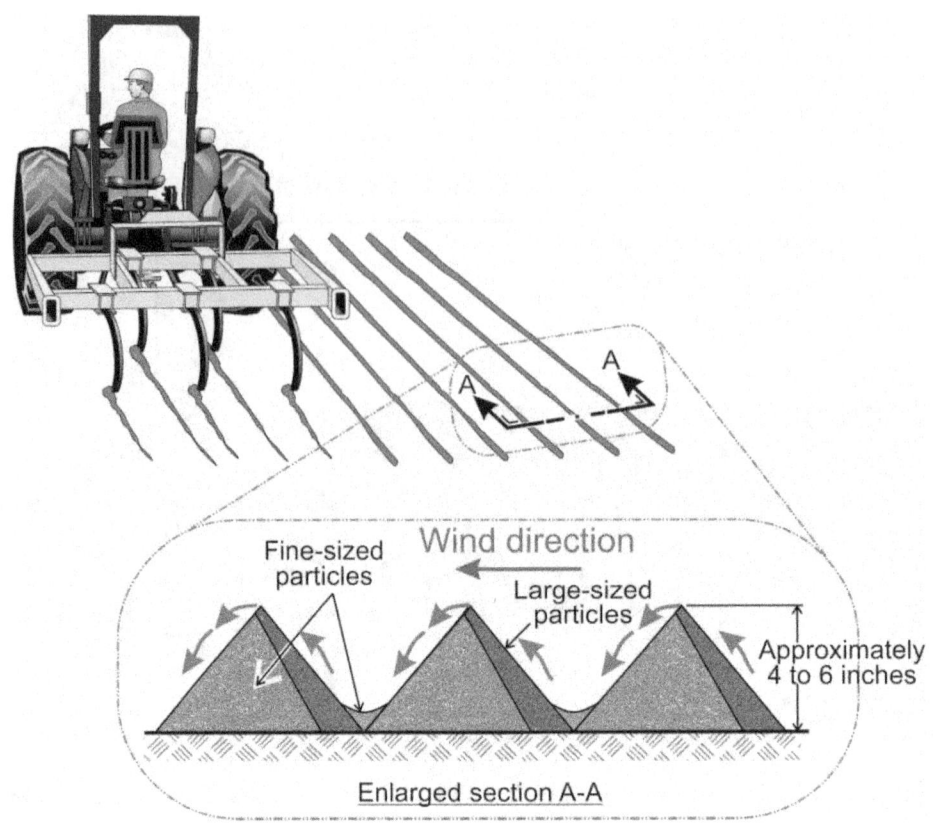

Figure 10.12. The armoring process for tillage rows. Blue arrow denotes wind direction perpendicular to tillage rows. Magenta arrows depict fine material being eroded from the windward side of ridges leaving large-sized particles, with the fine material being carried over ridge tops and deposited in the troughs on the leeward side of the ridges.

Another method for reducing wind erosion is the use of gravel blankets. The use of fine (0.02- to 0.25-inch), medium (0.1- to 0.5-inch), and coarse gravel (0.25- to 1.5-inch) applied at 20, 50, and 100 tons per acre, respectively, results in insignificant amounts of wind erosion (<25 pounds per acre). The smaller the diameter of the gravel, the smaller the amount of gravel required to prevent wind erosion. However, this only applies to a diameter down to 2 mm. Once the gravel diameter is less than 2 mm, wind erosion will increase, as particles <2 mm in the gravel will make the surface more erodible [Chepil et al. 1963].

Water

Water is an obvious method of dust control for stockpiles and open areas that has been recommended for use at mine sites. The water mixes with the material to create a crust which can be resistant to wind erosion. This crust generally requires a higher wind velocity to erode material. However, once any erodible material has scoured and the crust has worn through, then lower wind velocities may potentially cause erosion [Chepil 1957].

Cementation of the soil also affects soil erodibility and can be used for dust control of both stockpiles and open areas. Cementation occurs from alternate wetting and drying of the soil, forming a surface crust generally 1/16-inch in thickness. The cementation of the soil surface becomes compact and mechanically stable, thus becoming more resistant to wind erosion. The soil must have an appropriate amount of fine material in order to bind the larger materials to form the crust. However, the degree of cementation is hard to detect, making it difficult to measure the amount of cementation required for wind resistance, i.e., the number of wet/dry cycles required to prevent erosion [Chepil 1958].

The amount of moisture in the soil also affects the erodibility of soil. Basically, as the moisture is increased, the erodibility of the soil decreases [Chepil 1956]. This is important for dust control of open areas. However, it has been shown that even soils with high moisture contents can be susceptible to wind erosion. With loose soil, with a depth of 5 cm, the erodibility of a moist soil can be similar to that of a dry soil if the wind speed applied to the moist soil is 15–20 mph more than that applied to the dry soil [Cowherd and Grelinger 1996]. Therefore, while wetting is a good dust control measure, the area must still be monitored to maintain the proper amount of moisture to prevent erosion.

Coatings

Coatings as a dust control method generally include water or a myriad of chemical stabilizers in commercial production. The best coatings that prevent wind erosion form a crust on the surface of the material or soil. Additionally, the best coatings are nontoxic to plants, water permeable, and allow the material to be handled easily. The drawbacks to coatings are that they are expensive, and once the crust created by the coating is penetrated, erosion readily resumes [Diouf et al. 1990; Armbrust and Dickerson 1971; Lyles et al. 1969].

Mixing bentonite clay with soil is a dust control for open areas. Water is applied, allowing the soil-bentonite mixture to form a crust on only the top of the mixture. Beneath the crust the material is unconsolidated, as water is not able to penetrate the crust. Bentonite concentrations that are too high leave easily erodible sand particles (<1.0 mm diameter), while lower bentonite concentrations are desirable as the surface becomes tightly packed. However, it should be noted that a sand-only soil mixture with water application can resist erosion better than the soil-bentonite mixture, at least until an abrader is introduced. An abrader is an implement used to scrape or scratch the soil surface. The soil-bentonite mixture is the better choice if the surface will be abraded, because it will be more resistant to wind erosion than the sand-only soil mixture [Diouf et al. 1990].

Recently, as a dust control for open areas, a mineral operation in Wyoming applied a Guar Gum polymer to a tailings pond surface at a trona mine (sodium carbonate and sodium bicarbonate mixture). Guar Gum is extracted from the Guar Bean which can be fed to cattle. Industrial uses are in the oil industry to facilitate drilling and prevent fluid loss. It also has use in the minerals industry as a flocculant. At the trona operation in Wyoming, when the Guar Gum solution contacted the sodium bicarbonate strata of the tailings pond surface, bonds were formed which, once dry, created a crust which made the material resistant to wind erosion. This crust, because of the polymer involved, was high-strength, long-lived, and was resistant to microbial activity [Fuller and Marsden 2004].

A current method of stockpile formation for long-term storage is to build the stockpile and compact it using bulldozers (Figure 10.13). The compaction helps to prevent wind erosion. Additionally, coatings or binders can be used to aid in dust control. Wax mixtures (hot-melt formulations) and polyethylene and polypropylene latex emulsions have been found to be useful in past applications for coal stockpiles and could be adapted to be applied to other mineral stockpiles. The latex emulsion coatings are thinner and more water-permeable than the hot-melt formulations. The latex emulsion coatings also require a surface preparation of spreading fines over the entire stockpile prior to application. However, all coatings, both latex emulsion and hot-melt formulations, are significantly better at dust control over stockpiles with or without compaction (with the compacted stockpile better at dust control than stockpile without compaction).

One disadvantage to latex emulsion coatings for dust control is that this type of coating must be applied at temperatures >50°F, whereas hot-melt formulations can be applied in any weather condition [Kromrey et al. 1978]. Acrylic resins or polymers can also be used, but require a three-day curing period to form a crust on the surface of the stockpile. These coatings are very wind resistant to erosion, except if the stockpile surface is disturbed through settling or cracking, in which case wind erosion can resume as occurs with an untreated stockpile [Smitham and Nicol 1991].

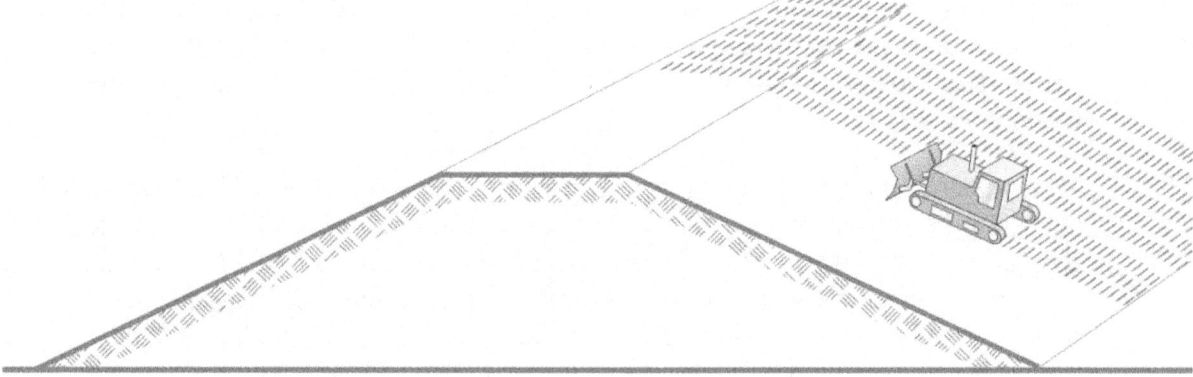

Figure 10.13. Dozer being used to compact stockpile for long-term storage.

Micro-foam is another method for controlling dust from fugitive dust sources, particularly load-out stockpiles. Typical foams used in firefighting (large 5-mm bubbles) are not effective at dust control, unlike micro-foam (small <100-µm bubbles). Micro-foam is stated to be better than water for dust suppression, because the water droplets, to be effective for dust control, must be similar in size to the dust particles which micro-foam can replicate. In addition, the velocity of the water must be high to break its surface tension upon contact with the dust particle. These dust particles penetrate the foam bubble, causing the bubble to break and wet the particles. Large particles are not a problem, as the small micro-foam bubbles wet the larger particles without affecting the bubble. Micro-foam can be injected at material transfer points in order to obtain optimum dust control, requiring approximately 0.4 gallons of water for each ton of bulk material treated for dust control [Eslinger 1997].

Wind Barriers

Wind barriers for wind erosion of open areas generally consist of vegetation (grasses, shrubs, trees, etc.). This vegetation can be grown on the edges of fields to attempt to stop the erosion, or as crop strips through the field [Woodruff et al. 1977]. Collection efficiencies of different types of plants and their differing optimal spacings, for <10 µm particles, can vary from 35 to 80 percent [Hagen and Skidmore 1977].

Vegetation can also be grown on topsoil stockpiles as they are most amenable to being vegetated (Figure 10.14). The topsoil is generally seeded after placement of the stockpile and all work activity has ceased. The vegetation creates a barrier preventing the high wind velocities from contacting the stockpile surface, thereby preventing entrainment of topsoil particles. The vegetation also prevents water erosion of the topsoil from the stockpile. Most materials other than topsoil that are stockpiled are not amenable to vegetation and must be watered or coated.

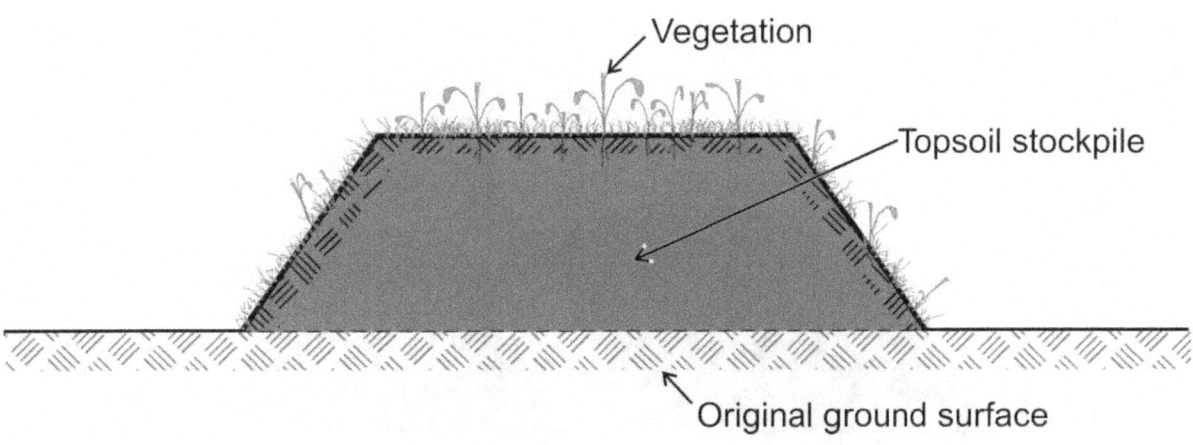

Figure 10.14. Example of a vegetated topsoil stockpile.

Solid board fences, snow fences, burlap fences, etc. can be used to control wind erosion of open areas. The fences should be 3 to 5 feet in height to be effective [Woodruff et al. 1977]. These fences are used to trap suspended dust. For use near roadways, optimal placement for maximum trapping efficiency is approximately 10–20 barrier heights upwind of the road (Figure 10.15). Additionally, the barrier material should have a porosity of approximately 30 to 40 percent for best performance [Hagen and Skidmore 1977]. Barriers or wind fences that are oriented perpendicular to the prevailing wind direction reduce saltation at the soil surface, and reduce the amount of collected airborne particles downwind of the fences by 90 percent [Grantz et al. 1998]. Barriers can also be used for stockpiles, but the barriers must be built to enclose the entire stockpile, making the barrier quite large, but effective (Figure 10.16). The stockpile barriers are generally constructed of common building materials.

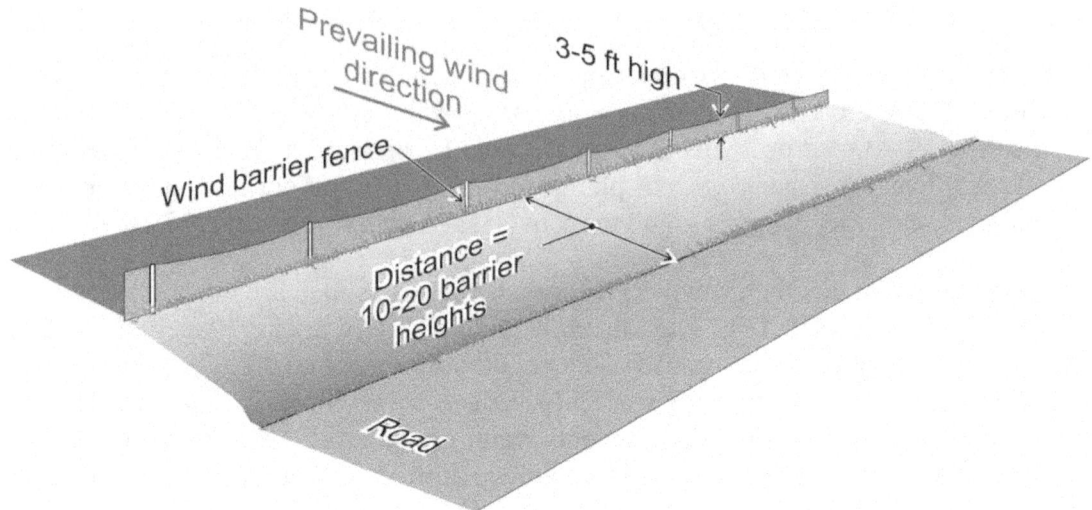

Figure 10.15. Demonstration of how fences are used as wind barriers.

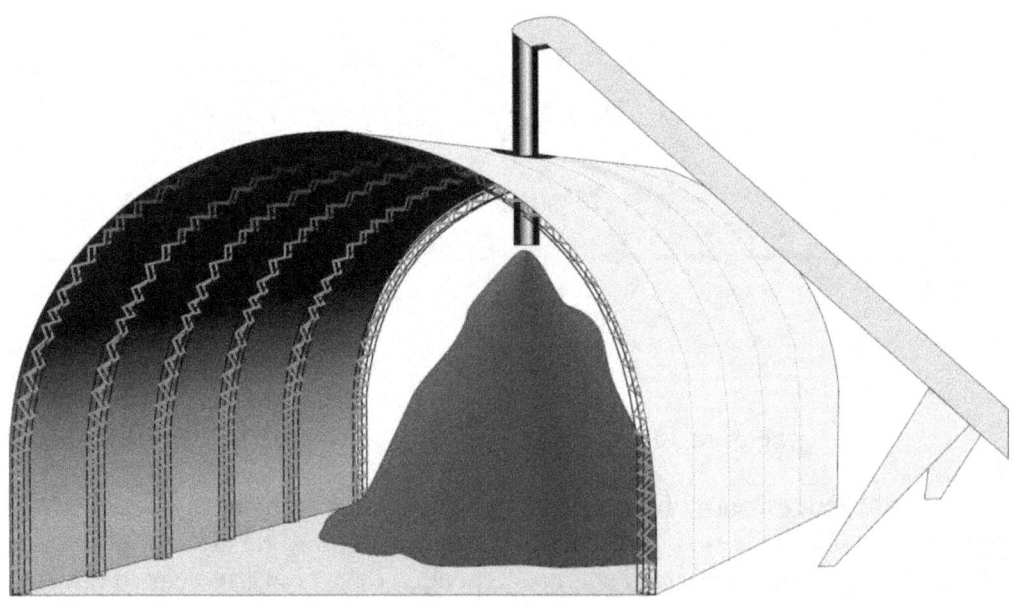

Figure 10.16. Barrier built to enclose a stockpile.

Other types of barriers that control wind erosion from open areas are dispersed barriers and earthen banks. Widely dispersed barriers can be effective in reducing the amount of windblown soil up to a height of 6 feet [Grantz et al. 1998]. Figure 10.17 illustrates the widely dispersed barriers using shrubs that were placed on a 30 x 7.5 foot rectangular grid and were protected from herbivores using cones having a base diameter of 8 inches, top diameter of 4 inches, and height of 2 feet. The cones were anchored to the ground using stucco wire netting. Earthen banks with heights of 2 feet are also effective for preventing wind erosion from open areas [Woodruff et al. 1977].

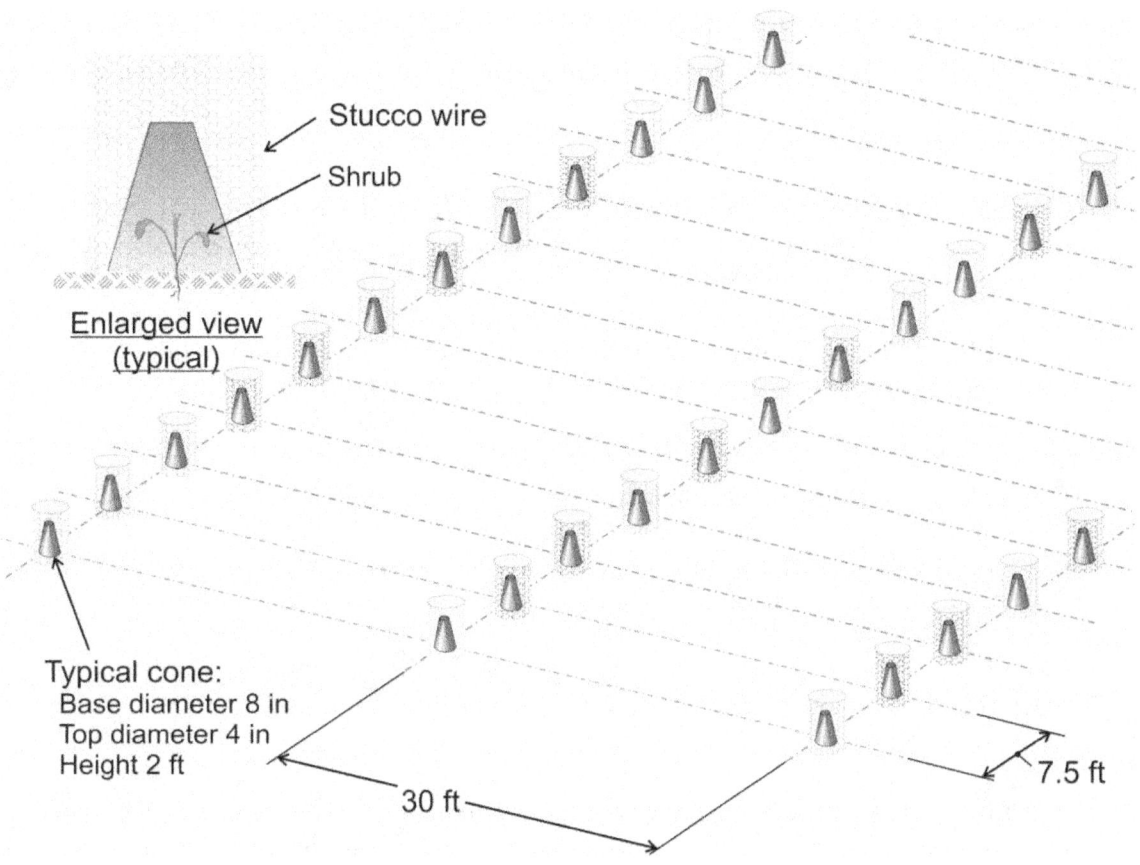

Figure 10.17. Layout demonstrating dispersed wind barriers.

REFERENCES

Amponsah-Decosta F, Annegarn HJ [Jan.-March 1998]. Assessment of fugitive dust emissions from an opencast coal mine. J of the Mine Vent Soc of South Africa 51(1):5–11.

Armbrust DV, Dickerson JD [1971]. Temporary wind erosion control: cost effectiveness of 34 commercial materials. J of Soil and Water Cons 26(4):154–157.

Bagnold RA [1960]. The physics of blown sand and desert dunes. London: Methuen & Co. LTD.

Blue Skies Campaign, Arizona [2010]. Quick reference dust control guide. Arizona Department of Transportation, Air Quality Planning, Research, Final Project Reports, Dust Control Training Templates BlueSkies Summary Sheet. [http://www.azdot.gov/mpd/air_quality/pdf/Blueskies.pdf]. Date accessed: December 3, 2010.

Caterpillar, Inc. [2008]. 785 mining truck specifications. [http://www.cat.com/cda/layout?m=169215&x=7&f=227128]. Date accessed: October 12, 2008.

Chekan GJ, Cecala AB, Colinet JF [2003]. An evaluation of cab filtration and pressurization systems: two case studies. Proceedings of the Environment, Safety & Health Forum and Expo, Alexandria, VA: National Stone, Sand, and Gravel Association, pp. 114-129.

Chekan GJ, Colinet JF [2003]. Retrofit options for better dust control—cab filtration, pressurization systems prove effective in reducing silica dust exposures in older trucks. Aggregates Manager *8*(9):9–12.

Chepil WS [1956]. Influence of moisture on erodibility of soil by wind. Soil Sci Soc of Amer Proc *20*(2):299–292.

Chepil WS [1957]. Erosion of soil by wind. In: 1957 Yearbook of Agriculture, pp. 308–314. U.S. Department of Agriculture.

Chepil WS [1958]. Soil conditions that influence wind erosion. U.S. Department of Agriculture Technical Bulletin 1195.

Chepil WS, Woodruff NP [1955]. How to reduce dust storms. Agricultural Experiment Station Circular 318. Kansas State College of Agriculture and Applied Science, Manhattan, Kansas.

Chepil WS, Woodruff NP, Siddoway FH, Fryrear DW, Armbrust DV [1963]. Vegetative and nonvegetative materials to control wind and water erosion. Soil Sci Soc of Amer Proc *27*(1):86–89.

Cole CF, Zapert JG [1995]. Air quality dispersion model validation at three stone quarries. Prepared for National Stone Association, Washington, D.C.

Countess Environmental [2006]. WRAP fugitive dust handbook. WGA Contract #: 30204–111. Western Governors Association, Denver, Colorado.

Cowherd C, Grelinger MA [1996]. High-wind failure of soil moisture as a wind erosion control. Air and Waste Management Association, 89th Annual Meeting & Exhibition, Nashville, Tennessee. Paper No. 96-TP40.04. Air & Waste Management Association.

Cowherd C, Grelinger MA [1997]. Dust control for prevention of vehicle accidents. Proceedings of the 1997 Air & Waste Management Association's 90th Annual Meeting and Exhibition, Toronto, Ontario, Canada.

Cowherd C, Mulesky GE, Kinsey JS [1988]. Control of open fugitive dust sources. Environmental Protection Agency Contract No. 68–02–4395, Midwest Research Institute Project No. 8985–14, Research Triangle Park, North Carolina: U.S. EPA, Office of Air & Radiation, Office of Air Quality Planning and Standards.

Diouf B, Skidmore EL, Layton JB, Hagen LJ [1990]. Stabilizing fine sand by adding clay: Laboratory wind tunnel study. Soil Tech *3*:21–31.

EPA [1998]. Section 13.2.2 Unpaved roads. In: Compilation of Air Pollution Emissions Factors, Vol. I: Stationary Point and Area Sources, 5th Edition AP-42. Research Triangle Park, North Carolina: U.S. Environmental Protection Agency, Office of Air Quality Planning and Standards, Emission Factor and Inventory Group.

Eslinger M [1997]. Controlling fugitive dust emissions with a foam suppression system. Cer Ind *147*(7):S3–S5.

Fryrear DW, Skidmore EL [1985]. Chapter 24, methods for controlling wind erosion. Follett RF, Stewart BA, eds. In: Soil Erosion and Crop Productivity. ASA-CSSA-SSA, Madison, Wisconsin. pp. 443–457.

Fuller J, Marsden L [2004]. Practical dust control agent and applications for alkaline ponds and playas. SME Annual Meeting, Denver, Colorado, 2004 SME Preprint #04–89. Littleton, Colorado.

Gillies JA, Watson JG, Rogers CF, DuBois D, Chow JC [1999]. Long-term effectiveness of dust suppressants to reduce PM10 emissions from unpaved roads. J of Air & Waste Man Assoc *49*:3–16.

Grantz DA, Vaughn DL, Farber RJ, Kim B, Ashbaugh L, VanCuren T, and Campbell R. [1998]. Wind barriers suppress fugitive dust and soil-derived airborne particulates in arid regions. J of Environ Qual *27*(4):946–952.

Hagen LJ, Armbrust DV [1992]. Aerodynamic roughness and saltation trapping efficiency of tillage ridge. Trans of the Am Soc of Agric Engrs *35*(4):1179–1184.

Hagen LJ, Skidmore EL [1977]. Wind erosion and visibility problems. Trans of the Am Soc of Agric Engrs *20*(5):898–903.

James DE, Piechota TC [2008]. Development of improved water spray patterns from construction water trucks. Presentation at Water Smart Innovations Conference and Exposition, Las Vegas, Nevada.

Kromrey RV, Scheffee RS, DePasquale JA, Valentine RS [1978]. Development of coatings for protection of coal during transport and storage. NTIS COO–4632–2. Prepared for U.S. Department of Energy, Division of Environmental Control Technology, Atlantic Research Corporation, Alexandria, Virginia.

Lyles L, Armbrust DV, Dickerson JD, Woodruff NP [1969]. Spray-on adhesives for temporary wind erosion control. J of Soil and Water Cons *24*(5):190–193.

Lyles L, Schrandt RL, Schmeidler F [1974]. How aerodynamic roughness elements control sand movement. Trans of the Amer Soc of Agric Engrs *17*(1):134–139.

Merritt FS, Loftin MK, Ricketts JT [1996]. Standard handbook for civil engineers, 4th ed. New York: McGraw-Hill.

Midwest Research Institute [1981]. Dust control for haul roads. U.S. Department of the Interior, Bureau of Mines Open File Report 130–81.

Moosmüller H, Varma R, Arnot WP, Kuhns HD, Etyemezian V, Gillies JA [2005]. Scattering cross-section emission factors for visibility and radiative transfer applications: military vehicles traveling on unpaved roads. J of the Air & Waste Man Assoc 55:1743–1750.

Mulesky GE, Cowherd C [2001]. Particulate emission measurements from controlled construction activities. U.S. Environmental Protection Agency, Office of Research and Development, Washington, D.C.

National Stone Association [1991]. The aggregate handbook. Barksdale RD, editor. National Stone Association, Washington, D.C.

Organiscak JA, Page SJ [1999]. Field assessment of control techniques and long-term dust variability for surface coal mine rock drills and bull dozers. Int J of Surf Min, Recl, & Environ. 13(4):165–172.

Organiscak JA, Reed WR [2004]. Characteristics of fugitive dust generated from unpaved mine haulage roads. Int J of Surf Min, Recl, & Environ 18(4):236–252.

Reed WR, Organiscak JA [2005]. The evaluation of dust exposure to truck drivers following the lead haul truck. Trans of the Soc of Min, Metal, and Expl, Inc., 318:147–153.

Reed WR, Westman EC, Haycocks C [2001]. An improved model for estimating particulate emissions from surface mining operations in the Eastern United States. Securing the Future, International Conference on Mining and the Environment, Proceedings. Skelleftea, Sweden: The Swedish Mining Association, pp. 693–702.

Skorseth K, Selim AA [2000]. Gravel roads, maintenance and design manual. South Dakota Local Transportation Assistance Program, U.S. Department of Transportation, Federal Highway Administration.

Smitham JB, Nicol SK [1991]. Physico-chemical principles controlling the emissions of dust from coal stockpiles. Powder Tech 64(3):259–270.

USBM [1977]. Design of surface mine haulage roads—A manual. By Kaufman WW, Ault JC. U.S. Department of the Interior, Bureau of Mines Information Circular 8758.

USBM [1983]. Cost effectiveness of dust control used on unpaved haul roads, Vol. 1 results, analysis, and conclusions. By Rosbury KD, Zimmer RA. U.S. Department of the Interior, Bureau of Mines Open File Report 106–85.

USBM [1987]. Fugitive dust control for haulage roads and tailings. By Olson KS, Veith DL. U.S. Department of the Interior, Bureau of Mines Report of Investigations 9069.

Watson JG, Rogers CF, Chow JC, DuBois D, Gillies JA, Derby J, Moosmüller H [1996]. Effectiveness demonstration of fugitive dust control methods for public unpaved roads and unpaved shoulders on unpaved roads. Final Report, DRI Document No. 685–5200.1F1, Prepared for California Regional Particulate Air Quality Study, Sacramento, California: California Air Resources Board.

Woodruff NP [1966]. Wind erosion mechanics and control. Proc First Pan Amer Cong of Soil Cons, Sao Paulo, Brazil, pp. 253–262.

Woodruff NP, Lyles L, Siddoway FH, Fryrear DW [1977]. How to control wind erosion. U.S. Department of Agriculture, Agricultural Research Service, Agriculture Information Bulletin No. 354.

Zingg AW [1951]. Evaluation of the erodibility of field surfaces with a portable wind tunnel. Soil Sci Soc of Amer Proc, pp. 11–17.

GLOSSARY

GLOSSARY

Abrader. An implement used to scrape or scratch a surface.

Accelerated silicosis. A form of silicosis resulting from exposure to high concentrations of crystalline silica and developing 5–10 years after the initial exposure. While accelerated silicosis presents with the nodular characteristics that typify chronic silicosis, accelerated silicosis develops only 5–10 years after the initial exposure and generally progresses more rapidly than chronic silicosis.

Access road. A road that is generally used for utility or maintenance purposes. This is usually a secondary road, which is smaller in design than a haul road. An access road is not normally used for moving ore to and from the pit and crusher locations.

Acute silicosis. A form of silicosis resulting from exposure to unusually high concentrations of respirable crystalline silica dust. Symptoms develop within only a few weeks to 4–5 years after the initial exposure. In contrast to the more common nodular forms of silicosis, acute silicosis involves a flooding of the alveolar region of the lungs with abnormal fluid containing fats and proteins, which gives rise to an alternative term for the disease—silicoproteinosis. Acute silicosis is uniformly fatal, often within several months of diagnosis.

Aerodynamic diameter. The diameter of a spherical particle that has a density of 1,000 kg/m^3 (the standard density of a water droplet) and the same settling velocity as that of the particle in question. Although particles have irregular shapes, this is an expression of a given particle's behavior as if it were a perfect sphere.

Agglomeration. The act or process of gathering into a mass. Using wet spray systems, the intent is to have the droplets collide with the dust particles and agglomerate, causing them to fall from the air.

Air atomizing. Mixing compressed air with a liquid to break up the liquid into a controlled droplet size.

Airborne dust prevention. Dust control measures taken to prevent dust from becoming airborne.

Airborne dust suppression. Dust control measures taken to knock down dust already airborne by spraying water into the dust cloud and causing the particles to collide, agglomerate, and fall from the air.

Air quantity. The amount of air used in ventilating a process, which is a combination of the air velocity or speed and the area of the duct, measured in cubic feet of air per minute (cfm).

Air to cloth ratio. In dust collection systems, a measure of the volume of gas per minute per unit area of the bag collecting dust.

Air velocity. The speed of air travelling in an area or duct measured in feet per minute (ft/min).

Annulus. The open area of the drill hole between the drill steel and the wall of the drill hole.

Armoring. A process used in wind erosion where large particles are left behind as smaller particles erode. The large particles then act as an erosion barrier, preventing further wind erosion.

Articulated loading spout positioner. A mechanical device that transfers product to a loading spout, while also providing the capability of moving the loading spout from side-to-side as well as forward-and-back to ensure proper positioning of the spout discharge.

Atomization. The process of breaking up a liquid stream into droplets.

Autogenous grinding. Size reduction of materials accomplished by impact and compressive grinding of the material by large units of the material itself—i.e., no balls or rods are employed.

Axial-flow fan. A fan that moves air in a direction parallel to the axis of the rotation of the fan. Axial-flow fans types include propeller, tubeaxial, vaneaxial, and two-stage axial-flow.

Back-flushing. The action of reversing the flow of air in a system to remove contaminants from a filter.

Bagging. Loading of product into some type of bag to be shipped/delivered to customers.

Baghouse collector. In dust collection systems, a type of collector that captures the particulate in a gas stream by forcing the gas through filter bags.

Bag perforations. Vent holes integrated into bags of product to allow air to escape rapidly and to reduce bag failures (e.g., rupturing or exploding).

Bag valve. Sleeve placed inside 50- to 100-lb product bags to allow for the insertion of the fill nozzle during bag loading and the sealing of the bag when loading is completed.

Bailing airflow. Compressed air blown down the drill stem through the bit in order to flush the cuttings out of the hole.

Blowback. Product spewing out from the bag valve during bag filling.

Bonding strength. The ability of the material to adhere to itself and to the material to which it is applied.

Capture velocity. A measure of the required airflow necessary to seize dust released at a source and then pull this dust-laden air into a (capturing) hood. Capture velocity is measured in feet per minute (ft/min).

Capturing hood. A hood positioned as close to a dust source as possible to capture the dust-laden air and pull it into an exhaust ventilation system.

Carryback. Material that sticks or clings to a conveyor belt after passing over the head pulley.

Cartridge collector. In dust collection systems, a type of collector that captures particulate from a gas stream by forcing the gas through filter elements, which are arranged in a pleated configuration.

Centrifugal collector. In dust collection systems, a device that separates particulate from the air by centrifugal force. Also called a cyclone.

Centrifugal fan. A fan in which the airflow is drawn into the rotating impeller and discharged radially from the fan blade into the housing. Centrifugal fans include radial, backward, and forward blade type.

Chimney effect. The behavior of heated air or gas rising in a vertical passage, as in a chimney, due to its lower density compared to the surrounding air or gas.

Chronic silicosis. A slowly progressive, nodular form of silicosis that typically develops after 10–30 years of exposures to respirable crystalline silica dust.

Coanda effect. The tendency of a moving fluid to be attracted to a nearby surface.

Collaring. The preliminary step in drilling, forming the beginning of a drill hole, or collar.

Contact angle. The angle at which a liquid meets a solid surface.

Control efficiency. A metric usually expressed as a percentage used to compare the results of two or more outcomes.

Crown. The mid-point of a road raised such that water drains off either side of the road surface into a ditch. In cross-section, the shape of the road is depicted as an A-shape, with the highest point in the center of the road.

Cure. The hardening of a chemical additive and base material.

Cuttings. Material that is produced during the drilling process. Cuttings are usually a fine-sized material.

Cyclone. In dust collection systems, a conical device that uses centrifugal force to separate large- and small-diameter particles.

Deliquescent. The ability of a material to absorb moisture from the air and become liquid.

Depth loading. In dust collection systems, a method of collecting particulate on a fabric by maintaining a filter cake on the fabric to optimize collection efficiency. When depth loading is used on a particular fabric, dust is intentionally allowed to penetrate into the fabric to form a layer of dust on the bag.

Diffusion. The spread of dust particles from high concentration areas to lower concentration areas. Many times this is achieved in ventilation systems by dust-laden air being mixed with clean, dust-free air.

Drifter drill. A drill used in underground mining to drill horizontal holes.

Drill deck (drill table). A deck or table on the drill located 2–3 ft above the ground surface and adjacent to the drill operator's cab. This deck is used by the operator to gain access to the drill rods when needed to add more rods or to change drill bits.

Drill rods. Metal rods, also called "steels," used to connect the rotary head to the drill bit as drilling proceeds through the ground.

Dual-nozzle bagging system. A type of fill nozzle for filling 50- to 100-lb bags of product where the inner nozzle is used for filling and the outer nozzle is used to exhaust excess pressure from the bag.

Dust. Small solid particles, often created by the breaking up of larger particles.

Dust collector. An air cleaning device used to remove air-entrained particles.

Dust collector airflow. The quantity of air that flows through a dust collector.

Dust collector airflow to bailing airflow ratio. A ratio of dust collector airflow and bailing airflow, mathematically represented by the dust collector airflow divided by the bailing airflow.

Dust suppressants. Chemicals or substances applied to a surface to prevent dust emissions from adhering to that surface.

Elastic limit. The maximum stress that can be applied to a material without causing permanent deformation.

Electrostatic charge. The physical electrical property of a dust particle that will cause it to experience a force, either attracting or opposing, from an oppositely charged electrical component or device.

Electrostatic precipitator. A particulate control device that uses electrical forces to move particles from the gas stream to collection plates.

Elutriation. The process of separating lighter from heavier particles by flow between horizontal plates.

Enclosing hood. A hood that either partially or totally encloses an area to capture the dust source or dust-laden air and prevent it from flowing out into a mine or plant environment where it could contaminate workers.

End-dumping. A process of spreading material (commonly gravel or sand) directly from the end of the dump truck. The material flow is controlled by limiting the tailgate opening to allow for the desired flow of material. Also called "tailgating."

Entrainment. The process of particles being lifted into the air and carried with the airflow.

Epitropic fiber. A fiber whose surface includes embedded particles designed to modify the fiber properties, typically electrical conductivity. Commonly used epitropic fiber materials include carbon, graphite, and stainless steel.

Exhaust. In ventilation systems, to discharge air into the atmosphere using an energy source (typically a fan).

Exhaust ventilation system. An engineered system designed to draw air through a network of openings via a fan.

Flexible Intermediate Bulk Containers (FIBCs). Large product containers, normally 1,000 lbs and greater when filled, used to ship bulk material to customers. Also called "bulk bags," "mini-bulk bags," "semi-bulk bags," and "big bags."

Fogger spray. Compressed air and water forced through a nozzle to create a fogged-in area, resulting in a mist that aids in dust suppression.

Freeboard. The distance from the top of a dump truck bed down to the top of the material loaded into the bed.

Fugitive dust. Solid airborne particulate matter that escapes from any dust-generating source.

Gradation. The distribution of different particle sizes in aggregate material.

Gravity separator. In ventilation systems, a large chamber where the velocity of the air stream is drastically reduced in order to facilitate the vertical drop of particles. Also called a "drop-out box."

Haul road. A road used for hauling ore to and from pit and crusher locations at a surface mine. Hauling is generally conducted using large off-road haul trucks.

Hydraulic (airless) atomization. Controlling droplet size by forcing liquid through a known orifice diameter at a specific pressure.

Hydraulic flat fan nozzles. In wet spray systems, nozzles used to produce relatively large droplets over a wide range of flows and spray angles, normally used in narrow enclosed spaces.

Hydraulic full cone nozzles. In wet spray systems, nozzles used to produce a solid cone-shaped spray pattern with a round impact area that provides high velocity over a distance.

Hydraulic hollow cone nozzles. In wet spray systems, nozzles used to produce a circular ring spray pattern, typically producing smaller drops than other hydraulic nozzle types of the same flow rate.

Hydrogen embrittlement. A process where metal (high-strength steel) can become brittle and fracture due to exposure to a hydrogen source.

Hygroscopic. The ability to absorb moisture from the air.

Impact plate. When material is being conveyed, a device used to control the direction and speed of the material.

Impingement plate scrubber. A type of wet scrubber in which the dust-laden air passes upward through openings in perforated plates, which hold a layer of water.

Induction. Airflow or product movement that creates sufficient momentum energy to pull additional surrounding air into the airflow or product movement.

Inhalable dust. A fraction of airborne dust intended to represent those particles likely, when inhaled, to be deposited in the respiratory tract from the mouth and throat to the deep lung. The collection efficiency of an inhalable dust sampler varies by aerodynamic diameter of the particles—to demonstrate, about 50 percent of 100-micrometer particles (but nearly 100 percent of particles less than 1 micrometer) are collected.

Inlet loading. In ventilation systems, the amount of dust traveling to the collector.

Intake air. In ventilation systems, the clean air supplied into an area or space to replace existing (contaminated) air.

Laminar. Sometimes referred to as streamline flow, the occurrence of air flowing in parallel paths with no disruptions between the paths.

Local exhaust ventilation (LEV). A system used to capture dust and remove it from a mine or plant environment. The concept is to capture dust-laden air, pull this air into an exhaust hood, transport the dust-laden air through ductwork to some type of dust collection system, and duct the dust-free air to some type of exhaust fan, which creates negative pressure, resulting in the airflow for the entire system.

Makeup air. In ventilation systems, the clean air necessary to supply a specific ventilation application (e.g., a local exhaust ventilation system).

Mechanical shaker collector. In dust collection systems, a type of collector that uses mechanical shaking to remove the excess dust cake from the collection media.

Musculoskeletal Disorder (MSD). Health problems in the body's muscles, joints, tendons, ligaments, and nerves. When work-related, these problems are caused by the work process or environment.

Open structure building design. A processing structure or building designed without exterior solid walls in an effort to lower respirable dust and noise levels.

Overhead Air Supply Island System (OASIS). A filtration system located over a worker at a stationary position to filter dust-laden air from the work environment and deliver a clean envelope of air down over the worker.

Palletizing. In the mineral processing industry, when bags of product, normally ranging in weight between 50 and 100 lbs, are loaded onto a pallet for shipment to customers.

Percussion drilling. Rock penetration through rotation and down pressure with a pneumatic drill, which contains a piston that delivers hammer blows to the drill column or the drill bit.

Pitot tube. A pressure-measuring instrument normally made out of two concentric metal tubes that can be used in conjunction with a manometer to measure the static pressure (SP), total pressure (TP), and velocity pressure (VP) due to airflow in the duct.

Protection factor. A comparison of outside versus inside dust concentrations.

Push-pull ventilation system. The technique of using an air jet (blowing system) directed towards a dust source and receiving hood (exhaust system) to capture and move dust-laden air into the hood.

Respirable dust. A fraction of airborne dust intended to represent those particles likely to be deposited in the alveolar region of the lungs when inhaled. The collection efficiency of a respirable dust sampler varies by aerodynamic diameter of the particles—to demonstrate, only about 1 percent of 10-micrometer particles (but nearly 100 percent of particles less than 1 micrometer) are collected.

Reverse air collector. In dust collection systems, a type of collector, that uses a traveling manifold to distribute low-pressure cleaning air to the filter bags for reconditioning.

Reverse jet collector. In dust collection systems, a type of collector that uses timed blasts of compressed air for filter bag reconditioning. Also called a pulse jet collector.

Roof ventilators. Either axial or centrifugal fans placed on the roof of a building and used to pull air out of the structure and discharge it into the atmosphere.

"Rooster tail." Product spewing from the fill nozzle and bag valve as the bag is ejected from the bag filling machine.

Rotary drilling. Penetration through rock by a combination of rotation and high down pressures on a column of drill pipe with a roller drill bit attached to its end.

Saltation. The movement of particles (usually soil or sand) in a jumping or bouncing motion caused by the wind.

Scarify. To break up or loosen the ground surface.

Semi-autogenous. Size reduction of materials accomplished by impact and compressive grinding of the material by large units of the material itself, with the use of balls or rods to provide additional crushing force.

Silicosis. A progressive and incurable lung disease that belongs to a group of lung disorders called the pneumoconioses. Silicosis is caused by repeated inhalation of respirable crystalline silica (e.g., quartz) dust.

Skirting. On conveyors, a horizontal extension of the loading chute used to contain ore and dust within the transfer point. On loading spouts, strips of belting or fabric installed on the spout discharge end, creating a physical seal with the product pile to reduce dust liberated during bulk loading.

Spoon. When material is being conveyed, a curved plate that helps steer the material in the desired direction.

Spray tower scrubber. A type of scrubber that employs water, which falls counter-current through a rising dust-laden air stream, to remove dust particles.

Spreader box. An implement used in road construction to spread road material (typically gravel or sand) in a uniform layer. The implement, commonly in the shape of a box, is situated behind an end-dump truck with material from the truck dumped into the box for spreading into the uniform layer.

Static pressure (SP). In ventilation systems, a measure of the pressure in a ventilation duct relative to the atmospheric pressure. Static pressure can be either a positive or negative value. Static pressure summed together with velocity pressure (VP) provides the total pressure (TP) for a duct in a ventilation system.

Stoper drill. A drill used in underground mining to drill vertical holes overhead.

Sub. A short length (usually 2–3 ft) of drill collar assembly, made of the same drill steel material, which is placed between the drill bit and the drill steel.

Subbase. In road construction, the layer between the subgrade and the wearing surface.

Subgrade. In road construction, the underlying soil or rock that serves as the foundation for the road.

Surface loading. In dust collection systems, a method of collecting particulate on the surface of a fabric.

Surface roughness. In wind erosion, a qualitative measure of the roughness of a ground surface. The higher the surface roughness, the more wind erosion is inhibited.

Surfactants. Substances that, when dissolved in water, lower the surface tension of the water, allowing the water to "wet" the dust more easily.

Table bushing. A bushing used to seal the opening where the drill steel passes through the drill deck.

Tilt sensor. On bulk loading spouts, a vertically mounted sensor designed to prevent blockages during loading. As the product pile grows and moves the sensor off of vertical by a predetermined amount (e.g., 15 degrees), an electronic signal is transmitted to initiate the raising of the spout.

Total pressure (TP). In duct ventilation systems, the sum of static pressure (SP) and velocity pressure (VP).

Total structure ventilation system. The use of exhaust fans high on the outside walls or roof of a structure, bringing in clean outside air at the base of a building, which sweeps up through the structure to clear dust-prone areas.

Total suspended solids. An air quality term used to denote a particulate emission's entire particle size distribution.

Tri-cone roller bit. A rotary drill bit with three roller cones. As the drill bit rotates, the roller cones, which have carbide components, roll against the bottom of the drill hole, causing the material to break into small particles. This type of drill bit requires high down-pressure on the drill bit.

Velocity pressure (VP). In ventilation systems, the pressure required to accelerate air from being at rest to a particular velocity. Velocity pressure summed with static pressure (SP) provides the total pressure (TP) in a duct.

Venturi eductor. In pneumatic conveying systems, a device that converts blower output into suction to draw feed material into the conveying line.

Venturi effect. A reduction in static pressure resulting from a liquid or gas flowing through a constricted space.

Venturi scrubber. A type of wet scrubber consisting of a venturi-shaped inlet and a separator.

Water cartridges. A plastic or PVC bag filled with water and inserted into the blasthole along with the explosive, resulting in dust suppression during blasting.

Water separator sub. A short length of a drill collar assembly placed between the drill bit and the drill steel. A water separator sub uses inertia to remove the injected water from the bailing air.

Wearing surface. The top layer of the road surface which is exposed to traffic.

Wet cyclone scrubber. A type of scrubber that uses centrifugal forces to throw particles on the collector's wetted walls.

Wet drilling. In surface drilling, the injection of water along with air to flush the cuttings out of the hole.

Wet scrubber. An air pollution control device used for particulate collection. Wet scrubbers employ water or another liquid as the collection media.

Wet spray systems. A system of water sprays used to wet fines so that each dust particle's weight increases, thus decreasing the particle's ability to become airborne.

Windrow. A long linear pile of material.

www.ingramcontent.com/pod-product-compliance
Lightning Source LLC
Chambersburg PA
CBHW080236180526
45167CB00006B/2294